Undergraduate Texts in Physics

Undergraduate Texts in Physics (UTP) publishes authoritative texts covering topics encountered in a physics undergraduate syllabus. Each title in the series is suitable as an adopted text for undergraduate courses, typically containing practice problems, worked examples, chapter summaries, and suggestions for further reading. UTP titles should provide an exceptionally clear and concise treatment of a subject at undergraduate level, usually based on a successful lecture course. Core and elective subjects are considered for inclusion in UTP.

UTP books will be ideal candidates for course adoption, providing lecturers with a firm basis for development of lecture series, and students with an essential reference for their studies and beyond.

More information about this series at http://www.springer.com/series/15593

Giovanni Organtini

Physics Experiments with Arduino and Smartphones

 Springer

Giovanni Organtini
Sapienza Università di Roma
Rome, Italy

ISSN 2510-411X ISSN 2510-4128 (electronic)
Undergraduate Texts in Physics
ISBN 978-3-030-65139-8 ISBN 978-3-030-65140-4 (eBook)
https://doi.org/10.1007/978-3-030-65140-4

Cover by Massimo Buniy, Gabriele D'Andreta and Alessio Ria: "Across the Black Hole"Artwork made for the "Art & Science Across Italy" programme: a European project of the CREATIONS network (H2020) organised by INFN and CERN to promote scientific culture among young people, combining the languages of art and science. Thanks to Prof. Georgia Conti.

This Springer imprint is published by the registered company Springer Nature Switzerland AG
The registered company address is: Gewerbestrasse 11, 6330 Cham, Switzerland

Preface

I have been dreaming about writing this textbook for years. Finally, in late 2019, before the COVID-19 pandemic exploded, I met a far-sighted publisher who agreed to print it. I wrote three chapters of this book while Italy was in lockdown and courses at our University had been switched to remote learning, to the obvious detriment of laboratory courses. I was then asked to provide some help as to how these remote courses might be carried out and I came up with a few ideas that were met with scepticism at the beginning but indeed, resulted in complete success. Students were, in fact, learning even more than they might have with traditional courses. The reason, in my opinion, is simple: conducting experiments by yourself (by which I mean assembling all that is needed to perform the experiment, organising data collection and analysing the results through the creation of your own code) teaches a student much much more than they are likely to learn simply by pushing buttons to start an automatic device prepared by a teacher. Moreover, classical physics laboratory courses suffer from the misguided notion that the purpose of the experiments to be proposed to students is to see how closely they can follow the lecture material and imitate original equipment in their own constructions, thus making experiments look quite old-fashioned.

On the contrary, the purpose of laboratory courses should be to teach students how to design and conduct an experiment, interpret data and derive models. These skills must be developed using modern tools, in both hardware and software domains, such that students are prepared for practice in modern physics laboratories.

The idea behind this textbook is that laboratory practice is learnt by doing and is much less formal than theoretical physics courses. Often, the emphasis in these courses is on statistics, and they are full of mathematics that seldom catches the attention of students devoted to experimental physics. The main goal of the proposed experiments is to prove that a physical law is correct, but given that most of the physical laws can be formulated within a framework in which many approximations are done implicitly, this goal is rarely achieved. As a result, students get frustrated and lose confidence in their experimental abilities. Finally, in many cases, operations are tedious (repeat the same measurement many times over, compute complicated derivatives and solve difficult equations or systems of equations, etc.), without even the benefit of being useful for the development of the competencies. None of us compute averages or uncertainties about measurements using the techniques adopted in university courses. We rather use automatic computation, having learnt a programming language. That's why computer programming is taught in physics courses. However, too often, computers are not used in laboratory courses to derive results, in the mistaken belief that doing math manually is the only way to learn and that writing computer programs is a waste of time.

My textbook is thus organised such that the knowledge is acquired progressively, increasing from chapter to chapter, sometimes overriding (or, more accurately updating) the knowledge acquired up to a given point. Different topics (experiment design, data acquisition, statistical data analysis and their interpretation) are not discussed in dedicated chapters but are mixed in with an introduction to a modern programming language such as Python and complemented by a relatively detailed description of Arduino programming.

Sections are organised such that students (and teachers) less interested in programming can easily skip them to no detrimental effect or, at most, read them superficially. There is no need to conduct the experiments exactly as proposed: the descriptions are intended to be suggestions from which the reader can draw inspiration for his/her own design. Also, the suggested experiments are just examples that any teacher can use to then propose his/her own. The organisation of the volume is such that a teacher is also free to organise his/her own course in the way he/she prefers. Even the order in which the topics are presented does not need to be strictly followed, even if chapters are necessarily organised such that the initial ones present very basic data analysis that is progressively refined.

Primary ideas and important results are highlighted as side notes. All the content of the side notes is then summarised at the end of each chapter.

One final remark point: my personal belief is that laboratory courses do not necessarily have to follow theoretical courses. In fact, physics is done the opposite way: investigate new and unknown phenomena experimentally, formulate models and test them. As a consequence, it is perfectly viable (indeed, in my opinion, sometimes preferable) to perform experiments on topics not yet mastered by students. However, most physics courses are organised differently, and we agreed that it was a bit audacious to propose a radically different textbook. For this reason, the experiments proposed have to do with mechanics, traditionally the first topic taught in physics courses. There is no need to perform the exact same experiments. Many variants can be imagined and possible alternatives are proposed on the book's website.

Rome, Italy

Giovanni Organtini

Acknowledgements

This book would probably never have been written if it had not been for Morten Hjorth–Jensen, from the University of Oslo, who urged me to write it in recognition of its potential. The idea was born during a short stay in Oslo, where I went thanks to a very kind and useful invitation from Morten, whom I want to warmly thank for the opportunity he gave me to visit his wonderful country and to collaborate with a team of researchers and teachers open to innovation and didactic experimentation. I hope to repeat this experience. I also want to thank Marina Forlizzi, who believed in this project and promoted it at Springer.

I also want to thank Julien Bobroff and Frédéric Bouquet, from Université Paris–Saclay, whom I had the good fortune to meet a few years ago, almost by chance, and with whom I share a vision of the importance of conveying science through new languages, including art. It was a great pleasure to work with them and their collaborators, too.

I thank Martín Monteiro, from Universidad ORT in Montevideo, Cecilia Stajano from Fondazione Mondo Digitale and Wolfgang Christian of the AAPT for their contagious enthusiasm.

Let me thank all my students, too: from the very best to the least proficient. Their success, as well as their difficulties, more than my career, is what drives me to always study new forms of teaching.

Last, but not least, I thank Federica for her patience and support. I recognise that, sometimes, it can be difficult to be my wife, but I will be forever grateful to her for that.

Contents

Physics and Nature

Contents

© The Author(s), under exclusive license to Springer Nature
Switzerland AG 2021
G. Organtini, *Physics Experiments with Arduino and
Smartphones*, Undergraduate Texts in Physics,
https://doi.org/10.1007/978-3-030-65140-4_1

1

This chapter is devoted to the tentative definition of the subject of your studies: physics. We already know that trying to define precisely what physics is about is, to say the least, very difficult, if not impossible. However, it has clearly emerged that physics needs measurements: this is the subject of this book.

1.1 Physics and Other Sciences

It is not simple to define physics. It is usually defined as the science of natural phenomena, but physicists deal with artificial phenomena, too. We suggest that you read some literature on epistemology. In particular, the works of Popper, Kuhn, Feyerabend and Lakatos are recommended.

Physics is often identified as a science, namely, that devoted to the study of natural phenomena. However, biology, too, is considered a science, and it certainly concerns things that belong to the set of natural phenomena. On the other hand, many physicists (i.e., people with a degree in physics) work in fields that have nothing to do with Nature, like finance, and completely new branches of physics are currently emerging, like *econophysics*.

Merely trying to define science itself is not at all simple. There are plenty of philosophers who have tried their hand at finding a precise definition of it and they still do not agree with each other. Popper's (1902–1994) falsificationism [1] is probably the most widely known work about epistemology, i.e., the theory of knowledge, but it is not the only one. Thomas Kuhn (1922–1996) developed a model [2] of scientific knowledge involving the so-called scientific revolutions, introducing the concept of the *paradigm shift*, consisting of a radical change in the interpretation of basic concepts, as a driver in scientific progress. In "Against Method" [3], Feyerabend (1924–1994) presented his radical idea that there are no recognised rules in the scientific community. Lakatos (1922–1974), constrastingly, was firmly convinced that scientific progress can be modelled as a sequence of *research programmes* [4]: progressive actions driven by a *positive heuristics*, i.e., a self-corroborating approach to solving problems.

Given the above-mentioned ambiguities, anyone can try to figure out what constitutes science and what constitutes physics, depending on his/her own attitudes. Our opinion (in fact, it is more than an opinion, but this word has been very deliberately chosen, see below) is that, indeed, the scientific community does not follow formal rules, as suggested by Feyerabend, but there are, however, unspoken rules, often not very well defined, that cause scientific results to be accepted as modelled by Lakatos. Rarely, paradigm shifts happen that, however, must be corroborated by some positive heuristic in order to be successful, also as in the Lakatos model.

Physics is a science based on experimental data, i.e., the science of what can be measured. As long as something can be measured, the phenomena to which it gives rise can be investigated through the methods of physics. Measurements play a major role in our science: this is why it is so important that you learn what measuring means and how to perform and interpret measurements, even if you are going to become a theorist.

Before addressing the problem of measurement, we must still say a few words about physics in general and its relationships with other sciences. From the above discussion, it should be quite clear that physics is quite different from mathematics, itself considered a science. Mathematics is not at all an experimental science. Mathematicians freely choose a set of axioms, true by definition, and then derive consequences that strictly adhere to a (sometimes arbitrarily defined) set of rules. As a result, many types of mathematics can exist at the same time: everyone knows that one can define alternative geometries known as non-Euclidean geometries, for example. Physicists are not as free as mathematicians: they cannot establish the rules to which physics must adhere. They can only observe the Universe and try to describe it in the proper way, i.e., such that they are able to predict what will happen in the future or determine what happened in the past, irrespective of how far into the past or future we wish to go, which depends on how good we are at describing the systems under investigation.

In short, we cannot tell the Universe how to behave or, to put it differently, the Universe does not obey the laws of physics. It's the laws of physics that obey (in other words, are set by) the behaviour of the Universe.

This is why we used the word "opinion" above. There is a well-known quote in Italy that goes "mathematics is not an opinion", ascribed to Senator Filippo Mariotti (1833–1911), meaning that a statement cannot be discussed or subject to interpretations. According to the above considerations, mathematics is indeed an opinion (in the sense that mathematicians are free to have "opinions" about what has to be considered as true). A more correct quote would be "Physics is not an opinion", since, whatever a physicist's believe may be, he/she can never model phenomena in such a way that the results of the model do not agree with observations.

There is no physics without experimental data. We cannot conceive the physics of love or the physics of dreams, because we do not have tools to measure love and dreams.

Mathematics and physics are often considered very similar. In fact, they are profoundly different.

Planets do not obey Kepler's laws. It is just the opposite: Kepler's laws *obey* (meaning they are set by) the way in which planets move.

1

Truth, in physics, is that which models predict correctly. As such, it may change depending on the context and it may evolve with time. There is no absolute truth in this science.

Studying physics, you will learn that many of your common sense beliefs are wrong, in the sense that they cannot be used to build models able to predict (or retrodict) phenomena. There's nothing you can do about it. You must believe (working) physics models as representing the truth, whether you like it or not. It is worth noting that this often happens in classical physics, too, even if the majority of people consider it a peculiarity of modern physics.

The evolution of cosmology

It was once believed that our planet defined the center of the Universe and that everything else (the Sun, the Moon, the stars and the planets) revolved around it. Even if it may appear patently false to us nowadays, there are still good reasons to believe this. Looking at celestial bodies, we are induced to conclude that they move around us in a more or less circular path. After centuries of observations, Ptolemy (ca 100–178) recorded a set of rules in his work Almagest [5]. These rules can be taken as a primitive form of physical laws.

Looking carefully at the night sky, one can see that, in fact, not all the celestial bodies move following regular circular paths. Some of them appear to invert their motion and exhibit so-called apparent retrograde motion. To explain such motions, Ptolemy and his followers introduced the epicycles: a circular path around a point that, in turn, follow a circular path around Earth.

It took more than 1 000 years for humanity to realise, thanks to Nicolaus Copernicus (1473–1543), that, in fact, Earth was not at rest in the centre of the Universe [6]. In his cosmology, Copernicus showed that all the planets, including Earth, rotated around the Sun, which was taken to be fixed, still at the centre of the Universe. The idea must have seemed odd to his contemporaries, but today, the fact that Earth and other planets orbit around the Sun is widely accepted.

Later, it was discovered that even the Sun moves around the centre of the galaxy and that many galaxies move around within the Universe, which surely does not have a center near us, if there is a center at all.

Till the work [7] of Sir Isaac Newton (1642–1726), no one realised that it took a force to hold the Solar System together, effectively wiping out the initial theory [8] formulated by Johannes Kepler (1571–1630), according to which planets moved as observed because they were playing music for the pleasure of God.

Albert Einstein (1879–1955) ultimately found that Newtonian's gravitation is only a first order approximation of his general relativity theory [9].

From this long story made short, one can see that, from time to time, a *theory* may be superseded by another *theory*. As a consequence, the newest theory has to be considered to be true, while the superseded ones must be taken as false. However, even Ptolemy's theory is not false, in the sense that it correctly describes planetary motions, at least if the precision with which their position is measured is not too high. Only the interpretation of his data can be considered wrong. It is worth noting that, even when the interpretation of a set of data is *wrong*, as in the case of Newton's gravitational theory as compared with general relativity, it can be taken as correct for many practical purposes. If we must compute the motion of freely falling bodies on Earth or even the trajectory of an interplanetary spacecraft, we do not use Einstein's general relativity: Newtonian mechanics is precise enough to provide the correct results. Hence, even if superseded by a new theory, an old one continues to remain valid and *true* within the domains in which it was formulated.

Another important difference between mathematics and physics is that the second is inevitably less rigorous than the first. Sometimes, physical concepts or quantities are not as well defined as mathematical concepts. Indeed, there is no need in physics to be as rigorous as in mathematics. Many physicists still discuss what is the *true* or *correct* definition of heat. Indeed, there is no need to struggle to find it: as long as we can measure it and use heat to compute other physical quantities, heat is well enough defined. In order for a physical concept to be well defined, it is enough that we are able to describe, as precisely as possible, a procedure to measure it.

1.2 Measurements and the Laws of Physics

A measurement consists of a set of operations leading to a measure. Measures, *per se*, are not so important if they do not lead to a model that is usually expressed in mathematical form as an equation, often called a **physical law**, even if such a term can be misleading according to the above.

Equations represent a physical law only if they express the relationship between at least two **physical quantities**, i.e., something that can be measured. There are no physical laws about love, for example, since there is no way to measure love. One can say that love does not exist for physicists, in the sense that physicists cannot make statements about love with the same "authority" with which they make similar statements about heat. A physicist's statement about love is just an *opinion*.

Physical laws are equations expressing the relation observed on two or more physical quantities, i.e., things that can be measured.

The way in which a physical law is written matters. Two mathematically equivalent expressions may not have the same meaning when interpreted as physical laws.

An equation representing a physical law has little to do with an equation in mathematics, but rather represent the rules for manipulating it. Consider an equation like

$$a = \frac{F}{m}. \tag{1.1}$$

Interpreted as in mathematics, such an equation means that a number represented by a is always equal to the ratio of two numbers F and m and can be written equivalently as

$$F = ma \tag{1.2}$$

or

$$m = \frac{F}{a}. \tag{1.3}$$

Moreover, as long as $m \neq 0$, Eq. (1.1) always makes sense and even in the limit $m \to 0$ the value of a is still a number and the equation holds.

For a physicist, Eq. (1.1), has a completely different meaning. It says that the physical quantity on the left (the acceleration a of a body) depends on those on the right side of the equation (the intensity of the force F acting on the body whose mass is m). Even if Eq. (1.2) can be found in many physics textbooks, it does not hold the same meaning and, in some sense, is not as correct. Read as a physical law, Eq. (1.2) states that a force depends on the product of a mass and an acceleration, which makes no sense. That equation, as well as Eq. (1.3), holds only for the numbers representing the outcome of the measurement of the acceleration of a body of mass m subject to a force F.

Despite the fact that physicists write their laws as equations involving the equal operator $=$, measurements are always affected by uncertainties, and models are often approximated, so, usually, $=$ turns out to be, in fact, $\simeq$.

There are other differences, too. All the equations hold if we substitute the numbers obtained during a measurement, but only approximately. A physical law like (1.1) does not guarantee that measuring the acceleration a of a body with $m = 2$ subject to a force $F = 3$, will be exactly $\frac{3}{2} = 1.5$. First of all, we must specify the units of a, F and m, otherwise the values do not make sense. Moreover, any measurement is affected by some uncertainty, hence the result of a measurement of a will probably lead to a value different from 1.5 (irrespective of the units), even if we expect to obtain a number close to it. How close depends on the uncertainty of the measurement, which we must learn how to estimate. In the end, the meaning of the symbol $=$ is somewhat different with respect to what it means in mathematics. The equal symbol states that we *believe* that those quantities must depend on each other in the given way, but only if we neglect many effects that, in practice, cannot be eliminated, such that the numerical results provided by the measurements only holds approximately and $=$ becomes, in fact, $\simeq$.

Another difference is that those equations hold only within the framework in which they have been experimentally established, i.e., *Newtonian mechanics*. For bodies moving at high speed (where high means close to the speed of light), Eq. (1.1) does not hold anymore.

Physical laws are only valid within the framework within which they are derived.

1.3 The Process of Measurement

A physical quantity is anything that can be measured, i.e., something to which we can assign a value, usually (but not always) expressed as a number.

For example, the height of a person is a physical quantity since we can tell how tall a person is (e.g., 1.82 m). The color of his/her eyes can be regarded as a physical quantity too, since we can attach a characteristic to it, such as brown or blue. It turns out that, in fact, all of the characteristics expressing the *value* of a physical quantity can be expressed as numbers. For example, one can define a table in which each color corresponds to an integer. For this reason, we are going to treat all the quantities of potential interest for a physicist as numbers.

Measurements can always be represented by numbers.

Even if we can assign a value to any characteristic of anything, we are not going to define all of them as physical quantities. We define as such only those characteristics that are of potential interest within the framework of the topic we are investigating. Collecting the value of all the potentially interesting characteristics of a system under investigation is called *characterisation*.

Measuring a physical quantity means performing a set of activities aimed at attaching a value to it, such that its value depends as little as possible on the procedure, on who performs the measurement and on any other condition that can alter the result. A good measurement should give the most objective result possible. The activities leading to the measurement of one or more physical quantities are said to be an **experiment**, whose purpose is always to obtain (at the very least) a number to be published for the community.

The outcome of an experiment is always, at the very least, a number that results from a measurement.

1

We are going to use computers and other modern tools to conduct our experiments, so as to approximate what happens, in fact, in modern physics laboratories. Often, physicists use cutting-edge technologies. While technologies evolve with time, working principles do not. We concentrate on the latter, so you will be able to use the best available technologies when needed.

In what follows, we are going to formally develop a theory of measurement by engaging in measurement. Instead of giving definitions and applying them in the process of measuring something, we will try to characterise a process or a system, whereby we will find that things are not as simple as they might seem. We will fall into many pitfalls out of which we will have to climb by revising our naive beliefs about the measurement process and we will come to define, through experience, the correct way to conduct it.

At present, physicists perform measurements by collecting data using some automatic instrument, often driven by computers. Data analysis, moreover, can be equally well done electronically, using either spreadsheets or *ad hoc* designed computer programs. Together with our quest to learn how to perform measurements and interpret collected data, we should then learn how to perform measurements using modern data acquisition systems and how to collect and analyse them using computers. The ability to code is as important as the ability to solve an equation. Sometimes it is even more important.

The details of data acquisition can change a lot from system to system. In order to provide sufficiently general information we are going to use simple and cheap tools like smartphones and Arduino boards (in truth, smartphones are not that cheap, however, anyone can buy them for other purposes and, regarded as instruments, they can be considered as almost free). Of course, the considerations arising from the analysis of the proposed experiments will be completely general and will apply to any data acquisition system (abbreviated as DAQ), be it automatic or manual.

Summary

Physics is an experimental science. Physics theories are grounded in experimental data. There are no physical theories that are not based on them. As such, we can define physics as the science of things that can be measured.

Solid experimental observations have to be considered to be the *truth*, irrespective of the fact that they may appear unreasonable to us. However, their interpretation is subject to change with time. Still, they remain true in the framework in which they were obtained.

A theory is a set of experimentally grounded mathematical rules allowing us to make predictions about the results of a measurement. If the rules do not allow us to make predictions, their set is not a theory. Two theories resulting in the exact same prediction are the same theory formulated in a different formalism.

Taking measurements is the job of experimental physicists. Physical laws are the mathematical formulation of observations. In particular, they are expressed as equations relating two or more physical quantities, i.e., things that can be measured.

A physical law can be treated as mathematical equations as long as they are used to compute and to make predictions. However, they are quite different from equations. They state what we believe depends on what and, as a matter of fact, they must be considered as exact only in principle.

An experiment consists in taking measurements in conditions under control. The outcome of an experiment is always, at the very least, a number.

References

1. Popper Karl R (1959) The logic of scientific discovery. Hutchinson, London
2. Kuhn Thomas S (1962) The structure of scientific revolutions. University of Chicago Press, Chicago
3. Feyerabend Paul (1975) Against method. New Left Books, London
4. Lakatos Imre (1968) Falsification and the methodology of scientific research programmes. Cambridge University Press, Cambridge
5. Claudius Ptolemy. Almagest (100–178)
6. Nicolaus Copernicus (1543) De revolutionibus orbium coelestium, Nuremberg (1543)
7. Sir Isaac Newton. Philosophiæ Naturalis Principia Mathematica
8. Kepler J (1619) Harmonices Mundi, Linz
9. Einstein Albert (1916) The foundation of the general theory of relativity. Ann Phys 49(7):769–822

Establishing a System of Units

Contents

© The Author(s), under exclusive license to Springer Nature
Switzerland AG 2021
G. Organtini, *Physics Experiments with Arduino and
Smartphones*, Undergraduate Texts in Physics,
https://doi.org/10.1007/978-3-030-65140-4_2

2

When we try to take a measurement, i.e., to attach a number to a physical quantity, we need a **unit**. A unit is a standard to which the quantity compares. We always use units without many formalities. For example, the height of a person is often given in centimetres or feet, depending on the country; age is given in years (except for very young babies, for whom it is customary to give it in months); food is often sold in units of weight, etc. In this chapter, we analyse the problem of defining a unit properly and we learn the basics of the systems of units used in physics.

2.1 Measuring Light Intensity

The need to measure something comes from the need to write equations representing the relationships between physics quantities. Equation terms represent numbers and we need to express physics quantities with them. Physicists are always dealing with measurements: experimentalists take them and theorists analyse them to make mathematical models.

Most textbooks start with the definition of simple and well-known units such as length. In this book, we make a different choice: we want to start with a non-naive measurement. Analysing units that people are familiar with might be boring and, more importantly, could induce some to neglect one or another fundamental step in understanding the processes that lead to their definition. This is why we chose to introduce the problem of measuring something with an unusual quantity: light intensity.

It is well known that looking at a light that is standing on the opposite side of a (partially) transparent screen causes that light to appear less *intense*. This simple observation describes a physical phenomenon, the **absorption of light**, in qualitative terms. It implicitly defines a physical quantity (the intensity of the light) and subtends that there is a relationship between the intensity of the light and the characteristics of the screen. The aim of the physicist is to precisely define what is intended for light intensity, i.e., how to measure it, establish which of the characteristics of the screen it depends upon (thickness, shape, material, weight, etc.) and how. Eventually, the origin of the observed phenomena may be ascribed to other known phenomena, leading to a *unified explanation* of them.

The need for measurements comes from the need to express such a qualitative observation in a more formal and precise way, such that one can make predictions about what will happen with a different screen. It is not enough, for a physicist, to affirm that a thicker screen will lead to more intense absorption of the light. We must know how much light is absorbed by a screen with given characteristics. It is not just a matter of practicality (for example, it may be requested that we design a filter that limits the intensity of the light shining on a plant to a level that the plant can tolerate). Attaching numbers (and units) to quantities is needed to understand the processes in which that quantity is involved. This, ultimately, is the physicist's job.

2.2 Definition of Units

The simplest way to define light intensity is to find a standard sample (often called simply a *standard*) to be used as a **unit**, whose comparison with the system under investigation allows us to assign a number to it. The standard must be manifestly chosen among those that exhibit the same characteristic, and are thus **homogeneous** with the system to be characterised.

One way to measure something is to compare it with a standard.

To define a unit for light intensity, we need a source of light whose intensity is as stable as possible. Sunlight is not a good choice, since it changes with time, season, weather, etc. We might use artificial light, like that produced by a candle or a low wattage lamp. The light of a candle fluctuates a lot, thus it is a bit impractical. A good choice could be a smartphone's flashlight. We hardly see any difference in its intensity, at least if the smartphone is sufficiently charged. There can be differences from smartphone to smartphone, however, we can overcome this difficulty by establishing a specific model as the reference.

Units must be stable, reproducible and easy to use.

In order to perform a direct measurement of the light intensity of another light source, we need a way to compare it with the unit. Suppose that the intensity of the light to be measured is lower than that of the reference smartphone. In this case, it is enough to choose a set of standard *absorbers*, equal to each other, to be used as screens for the light of the reference smartphone. One can observe the light to be measured and compare it with the light of the reference smartphone transmitted by n such screens.

Direct measurements are made by comparing the quantity to be measured with a standard unit.

Screens can be cut out from a clear pocket for $A4$ sheets of paper. As long as the screens can be considered equal to each other, one can identify the number n of screens needed to observe the light of the reference smartphone as being identical to that emitted by the source under investigation and say that its light intensity is $I = n$.

The procedure for obtaining the measurement is part of the definition of a physical quantity.

However, we cannot require everyone to own the same smartphone as we do. So, while, e.g., Amelia claims that the intensity of the light is $I_A = n$, Ben may correctly report that the intensity of the same light is $I_B = m \neq n$. Man-

Different units may be needed to measure the same thing.

2

ifestly, the light intensity is the same, thus their measurements must be so as well, i.e.,

$$I_a = I_b . \tag{2.1}$$

The same quantity can have different measures, depending on the units chosen.

If the above were mathematical equations, the only possible solution would be $n = m$. However, these are not mere mathematical equations. They represent the relationships between physical quantities: the quantities are the same, but the ways in which we express them are not. A possible workaround is to write an equation like

$$n\,u_1 = m\,u_2 \tag{2.2}$$

that allows $n \neq m$, provided that there is a precise relationship between u_1 and u_2. The equation written above makes it clear that simply attaching a number to a measurement is not enough. We need to specify a unit, i.e., the standard to which the measured quantity had been compared. In the example above, Amelia used the u_1 unit, while Ben used the u_2 unit. The names assigned to those units are impractical. A better choice would be to use specific words invented for each of them. Often, the names of units come from the names of famous physicists of the past, as joule, newton, ohm, etc. Moreover, units can be indicated by using a symbol, for the sake of brevity (J, N, Ω). Adopting the same approach, let's call the unit represented by u_1 the **floyd** (in honour of the rock band Pink Floyd), to which we assign the symbol **fl**, and the one represented by u_2 the **stone** (symbol **st**, from the Rolling Stones). The equation

When quoting the result of a measurement, we must quote its units, too.

$$n\,\mathrm{fl} = m\,\mathrm{st} \tag{2.3}$$

Unit conversion consists in finding the number that represents the measure of something in a different unit. Conversion factors are easy to obtain by treating units as algebraic symbols.

is perfectly valid in both physics and mathematics. Given a measurement x in floyds, one can transform it into stones using a conversion factor obtained by treating units as algebraic symbols. Following the usual rules, one can write

$$x\,\mathrm{fl} = x\frac{m}{n}\,\mathrm{st} . \tag{2.4}$$

The ratio $\frac{m}{n}$ is the desired conversion factor that allows us to transform a light intensity x expressed in fl into units of st. It is enough to multiply x by the conversion factor.

For simplicity, we can construct tables of conversion factors where, for each unit u_1, we provide the corresponding value in another unit u_2 when the value in u_1 is one, e.g.,

$$1\,m = 100\,cm.\tag{2.5}$$

So, to transform a height of 1.82 m into centimetres, we can just multiply the number by 100, to obtain

$$1.82\,m = 182\,cm.\tag{2.6}$$

In fact, we can work as if we were substituting the symbol m with $\frac{100}{1}$ cm. Each time you are in doubt, just write the equation stating the equivalence and treat units as algebraic symbols.

Exercise 1
   A slim, elegant TV is on sale in an online shop for 449 USD. The data sheet says it has a screen size of 58″ and its aspect ratio is 16 : 9. Before buying it, you must check whether there is enough space on the wall of your room. Moreover, you want to compare the price of this TV with another on sale at a physical shop whose price is quoted in euros. Compute the width of the TV and its price in euros.

   The size of a TV screen is given in **inches** (symbol ″), a unit commonly used in countries formerly under the influence of the British Empire, and represents the length of its diagonal. One inch corresponds to 2.54 cm, i.e.,

$$1'' = 2.54\,cm.\tag{2.7}$$

In this case, it is very easy to find the size of the TV in cm as $d = 58 \times 2.54 = 147.32$ cm. It is bit less straightforward to express the price in another currency. The exchange rate is usually given as EUR/USD ratio. Assuming that, at the time of the purchase, the rate is EUR/USD= 1.12, treating the units as symbols, we can invert both members to obtain

$$\frac{USD}{EUR} = \frac{1}{1.12}\tag{2.8}$$

such that

2

$$449 \, \text{USD} = 449 \frac{1}{1.12} \, \text{EUR} \simeq 400.89 \, \text{EUR} \,. \tag{2.9}$$

Exercise 2

Cosmetics are often sold in bottles, whose content is expressed in **fluid ounces** (fl. oz.). A *cologne* bottle contains 1.7 fl.oz. and costs \$ 156.25. How much does the cologne cost per ml, in euros?

Units used in the Anglo-Saxon world are usually expressed as fractions. For fluids,

$$\frac{1}{8} \, \text{fl.oz.} = 3.70 \, \text{ml} \,. \tag{2.10}$$

We can get the conversion factor from this equivalence as

$$1 \, \text{fl.oz.} = 8 \times 3.70 \, \text{ml} = 29.6 \, \text{ml} \,, \tag{2.11}$$

such that the bottle's content is $1.7 \times 29.6 \, \text{ml} = 50.32 \, \text{ml}$. It thus costs

$$\frac{156.25}{50.32} \frac{\text{USD}}{\text{ml}} \simeq 3.11 \frac{\text{USD}}{\text{ml}} = 3.11 \frac{\text{USD}}{\text{ml}}$$
$$\times 1.12 \frac{\text{EUR}}{\text{USD}} = 3.48 \frac{\text{EUR}}{\text{ml}} \tag{2.12}$$

For historical reasons, there is one notable exception to the above rule. Temperatures can be measured in units such that conversion factors are not as simple to obtain.

Temperatures are measured in degrees Celsius (°C) in most countries, except in few Anglo-Saxons ones, where they are measured in degrees Fahrenheit (°F). The following equation holds:

$$T(°\text{F}) = \frac{9}{5} T(°\text{C}) + 32 \,, \tag{2.13}$$

such that $0°\text{C} = 32°\text{F}$ and $100°\text{C} = 212°\text{F}$ (correspondingly, $0°\text{F} \simeq -18°\text{C}$ and $100°\text{F} \simeq 38°\text{C}$). The reason for this bizarre conversion is historical (see box).

> **Temperature scales**
>
> The German physicist Gabriel Fahrenheit proposed a temperature scale in which the coldest temperature that he could achieve in his laboratory was defined as the zero of the scale, while the temperature of a human body was taken to be 100 degrees. It turns out that, indeed, this temperature corresponds to about 38°C, so either Fahrenheit had a fever or he was wrong in taking his measurements. One argument put forward was that he used the blood of his horse to define such a temperature, a horse's blood being a bit hotter with respect to that of a human. According to Encyclopedia Britannica, the reason for the discrepancy has to be ascribed to a redefinition of the scale, after Fahrenheit's death, made for convenience (so that the conversion between Celsius and Fahrenheit would be easier).
>
> Anders Celsius's proposal was in favour of a more objective scale [1], in which $T = 0°$ was taken as the temperature of boiling water, while $T = 100°$ was the temperature of melting ice. Note that this scale uses the opposite convention with respect to that of Fahrenheit: the higher the temperature, the colder the system.
>
> Jean-Pierre Cristin subsequently proposed [2] inverting the Celsius scale and called it the Centigrade scale. Nowadays, Celsius and Centigrade are, in practice, synonyms.

Converting between units is made extremely simple through use of the Google search engine (▶ https://www. google.com/). Simply typing the requested conversion into the search input box will result in the proper value, e.g., typing `3.2 miles to km` shows something like

Conversions between units are easy with Google.

Some computer operating systems also include such a feature. Note that having an easy-to-use tool for effecting conversions does not mean that you can ignore how to do them yourself. This is a general rule: throughout this book, we make use of various automatic tools. However, we always give details about how these tools are implemented. Physicists need to know how their tools work.

Even if tools exist to provide an easy answer to problems, you need to learn how to solve them without their help.

2

2.3 Systems of Units

From the example above, it is clear that defining a unit is not as straightforward as it may appear at first sight. The definition of the units must be stable and independent on as many manufacturing details as possible, however, it is always subject to improvement and, if needed, can be redefined. Units must be agreed upon by everybody working in the field and their definition has been assigned to an international body called the *Bureau International des Poids et Measures* (BIPM), based in Sèvres (France) following the first adoption of the **metric system** in 1799, after the French Revolution. The metric system was first introduced in order to define units based on natural dimensions. For example, the unit of **length**, the meter (symbol: m), was initially defined as one ten-millionth of the distance between the equator and the North Pole, measured along a great circle.

The BIPM (Bureaus International des Poids et Measures) is in charge of the precise definition of units.

Since then, researchers have tried to define fundamental units in terms of universal constants. The meter, for example, is now defined as "the length of the path travelled by light in vacuum during a time interval with duration of $1/299\,792\,458$ of a second", as stated on the BIPM website at ▶ https://www.bipm.org/.

If possible, units are defined in terms of fundamental constants.

The **International System** of units, abbreviated as SI (from its French name Système International), comprises seven fundamental units, namely, the second (s) for time, the meter (m) for length, the kilogram (kg) for mass, the ampère (A) for electrical current, the kelvin (K) for temperature, the mole (mol) for the quantity of matter and the candela (cd) for luminous intensity. It is worth noting that the symbol for the second is just s, not sec, as is often found, and that the unit for temperature is the kelvin, not the "degree kelvin".

The International System of units (SI) is the main system of units and comprises fundamental units for time, length, mass, electrical current, temperature, quantity of matter and light intensity.

Fundamental units are units arbitrarily chosen as such, whose definition does not rely on other units. We stress that the choice of what unit is fundamental is completely arbitrary and, in fact, other systems of units exist in which units that are fundamental in the SI are not so in those systems. For example, in the **natural units** system, the speed of light c is taken as a fundamental unit and its value is chosen to be $c = 1$. In such a system, lengths are measured as the time needed for light to cover the corresponding distance and are given in units of time. The light year, for example, used in astronomy, is defined as the distance travelled by light in one year. In this system of units, $1\,\text{m} = 1/299\,792\,458\,\text{s} = 3.335640952 \times 10^{-9}\,\text{s}$. Following

Besides SI, there are other systems of units such as the system of natural units, the cgs, the imperial and the US system of units, as well as the Gaussian system.

the Einstein relation $E_0 = mc^2$, masses can be measured in units of energy in this system. Since $c = 1$, $E_0 = m$. Using natural units, it is easy to remember the masses of electrons and protons that are, respectively, 0.5 MeV and 1 GeV (see Table 2.1 for prefixes). To obtain the masses in SI units, it is enough to divide these numbers by c^2 and convert eV into J (the unit of energy in SI), knowing that $1\,\text{eV} = 1.6 \times 10^{-19}\,\text{J}$.

The imperial units are popular in countries like the USA, the United Kingdom and countries formerly under the authority of the British Empire. In this system, lengths are measured in feet (ft), corresponding to 12 inches (in) and 0.3048 m.

Imperial units are popular in Anglo-Saxon countries.

Duodecimal and other numeral systems

There are a few systems of units that are based on the duodecimal system, i.e., a base-12 system, in contrast to SI and other systems, which are based on base-10 numeral system.

The most common way to express numbers involves positional notation, in which each digit in the number has a *weight* depending on its position. The base of a positional system is the number of digits defined in that system, with each digit having a value that corresponds to the value represented by the digit itself times an increasing power of the base from right to left. The decimal point in a number separates negative powers from positive powers. For example, the number 8 237.43 in a decimal system, i.e., a system in base-10, corresponds to

$$8 \times 10^3 + 2 \times 10^2 + 3 \times 10^1 + 7 \times 10^0 + 4$$
$$\times 10^{-1} + 3 \times 10^{-2}. \tag{2.14}$$

The same number can be expressed in other positional systems. For example, in computer science, hexadecimal (base-16) and binary (base-2) are popular. The latter is used because it is easy to build devices that, having two states, can represent any of the two digits (0 and 1) existing in this base, and it is then possible to represent, e.g., numbers in memories made of capacitors (charged or empty), transistors (in conduction or in a state of interdiction), switches (open or closed), etc.. Hexadecimal is used because it allows for a short notation that can be easily turned into binary, and vice versa, thanks to the fact that 16 is a power of 2.

In base-16, the number above reads as 202D.6E. In fact, the digits of the hexadecimal system are the same as those used in the decimal one, to

2

which we add the first six letters of the alphabet, such that A=10, B=11, etc. The value of the above numeral (i.e., the symbol used to write a number) is

$$2 \times 16^3 + 0 \times 16^2 + 2 \times 16^1 + 13 \times 16^0 + 6$$
$$\times 16^{-1} + 14 \times 16^{-2} \simeq 8\,432.42\,.$$

It is worth noting that a rational number in a positional system can be irrational in a different system and the number of digits after the decimal point of a number can be different in a different system. The same number in base-2 can be obtained in the same way. Converting it from base-16 is easy, as each digit in the hexadecimal system corresponds to four digits in base-2. The first digit in binary, 2, is written as $10\ (1 \times 2^1 + 0 \times 2^0)$, equivalent to 0010 (the trailing zeros are called non-significant), D= 13 = 1101, 6 = 0110 and E= 14 = 1110, such that

$$202D.6E = 0010\,0000\,0010\,1101.0110\,1110\,.$$

Historically, in the past, people used numerals expressed in different notations. A rather common base is 12, upon which many units of length are based. This system is called the duodecimal system.

We still use it to measure time, indeed. The reason being that the number of divisors of 12 (2, 3, 4, 6) is higher than that of 10 (2 and 5) and it is simpler to express numbers as fractions of the units. This is the reason why in the USA, where a system derived from the imperial one is used, most road signs report distances as fractions of a mile (a unit of length corresponding to 1.609 km).

Multiples and submultiples of a unit are expressed by prefixing the symbol of the unit.

Measurements can be expressed in terms of the units defined in the chosen system or in one of its multiples and submultiples, represented by a prefix added to the unit symbol. Table 2.1 shows the most common multiples and submultiples, the corresponding symbol and how it is pronounced. For example, a centimeter, whose symbol is cm, corresponds to 10^{-2} m, while 1 MW (megawatt) corresponds to 10^6 W and 1 m in natural units is about 3.3 ns.

Measurements can also be expressed as a combination of units. For example, speed indicates how fast you are

□ **Tab. 2.1**   Most common multiples and submultiples , along with prefixes and names

multiple	symbol	prefix name	submultiple	symbol	prefix name
10^{15}	P	peta	10^{-1}	d	deci
10^{12}	T	tera	10^{-2}	c	centi
10^{9}	G	giga	10^{-3}	m	milli
10^{6}	M	mega	10^{-6}	μ	micro
10^{3}	k	kilo	10^{-9}	n	nano
10^{2}	h	hecto	10^{-12}	p	pico
10^{1}	da	deca	10^{-15}	f	femto

traveling as it is defined as the length of the trajectory per unit time, i.e.,

$$v = \frac{\Delta x}{\Delta t} \qquad (2.15)$$

(the Δ is used when measurements are obtained as differences: $\Delta x = x_2 - x_1$ and $\Delta t = t_2 - t_1$). As a result, measuring lengths in m and times in s, the units for speed are m/s (meters per second). The unit of speed is said to be a derived unit.

Derived units are units obtained as the product of powers of base units.

Summary

When quoting the value of a physical quantity, its unit must be quoted, too.

A unit is a standard, adopted worldwide, to which all measurements can be compared. It must exhibit the same characteristic as the subject of the measurement, e.g., a unit length must have a length.

Measures in one unit can be translated into another unit through conversion. Conversion factors can be found by treating units as algebraic symbols in equations.

Despite the existence of tools that can provide answers to more or less complex problems, it is of capital importance that you learn how to solve these problems without

2

using them. Only when you master the solution can you start using them regularly.

A system of units comprises a set of units arbitrarily chosen as base or fundamental units.

The Bureau International des Poids et Measures (BIPM - ▶ https://www.bipm.org/) is the organism responsible for the definition and maintenance of the International System of units (SI).

Fundamental units in SI are the second (s) for time, the meter (m) for length, the kilogram (kg) for mass, the ampère (A) for electrical current, the kelvin (K) for temperature, the mole (mol) for the quantity of matter and the candela (cd) for luminous intensity. Visit the website for definitions.

Units that are fundamental in one system may not be so in another system. For example, lengths are measured in units of time in the natural system of units in which the speed of light $c = 1$.

Units whose definition is based on combinations of fundamental units are said to be derived. For example, speed is measured in units of length per unit of time. In SI, its units are m/s.

References

1. Celsius, Anders (1742) "Observationer om twänne beständiga grader på en thermometer" (Observations about two stable degrees on a thermometer), Proc. of the Royal Swedish Academy of Sciences, 3, pagg. 171–180
2. Fournet, M.J., "Sur l'invention du thermomètre centigrade a mercure", Not. lue a la Societé d'Agriculture de Lyon, 4 juillet 1845, pagg. 1–17

Collecting Data

Contents

3

It is now time to start taking some serious measurements. We will avoid boring you with trivial measurements that nobody is interested in. Apart from a few very basic ones, most modern apparatuses for taking measurements are based on electromagnetic phenomena and usually display the results on digital screens. Advanced experiments profit from the ability of computers to interface with external devices so as to control data acquisition, filter that data and provide intermediate results before storing the raw data on files for further analysis. As you are still not used to taking measurements, the problems with which you are presented at this point should be simple ones, however, we will mimic more complex experiments using instruments that work similarly to professional ones.

3.1 Instruments

To take measurements we need instruments.

Instruments are the tools for taking measurements. The simplest instrument is a tool exhibiting the same characteristic as the subject of the measurement with which it is directly compared. A ruler is an example of such an instrument. It has a length, and the length of anything else can be obtained by directly comparing its length with that of the ruler. In fact, rulers are graduated instruments, i.e., instruments upon which a scale is impressed such that we can directly compare the length of something with a fraction of its full length.

Instruments can be graduated.

A caliper with a Vernier scale.

Sometimes, graduated instruments are equipped with a Vernier scale: a device that facilitates the interpolation of fractions of the interval between two adjacent marks on the instrument scale. Technology made the usage of Vernier scales almost obsolete, and this device can only be found on rather old tools (still, they are widely used).

Instruments often respond to certain stimuli by producing signals whose measurement depends on the intensity of the stimulus. In that case, we measure a different quantity with respect to the one we are interested in, which, however, is proportional to the latter. The graduated instrument in Fig. 3.1 can be used to measure the mass of some flour. Since the mass is proportional to its volume, a measurement of volume V is turned into a measurement of mass m just by multiplying the volume by its density, defined as

$$\rho = \frac{m}{V}.$$

(3.1)

□ Fig. 3.1 A *pint* is an instrument for directly measuring volumes containing 0.473 l of a liquid. The jug is a graduated instrument when we read the red scale, since it reports the fractions of its volume. It works as a calibrated instrument if we read the blue scale

When the instrument measures the quantity a to provide the value of quantity b as a function of a, the instrument is said to be calibrated. The process of finding the parameters **p** needed to transform a into b is called calibration.

For example, temperatures are often measured with a device called a platinum resistance thermometer (PRT). As the name suggests, these thermometers are made of platinum, whose electrical resistance depends on the temperature. Assuming we have an instrument able to provide the measurement of the resistance in the appropriate unit (ohm, symbol Ω), the following relation between temperature and resistance holds for the very common PT100 sensor:

Instruments are calibrated when their scale reports a different unit with respect to the one we read.

$$R(T) \simeq R(0) \left(1 + AT + BT^2\right), \tag{3.2}$$

where $R(T)$ is the resistance at temperature T, given in degrees Celsius, and A and B are constants, whose values, for resistance given in ohm, are $A = 3.9083 \times 10^{-3}$ and $B = -5.775 \times 10^{-7}$. Measuring $R(T)$, one can obtain T as

$$T = \frac{1}{2B^2} \left(\sqrt{\frac{A^2 R(0) + 4B^2 \left(R(T) - R(0)\right)}{R(0)}} - A \right). \tag{3.3}$$

3

Here, the right member of the equation is $f\,(\mathbf{p},\,T)$, while $\mathbf{p} = (A,\,B,\,R(0))$.

In the following we use **smartphones** and **Arduino** boards as instruments. Smartphones are capable of taking many types of measurement, thanks to the various sensors often on board. Arduino is a programmable microprocessor to which we can attach various sensors.

> Smartphones and Arduino boards are very effective instruments to conduct physical experiments.

3.2 Smartphones

All smartphones are equipped with a microphone for making phone calls, a camera to take pictures and an accelerometer for determining their orientation. Many of them include a light sensor to measure the environmental illumination (to adjust the display luminosity) and a magnetometer (to work as a compass in navigation systems). In the latter case, they also have a GPS sensor to determine their position on Earth. A few smartphones are equipped with a gyroscope, used to play games and for augmented and virtual reality applications. Some also have a pressure sensor.

Most of these sensors can be exploited to conduct physical experiments.

> Smartphones are rich in sensors. At the very least, they have a microphone, a camera and an accelerometer.

> PHYPHOX is an award-winning App for exploiting sensors within a smartphone.

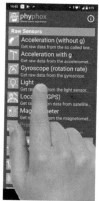

For them, we need data coming directly from the sensors and, to this aim, we can use an App called PHYPHOX (acronym for *Physics Phone eXperiment*), developed by a team of physicists from the RWTH Aachen University, lead by Sebastian Staacks. The App is available for free and can be downloaded and installed for the two major operating systems: Android and iOS. Once installed, PHYPHOX looks for sensors in your phone and configures it such that you can start taking measurements with them.

Missing sensors are shown as shaded in the list of available sensors (see picture).

Choosing among available sensors starts a dedicated application that collects data from the phone upon request. Once data are collected, they can be exported in the form of a table via various methods, including e-mailing it to an address or saving it directly to the cloud, as with Google Drive or Dropbox.

Collected data can be saved in a convenient format for further analysis.

There exist other Apps dedicated to the same task. As of this writing the only other App we feel like recommending is PHYSICS TOOLBOX. In the following, all the experiments are proposed using PHYPHOX, however, they can be easily done using the latter, too.

3.3 Arduino

The term "Arduino" is often used to identify an electronic board with a programmable microcontroller on board and a series of input and output interfaces (I/O) of various kinds (digital, analogue and serial) that allows for communication with external devices. In fact, it is much more than that. It is what computer scientists call an *ecosystem*, of which the board is just a component. The ecosystem is made up of numerous external devices such as sensors (devices sensitive to some physical or chemical phenomenon) and **actuators** (devices intended to perform actions, such as activating an engine, turning on an LED, producing sound with a speaker or displaying something on an LCD screen). A *software* platform (IDE: *Integrated Development Environment*) needed for its programming is part of the ecosystem. It can be downloaded for free from the Arduino website and exists for the most popular operating systems. Its installation is simple and guided. There is also a completely online version of the development tool that does not require installation, but does require an Internet connection. The whole community of users who share their experiences with others is part of the ecosystem, which has made Arduino a success on a planetary scale.

Arduino is an ecosystem consisting of a board with a programmable microprocessor, numerous external devices and a tool for programming.

Arduino can be programmed with a dedicated language, very similar to the C-language, via an open source platform.

Its Italian origin is revealed by the fact that most Arduino-based projects have an Italian-sounding name.

Arduino UNO (Fig. 3.2) is the most common and looks like a module with an approximate rectangular shape of about 69×53 mm^2. Pins, i.e., connectors that allow the microprocessor to interface with the outside world, are aligned in two rows on the long sides of the board. A USB

All the examples in the following are described as if we used the Arduino UNO board. However, all the Arduino flavours can be used.

3

□ **Fig. 3.2** The Arduino UNO board. The processor is the big black box on its surface. I/O ports are aligned on the top and bottom rows of connectors. On the left side, the USB connector and the external power connector can be seen

connector allows for communication with a computer. If not powered through the USB connection, Arduino can be connected to any voltage generator between 9 and 12 V.

Its form factor has been carefully designed such that it is small enough to fit in many environments, while, at the same time, not so small as to make it difficult to manipulate.

The genesis of Arduino

The name Arduino comes from the name of the bar where Massimo Banzi and David Cuartielles, the two main inventors of the project, used to hang out when they were working at the Interaction Design Institute at Ivrea.

In turn, Arduino was the name of a nobleman of Ivrea, who later became King of Italy in 1002. At that time, Italy was a constituent of the Holy Roman Empire.

Ivrea, a UNESCO World Heritage Site, is renowned for being one of the most innovative Italian districts, distinguished by attention to both performance and design. It was here Adriano Olivetti developed the first personal computer, called Programma 101.

A plethora of different boards are sold, with various characteristics in terms of computing power, memory, speed, etc. Since they all share the same programming language and the same architecture, in the following, the version to which we refer is the most common and least expensive board: Arduino UNO. Every flavour, however, is equivalent to it for the purposes of this book.

3.4   Open Source Makes It Better

Both Arduino and PHYPHOX are open source projects. In the case of Arduino, even the hardware is released under an open source license. Check out the documentation section of each product on the Arduino website to obtain the production files.

The open source movement originated in the software realm. The source code of open source software, also called *free software*, can be inspected, modified and redistributed. This practice has been extended to hardware, in which the construction plans of a device can also be inspected, modified and redistributed. A frequent misconception about open source is that products released under this license are free, in the sense that users do not have to pay to use them (whatever "to use" means). The best definition of what it means for software to be free is given on the GNU project website:

> » "Free software" means software that respects users' freedom and community. Roughly, it means that the users have the freedom to run, copy, distribute, study, change and improve the software. Thus, "free software" is a matter of liberty, not price. To understand the concept, you should think of "free" as in "free speech", not as in "free beer".

The most common open source licenses are the General Public License or GPL (▶ https://www.gnu.org/) and the Creative Commons or CC (▶ https://creativecommons.org/).

Open source is a great opportunity for learning. Copying source code is encouraged in this course, provided you understand what you are doing and you learn from it. Using the techniques developed in this book, you are going to learn much more than just how to conduct good physical experiments. Our goal is to develop your skills in computing and programming, critical thinking, design and communication.

In learning about the Arduino programming language, you are going to learn some C and C++ languages. We use Python for offline data analysis, as its popularity is already high and still increasing. However, we will not go into too many details. The idea here is that you must learn about "programming", in whatever language. You are not simply supposed to learn a programming language. Being a good programmer is important. The language in which you are proficient is much less important. Speaking is not just a matter of pronouncing words. In order to speak, you must be able to think and translate concepts into the right words in the right order, such that the meaning of

Learning how to perform measurements using Arduino and smartphones will greatly improve your skills: you will learn about computer programming, design, critical thinking and communication.

3

your thoughts is reflected in the commonly agreed upon meaning of the pronounced words. As in spoken languages, once you know how to speak, it is not so difficult to learn a different language.

3.5 Measuring Light Intensity with a Smartphone

In order to perform our first measurement, we start PHY-PHOX and click on the LIGHT icon. The display shows the GRAPH tab of the "Light" experiment.

Each experiment has an icon to start data acquisition (a triangle-shaped arrow) and one to discard data (a bin). In SI, light intensity is measured in "candela" (cd). The flux of light emitted by a source in a solid angle is measured in "lumen" (lm). The flux per unit area is called illuminance and is measured in "lux" (lx).

The row in orange is common to all experiments. It includes a triangle-shaped arrow, a bin-like icon and three dots aligned in a vertical line. The arrow is used to start data acquisition. Clicking on the bin discards all the data collected so far. A menu with different options is shown when touching the dots.

The display is divided into tabs. The default one, "GRAPH", is shown underlined. It is intended to show a graph of the data as a function of time. The graph is displayed in the rectangle below, whose title is ILLUMINANCE, as a function of time. The "SIMPLE" tab shows the same data in numerical format.

In this experiment, times are measured in seconds, while **illuminance** is given in lux (symbol lx). Illuminance is defined as the luminous flux per unit area. In turn, the luminous flux is defined as the amount of light per solid angle radiated from a source. The luminous flux is measured in lumen (lm) and 1 lm corresponds to the light emitted by a source whose intensity is 1 cd (one of the base units of the SI) under a solid angle of 1 steradian. What is important to understand, here, is that illuminance is a measure of the light effectively impinging on the sensor: the larger the sensor, the higher the illuminance for the same source. As a result, we cannot compare illuminance measured with one detector with illuminance measured with another, unless we know the area of both.

Touching the triangle-shaped arrow on the top right corner, data acquisition begins and the data is shown in the graph as a function of time.

Start the experiment touching the triangle-shaped arrow and pause it clicking on ⏸. Then export data to your preferred service.

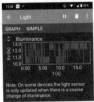

Clicking on SIMPLE shows the current reading as a number. If the light sensor is updated frequently enough, you can see the number changing (and, correspondingly, the same behaviour can be seen on the graph within the other view).

Data acquisition can be paused by touching the "pause" symbol ⏸. The data collected up to that point are still in the memory of the smartphone and can be saved as a two-column table containing time and illuminance.

Data can be saved in a variety of formats. Click on the vertically-aligned dots, such that a menu appears, and then click on "EXPORT DATA". The menu also contains few other lines. "EXPERIMENT INFO" gives access to a short text explaining how the experiment works. "SHARE SCREENSHOT" is self-explanatory, while "TIMED RUN" allows the user to program the start and duration of the data acquisition phase. When "ALLOW REMOTE ACCESS" is activated, the display can be replicated on a web page, opening up the address specified at the bottom of the display in the user's favorite web browser when the experiment begins. Note that the specific configuration of your network may make this option ineffective. The current configuration of the experiment can be saved for later use by using "SAVE EXPERIMENT STATE".

A few functions are available under the experiment menu, accessed by clicking on the three vertically aligned dots.

| Experiment info |
| Export Data |
| Share screenshot |
| Timed run ☐ |
| Allow remote access ☐ |
| Save experiment state |

3

Data can be exported in a variety of formats, such as tables.

Upon clicking "EXPORT DATA", a choice by the user is required concerning the format of the output. Data can be exported as an Excel file or as a CSV (comma separated value) text file. In the latter case, the separator between fields can be a comma, a tabulator or a semicolon. The user can also choose whether he/she wants the separator between the integer and the fractional part of numbers to be a point or a comma. Once the choice is made, the user has the opportunity to choose where he/she wish to be exported. Common options are sending them to an e-mail address or saving them to the cloud, e.g., Google Drive or Dropbox. The first line always contains a legend describing the data.

Time (s)	Illuminance (lx)			
0	12			
0.10876	13			
0.218672	12			
0.329632	15			
0.438514	14			
0.548012	15			
0.659466	14			
0.768024	15			
0.878746	14			

Usually the sampling rate of a smartphone is relatively high.

Saved data appear as shown in the figure. A quick look at the data gives a rough idea of the sampling rate of the smartphone, which, in the example, was about 10 Hz (the measurements follow each other at intervals of about 0.1 s). Though Excel is very convenient in many cases, we are going to use the CSV format, with a comma as the separator between fields and a decimal point as the separator between the integer and the fractional part of numbers. This way, we can play with Python to analyse data and learn about the latter.

Analog sensors have three leads: two are needed to connect to power supply, while the third is the output, converted into a voltage between 0 V and V_0, where V_0 is the voltage at which the sensor is powered.

3.6 Measuring Light Intensity with Arduino

Analog sensors are the simplest sensors to be used in conjunction with Arduino. Analog sensors are devices that produce a voltage proportional to the quantity to which they are sensitive. They must be powered by connecting them to a power source that provides a voltage usually between 3.3 V and 5 V. As a response, they provide an electrical signal between 0 and V_0, where V_0 is the voltage with which they are supplied.

All analog sensors have three leads: two for the power supply and one for the outcome. The power supply connectors are usually marked as GND (short for *ground*, a lead whose voltage is taken to be 0 V) and VCC (Voltage Continuous Current).

fritzing

☐ **Fig. 3.3** Connecting an analog sensor requires three wires: two for the power supply and one for the output signal

Among the Arduino pins, two serve to provide power to external sensors. They are marked 3.3v and 5v, respectively.

Sensors are connected to Arduino by means of dedicated cables, called jumpers. Jumpers end with Dupont connectors that can be either male or female.

A set of six Arduino pins, marked from A0 to A5, are internally connected to an ADC: an Analog to Digital Converter. This device digitises voltages up to 5 V and converts them into integer numbers that can be read by the processor onboard. The output of an analog sensor, usually marked as SIG or OUT, must be connected to one of these ports. For this project, any analog light sensor is fine.

Figure 3.3 shows a schematic of the connections needed. Schemas like this can be easily made using FRITZING, an open source software tool available at ► https://fritzing. org. The project is open source, so you are free to download the source code and *compile* it yourself, i.e., generate the executable application. You can also download the executable application from the website, ready to be used. In this case, you are asked to pay a small fee.

The GND and VCC leads of the sensor must be connected using jumpers to, respectively, the pins marked GND and 5v or 3.3v on Arduino, depending on the sensor.

Schematics can be produced using Fritzing, an open source tool.

3

Always use red jumpers for VCC and black ones for GND. This way, debugging your project will be easier.

Reduce the chance of damaging Arduino and other components by not connecting it to the computer until the last step.

When connecting a sensor to Arduino, the color of the wires is obviously irrelevant. However, following conventions helps to maintain a consistency among projects, as well as aiding in documenting and debugging them. We suggest getting into the habit of always using red and black wires for power, reserving the black color for the GND pin. The signal output can be connected to Arduino with a wire of a different color.

Once the system is ready to take measurements, connect Arduino to a computer via a USB cable and open its programming tool (IDE). Write the following program (called "sketch" in Arduino jargon) in the editor window.

```
void setup() {
    Serial.begin(9600);
}

void loop() {
    Serial.println(analogRead(A0));
}
```

Any program must be *compiled*, i.e., translated into machine language, before deploying it to the Arduino memory. The transfer happens via the USB cable.

Serial.begin() configures the serial communication channel between Arduino and the computer. Data can be inspected using the serial monitor.

After saving the program, *compile* it by clicking on the checkmark icon on the top left side of the window. This translates the program, written in the Arduino programming language, into machine code, consisting of binary coded, low level instructions for the processor.

Machine code must be transferred to the Arduino memory. This is done via the USB cable upon clicking on the small arrow icon, close to the one used to compile the program. The LED's on Arduino will start blinking and, as soon as the code has been loaded into the Arduino memory, it will begin to execute it.

The program configures the communication channel via the `Serial.begin(9600)` statement, then repeatedly executes `analogRead(A0)`, which is needed to obtain the voltage on the A0 port, and transfers its value to the computer by `Serial.println()`.

A tool called a "serial monitor" can be activated by clicking on the magnifying glass icon in the top right corner of the IDE or via its menu, so as to quickly inspect the output, consisting of an infinite series of integer numbers scrolling in the window.

In order to collect data onto a file on the computer disk, we can use the following Python code:

```
import serial
import time

port = '/dev/cu.usbmodem14101'
usb = serial.Serial(port)
f = open('illuminance.csv', 'w')
while True:
    arduino = usb.readline().rstrip()
    print(arduino.decode())
    f.write('{}, {}'.format(time.
        time(), arduino.decode()))
```

When running the Python code, the serial monitor must be off. In short, this program opens the communication channel with Arduino using serial.Serial(), then writes the data onto a file called illuminance.csv. The data are read using usb.readline().rstrip().

The USB port is identified by the string /dev/cu.usbmodem14101, typical on the macOS X operating system. On Linux, USB ports are usually called /dev/ttyUSB0, /dev/ttyUSB1, or similar, while on Microsoft Windows, they are called COM1, COM2, etc.

After collecting data for 20–30 s, one can stop the program by simultaneously pressing the Ctrl and C keys.

Python is an interpreted language: Python applications do not need to be compiled before execution. Indeed, each statement is translated into machine language at run time.

3.7 Understanding Arduino Programming

Any Arduino sketch must contain at least two "functions": blocks of code enclosed in braces. Their names are setup() and loop(). The user is responsible for defining them. This is done by writing the "type" in front of the name, and providing the list of actions to be taken by the function enclosed in parenthesis.

As in C or C++, there are no strict rules about how to write statements in the Arduino language. Nevertheless, it is a good habit to be considerate when writing code so as to make its maintenance easier. Strictly following conventions will help in getting you recognised as a good programmer. Usually, no more than one statement is present on each line of code, with very few exceptions. A newline is started after an open curly bracket, while closed ones

Arduino sketches contain a setup() function executed once at the beginning of the run, and a function loop() executed repeatedly without cessation.

Always adhere to a consistent formatting convention, never deviating from it.

3

are placed on a line by themselves. Use consistent indentation rules: add a few spaces on the left each time you open a curly bracket and align the closed one to the beginning of the line in which it opens. Many text editors automatically indent lines for you, but the programmer can always override the editor's behaviour. If lines are very long, it is customary to break them into individual lines.

When a new program is loaded into the Arduino memory, the execution of the program starts a single execution of the statements listed in `setup()`, after which `loop()` is executed repeatedly: it restarts upon exiting. The sketch begins again when the RESET button on Arduino is pressed or when it is powered up.

In serial communications, bits are sent over a line one after the other N times per second (baud).

`Serial.begin(9600)` in `setup()` is needed to configure the communication channel between Arduino and the computer to which it is connected via the USB cable. The number 9600 in parentheses represents the communication speed measured in "baud": a unit for transmission speed. Data travels over the USB cable as a series of bits, 0 and 1. Each bit is represented by a voltage level and is transmitted over the line in series, i.e., one bit after the other. The speed of serial communication is given by the maximum number of voltage changes over the line per unit time. 9600 baud is, in practice, equivalent to 9600 bits per second and is a relatively low speed. Modern computers can handle speeds up to more than 115200 baud, which is the current maximum speed tolerable by Arduino.

The Arduino ADC has 10 bits and transforms any voltage between 0 V and 5 V into an integer between 0 and $2^{10} - 1 = 1\,023$. Digitised data can be retrieved with `analogRead()`.

Data can be collected from the ADC's with `analogRead(A0)`, where A0 is the analog port identifier to be read. Executing `analogRead()` **returns** an integer number between 0 and 1 023. Such a number n is proportional to the voltage V_{pin} present on the given pin (A0), i.e.,

$$n = \alpha V_{pin}, \tag{3.4}$$

where $\alpha = \frac{1\,023}{5\,V}$. You can imagine some statements as a sort of mathematical function $y = f(x)$. In this case, `analogRead(A0)` represents $f(x)$, and the data returned is y. Putting this statement inside the parentheses of `Serial.println()` causes its returned value to be transferred over the serial line to the computer; a newline character is added at the end of transmission.

When a computer is connected to the other end of the USB cable, it can extract characters from it, one after the other. They can be shown on the display, as happens with the serial monitor. The newline character causes the computer to advance one line, such that the next data appears on the line below the previous one. Using

```
Serial.print(analogRead(A0));
```

without the `ln` after `print`, the data are shown on the same line and it is impossible to separate the different readings.

Despite the name, the `println()` method of the class `Serial` only sends bytes over the serial line. To actually print data, a computer should be used.

3.8   Python Data Collection

The `import` keyword requires Python to load the corresponding module: a collection of software intended to simplify certain operations (also known as a library). The `serial` module is needed to communicate with serial ports (the USB port is a serial port; in fact, USB stands for Universal Serial Bus). The `time` module helps in manipulating times.

Python's core functions can be extended by importing external modules.

In contrast to C or C++, Python requires statements to be written one per line, even if a single statement can be broken up and distributed on more than one line so as to improve readability. Indenting is not an option: it is mandatory under certain conditions, as seen below for the `while` statement.

To establish communication with the port, the statement `serial.Serial()` is used. The statement requires an argument representing the port identifier. The identifier differs from computer to computer and from port to port, however, it is usually in the form `/dev/cu.usbmode mxxx` on macOS X, `/dev/ttyUSBxxx` on Linux and `COMxxx` on Microsoft Windows. Here, xxx is a number. Strings are written in single ' or double " quotes. `serial.Serial()` returns an object to which we give the name usb via the = operator.

The `serial` module allows for connections to the USB ports.

The object `f` represents a file. It is returned by the `open()` function, which requires two string arguments: the name of the file to be opened and its mode of operation. A file can be opened in write (`'w'`), read (`'r'`) or append (`'a'`) mode. When a file is opened in write mode, if it does not previously exist, it is created, while if it does already exist, the next writing operation will overwrite its content. In read mode, data can only be read from the file,

Files must be opened prior to being used. We specify how we are going to use them.

3

Iterations are realised using the `while` statement. It repeats the following indented lines as long as the associated logical expression is true.

In OOP, a program is made of objects belonging to classes, interacting via their methods. Methods allow for accessing and altering the state of an object. The state of any object of the `serial` class, for example, comprises data in the channel, which can be retrieved by the `readline()` method.

while in append mode, if the file previously exists, new data will be appended to the end of that which is already there.

`while` is used to build an iterating structure. With it, we can repeat the same set of instructions as long as a condition, placed after `while`, is true. In the example, the condition is, by definition, always true (`True`). The colon marks the start of the instructions to be repeated, which must be written with indentation, i.e., with leading whitespaces corresponding to a tabulator sign (`tab`). All the following indented lines contain statements that will be repeated without cessation.

The first operation done in the while loop is extracting data from the serial line, interpreting these data as characters that build a string and assigning that string to a variable whose name is `arduino`.

Python is an "Object Oriented Programming" (OOP) language. In OOP, "objects" belonging to a "class" have the ability to perform actions by means of "methods". Objects of the same class share the same methods, and thus exhibit the same behaviour. In our application, `usb` is an object of the class `serial`. The method `readline()` of this class returns a sequence of bytes, i.e., a series of 8 bits, extracted from a serial line. You can imagine the USB port as a stockroom where pieces of data enter from outside one by one, maintaining their order. They can then be extracted from the stockroom by a processor that works according to what computer scientists call a FIFO (first input, first output). With `usb.readline()`, we extract all the characters up to the EOL (end-of-line) character, currently represented in the FIFO by the `usb` object, in the order in which they entered. This operation returns an ordered set of bytes called an array. An array can be regarded as a list whose elements all belong to the same type. In the following, we often refer to homogeneous lists as arrays, while the term list is used for more general collections. The last byte, interpreted as an 8-bit integer, corresponds to 10 (0×0A in hexadecimal) and corresponds to a newline character. If printed, this character causes the current line to scroll up, such that the next characters will be printed on the next line. `rstrip()` is a method of the `String` class. It returns the object to which it is applied, except for the last character. `arduino`, this way, is a `String` object containing all the characters read from the USB port, except for the last newline one.

With print(), we print the string to the display associated with the running application. The method decode() forces the system to interpret the list of bytes as a character string.

Finally, we write both the current time and the content of the arduino string to the file, previously opened, represented by f. The write() method of the file object takes, as its argument, the string to be written. Here, the string is represented by '{}, {}'. Applying the method format() to the string object, each pair of braces is substituted with the actual values of the arguments of format(). In the example, time.time() returns the current time expressed as the number of seconds elapsed since the "epoch". The latter corresponds, by convention, to January 1, 1970 at 00:00:00 UTC (Universal Time Coordinated). The second argument of the format() method is the string content, read from the USB channel. As a result, time is substituted for the first pair of braces, while the number provided by Arduino as a result of the measurement of the light intensity is substituted for the second pair. In the end, a pair of numbers, separated by a comma, is written on the output file, as follows:

> Data of any type can be printed using print(). The same data can be interpreted in a variety of ways. For example, an array of bytes can be interpreted as a character string.

> The epoch is, by convention, the start of the computer age and corresponds to January 1, 1970, at 00:00:00 UTC.

```
1581615606.776129, 36
1581615606.7800908, 37
1581615606.784203, 37
1581615606.788279, 36
1581615606.7923782, 36
1581615606.79651, 36
1581615606.800616, 37
```

The first number in the first line corresponds to the time of the measurement. It happened about 1 581 615 607 seconds after the epoch. Using any epoch converter online, you can easily see that such a measurement was carried out on February, 16, 2020, at 10:35:13 UTC (or GMT, standing for Greenwich Mean Time, which coincides with UTC). The number after the comma represents the measured intensity of the light in arbitrary units, i.e., as an integer proportional to the actual value.

Since there are no other lines in the program, those after the while statement are repeated indefinitely, until the program is forced to stop.

3

Summary

Graduated instruments have a scale upon which marks are engraved, allowing for a direct comparison of a quantity with the instrument itself. If the quantity to be measured is not homogeneous with the instrument characteristics giving the response, the instrument is said to be calibrated.

Automatic data acquisition allows us to take several measurements in short amounts of time. Smartphones and Arduino boards are useful tools.

Both PHYPHOX and Arduino are open source projects.

phyphox

Smartphones' sensors can be exploited using PHYPHOX: a free App available for both Android and iOS operating systems that gives access to all the sensors found in a smartphone. All smartphones have at least a microphone, a camera and an accelerometer. Many have a GPS sensor, a gyroscope, a magnetometer and/or a pressure sensor.

Data from PHYPHOX can be exported in various formats, including Excel tables.

Light intensity can be directly measured with a smartphone, thanks to the light sensor. The values are given in lux, a measurement of the light intensity per unit area. PHYPHOX performs as many measurements as possible and reports them as a function of time.

Arduino

Arduino is a versatile programmable board with a microprocessor connected to several I/O ports. It can be connected to sensors and actuators and communicates with a computer via a USB cable.

Light sensors exist for Arduino as analog sensors. Once powered, analog sensors provide a voltage on the signal lead proportional to the quantity to which they are sensitive. The voltage can be digitised, thanks to Arduino's analog pins, and transferred to a computer to store its value in a file for further analysis. In this case, light intensity is given in arbitrary units.

When writing code, always adhere to a consistent formatting convention and never deviate from it, even if the programming language allows it.

An Arduino program consists in executing the statements in the `setup()` function once, then repeating those listed in `loop()` indefinitely. A program is started each time Arduino is switched on, is loaded with a new program, is reset or when tools like the serial monitor and the serial plotter are started.

The `analogRead()` statement returns an integer number n between 0 and 1 023 proportional to the voltage V present on the pin given as an argument, such that $V = n\frac{5}{1023}$.

Data can be sent over a serial line using `Serial.println()`, after it has been properly configured with `Serial.begin()`.

Python

Data extracted from the USB channel can be inspected using the serial monitor and collected using a Python script. The latter opens the communication channel, then starts reading characters in a FIFO (First Input First Output) structure prior to storing them in a file that has been previously opened.

Python is an Object Oriented Programming language. A program is made of objects belonging to classes that interact with each other. Methods are used to access or alter an objects' state. The latter is a collection of data characterising the object. Methods are similar to functions in procedural programming and apply to objects whose name is specified before a dot, as in `serial.readline()`. Methods can be concatenated like `serial.readline().rstrip()`. In this case, the `rstrip()` method is applied to the object returned by the `readline()` method applied to the `serial` object.

On computers, time is measured in units of seconds elapsed since the epoch, corresponding to the conventional date of January, 1, 1970, at 00:00:00 UTC.

Uncertainties

Contents

© The Author(s), under exclusive license to Springer Nature
Switzerland AG 2021
G. Organtini, *Physics Experiments with Arduino and
Smartphones*, Undergraduate Texts in Physics,
https://doi.org/10.1007/978-3-030-65140-4_4

4

In this chapter, we conduct an initial analysis of data collected using both PHYPHOX and Arduino. We see that, despite the subject of the experiment being the same, so that we expect the same values for all of the physical quantities we can collect, the numbers we get from the instruments are both different and differently distributed. We learn that each experiment results in numbers with an associated uncertainty, which depends on the experiment's setup.

4.1 Data Analysis

Collected data are observed to fluctuate during the measurement. Fluctuations can be ascribed to tiny random effects that influence the result of the measurement.

The first characteristic we notice is that collected data are not constant, as we may naively expect. They may not be constant for a variety of reasons. Indeed, the light we are measuring can be intrinsically variable. However, the process of measuring it involves a set of processes, each of which is subject to influence from many unwanted effects, which may lead to fluctuations. Even if we believe that the light is, in fact, constant, it must travel to the sensor, and in this travel, it may partly scatter with dust. Even microscopic variations in the pressure, temperature or speed of the air in front of the sensor may change the way in which it reacts to light. Also, the voltage used to feed the sensor may not be entirely stable, causing fluctuations in its response.

Despite all our efforts, we can never get rid of fluctuations. If they are not observed, it is only because they are smaller than the instrument's resolution: the smallest unit we can appreciate.

We can try to reduce the degree to which these sources of fluctuation affect the proceedings, however, it turns out that it is impossible to remove them completely.

In any case, the readout of any instrument will always be a number with a limited number of digits, so, in the end, the precision of our reading will be limited. Thus, at the very least, measurements are affected by a reading uncertainty that corresponds to the smallest division of the instrument scale. Even if you are tempted to interpolate between two adjacent marks on a scale, you are not usually authorised to do so, since marks are imprinted by the manufacturer according to the instrument's resolution, corresponding to the smallest unit the instrument can appreciate.

According to quantum mechanics, there is an intrinsic limit to the resolution we can achieve.

Even in principle, a measurement cannot be taken with infinite precision. In classical mechanics, physical quantities are considered as continuous and real. In fact, this is a prejudice. We believe that length, in fact, is a continuous function and that, in principle, it is always possible to divide something with a given length into two parts indefinitely, such that it makes sense to write

$$\ell = \lim_{n \to \infty} \frac{L}{2^n}, \qquad (4.1)$$

where L is a distance. Though we assume that, we must know that such an assumption is valid only approximately, in the sense that the smallest length we could imagine measuring is so small that, in practice, it can be considered as null, but, in fact, it is not. However, that can only be true in classical mechanics. Quantum mechanics taught us that lengths cannot be as small as we wish. The microscopic world (and not only the microscopic world) is discrete. According to quantum mechanics, lengths can only be measured with an **uncertainty**

$$\Delta x \geq \frac{\hbar}{2\Delta p}, \qquad (4.2)$$

where $\hbar = \frac{h}{2\pi}$ is a constant. $h \simeq 6.63 \times 10^{-34}\, \text{m}^2\text{kg}\,\text{s}^{-1}$ is called the Planck constant, after Max Planck (1858–1947), and Δp is the uncertainty with which we can measure momentum.

Uncertainty, then, is not simply accidental and the result of our limited technology. It is an essential ingredient of Nature, though, in most cases, we can think to physical quantities as we would in classical mechanics: as continuous functions of real variables. In what follows, we make such an assumption, for simplicity. However, we should never forget that every result is only a low energy approximation of the reality.

> In this book, we assume that quantities are continuous functions of real variables, even if this may not be rigorously true.

In summary, experimental data are always affected by an uncertainty. As a consequence, we need to find a way to express the result of a measurement that is not a useless, sometimes very long list of numbers.

It can be useful to look at how data distribute, making a "histogram". In a histogram, we plot the number of events as a function of the result of the measurement. For example, in the Arduino data, we have 2, 2 099, 9 609, 859 and 5 occurrences, respectively, of 34, 35, 36, 37 and 38. Its graphical representation is given on the left side of Fig. 4.1. The right plot shows the histogram of data taken with the smartphone.

> A histogram is a plot of the number of events occurring for each result of the measurement or group of them.

4

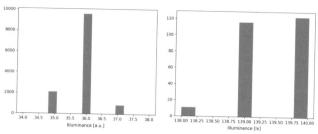

□ **Fig. 4.1** Histograms of data collected using Arduino (left) and PHYPHOX (right)

Instead of plotting the absolute number of events, in a histogram, we can show the frequencies, defined as

$$f_i = \frac{n_i}{N} \, , \tag{4.3}$$

where n_i is the number of events belonging to class (or bin) i and N is the total number of events $N = \sum_i n_i$. From the latter equation, it follows that

$$\sum f_i = \frac{1}{N} \sum_i n_i = 1 \, . \tag{4.4}$$

Data distribute around a few values. A good summary of the data is an index of position called the mean. The mean is computed as the arithmetic average of data.

Clearly, data tend to concentrate in one place that, for Arduino, is close to 36. An estimate of the position at which data are thickened is given by the mean, defined as

$$\langle x \rangle = \frac{1}{N} \sum_{i=1}^{N} x_i \, , \tag{4.5}$$

where x_i are the single measurements and N their number. The mean is often called the average. The two terms are used as synonyms, even if they are not rigorously such.

It does not make sense to express the mean with a long string of digits. We truncate it according to a rounded value of the uncertainty.

For our data, the mean is 35.90186098298076.... Does it make sense to specify all these digits? Indeed, what we learn from our first experiment is that the value of the illuminance is known with some uncertainty that, in this case, is on the order of ±1, since data fluctuate mostly between 35 and 37. A better way to express our knowledge is to write the illuminance $\mathcal{L}$ as

$$\mathcal{L}_A = 36 \pm 1 \text{ a.u.} \tag{4.6}$$

(the subscript A stands for Arduino), Looking at PHY-PHOX data, it seems more difficult to identify the *center* and the width of the distribution. Using the same definition as above, we find

$$\mathcal{L}_p = 139 \text{ lx}, \tag{4.7}$$

which, in fact, can be taken as an index of position of the distribution, even if it is unbalanced towards the right. It is possible to define other indices like the median, defined as the value such that exactly half of the data is on its left, with just as many on its right. The mode is the most frequent value in a distribution. In the following, we take the mean as the index of position of our data.

> Besides the position of the data, we need to quote their dispersion.

What about dispersion? Can we take it as ± 1, too? Well, the PHYPHOX distribution seems a bit *wider* than that of Arduino, not only because of the scale factor (35 vs. 139), but also because the average position appears more uncertain in the first case. We should find a better way to express the width of the distribution or, better yet, its index of dispersion.

A possible, natural definition is that the index of dispersion could be the average distance between individual measurements and their average, i.e.,

$$\Delta = \frac{1}{N} \sum_{i=1}^{N} |x_i - \langle x \rangle|. \tag{4.8}$$

Despite this definition seeming very reasonable, it is not as good as it may appear, because of the absolute value operator $|\dots|$. As a matter of fact, it is very impractical, because it forces us to evaluate the sign for each single term in the sum. However, the distance squared is always positive and we can define

> When the absolute value of any quantity z is needed in a formula, better to compute the square of z and get its square root.

$$\sigma^2 = \frac{1}{N} \sum_{i=1}^{N} (x_i - \langle x \rangle)^2 \tag{4.9}$$

4

as the square of the index of dispersion. We call this number the variance of the distribution. Still, this definition has a problem. If we take just one measurement, the variance is zero. This leads to the absurd situation in which it appears to be better to take just one measurement in order to obtain an infinitely precise measurement. To force us to take at least two measurements, we define the variance as

Equation (4.9) defines the variance of the population.

$$\sigma^2 = \frac{1}{N-1} \sum_{i=1}^{N} (x_i - \langle x \rangle)^2, \tag{4.10}$$

A modified variance with $N \to N-1$ is a better estimation of the dispersion. This is called the variance of the sample. We take its square root as the uncertainty and we call it the standard deviation.

such that, for $N = 1$, $\sigma^2 = \infty$. On the other hand, the distance $\sigma = \sqrt{\sigma^2}$ is still a measurement of the width of the distribution. It is just a bit wider than the first one, however, the difference tends to zero as N tends to infinity. σ is called the standard deviation. Computing it for our data, and taking it as the uncertainty, results in

$$\mathcal{L}_A = 35.9 \pm 0.5 \,\text{a.u.}$$
$$\mathcal{L}_p = 139.4 \pm 0.6 \,\text{lx}. \tag{4.11}$$

Stating that, e.g., $\mathcal{L}_A = 35.9 \pm 0.5$ a.u., means that, according to our measurements, the *true* illuminance value is close to 35.9; however, we do not know it exactly, but rather with an uncertainty of 0.5, i.e., it is comprised between $35.9 - 0.5 = 35.4$ and $35.9 + 0.5 = 36.4$. In turn, this does not mean that all the measurements lie within the interval [35.4, 36.4]. As a matter of fact, data also appear outside this interval.

Using the standard deviation as a measure of the uncertainty gives a probabilistic interpretation of it.

However, if we take another set of measurements, we should find that about 70% of them lie within the given interval in both cases. Statistics tell us that this fraction is expected to be close to 68%, when $N \to \infty$. Correspondingly, 95% of data lie within $\pm 2\sigma$ from the mean and 99.7% within $\pm 3\sigma$.

If we take the standard deviation as a measure of the dispersion of the data, we give it a statistical meaning. Manifestly, the more data, the better the estimation of the uncertainty. When quoting a result as above, we intend that repeating the measurements will lead to a result within $\pm\sigma$ around the mean with a probability of 68%. We are implying that fluctuations observed in the measurements are of a statistical nature, and must thus be understood using statistics.

Note that, when quoting the results, we truncate the standard deviation to one significant figure and quote the mean using a number of digits that matches those of the uncertainty. It makes no sense to quote an uncertainty of 0.4773289079373387.... if we cannot know a number precisely enough to distinguish it between 35.9 and 36.3, because of the uncertainty of the order of 0.4. Furthermore, we cannot distinguish them if they differ on the second significant figure. We thus keep the first two digits and approximate them to the closest value written with one digit. The closest number to 0.477... is 0.5. In general, we approximate the first non-null digit to the next integer if the following digit is between 5 and 9, while we keep it as such if the following digit is between 0 and 4.

> Numbers must be written with the proper number of significant figures. We keep just one significant digit for the uncertainty and write the mean accordingly.

It is worth noting that, if $\sigma = 0$, it does not mean that the value of the physical quantity is known with infinite precision. It only means that the reading uncertainty is larger than statistical fluctuations. Thus, the uncertainty must be quoted after it. When quoting the reading uncertainty, however, we provide different information with respect to that provided with the standard deviation. The interval defined by the standard deviation has an amplitude of about $\frac{2}{3}$ of the whole interval in which all the measurements fall. If we want to quote an uncertainty having more or less the same meaning when the reading uncertainty dominates, we need to divide the width of the resolution by three. For example, if we measure the width of an A4 sheet of paper to be 210 mm, we, in fact, intend that, in repeating the measurement, we will for sure find a number between 209.5 mm and 210.5 mm. Dividing the width of the interval (1 mm) by three, we can quote the result as

> A null standard deviation means that the resolution of the instrument is poor and the reading uncertainty dominates. In this case, we can quote the resolution divided by three as the uncertainty, giving an interpretation to this number similar to that of σ.

$$w = 210.0 \pm 0.3 \, \text{mm} . \tag{4.12}$$

This way, about 70% of the measurements fall within $\pm 1 - \sigma$ interval, even if their distribution is completely different. In fact, we are forced to believe that the probability of finding any value between 209.5 mm and 210.5 mm is the same and the distribution is said to be uniform.

4

4.2 Data Analysis with Python

pandas is a module for data analysis, numpy specialises in scientific computations and matplotlib is needed to make plots.

To conduct the above analysis, we can use Python. There are a variety of useful modules for data analysis, namely, pandas, a data structure and data analysis tool, numpy, a package for scientific computing, and matplotlib for plotting.

The following script has been used for the analysis above.

```
import sys
import pandas as pd
import numpy as np
import matplotlib.pyplot as plt

if len(sys.argv) <= 1:
    print('Usage: analyse.py [filename]')
else:
    filename = sys.argv[1]
    unit = 'a.u.'
    if len(sys.argv) > 2:
        unit = sys.argv[2]
    f = pd.read_csv(filename)
    data = f.T.values.tolist()
    h, bins, rect = plt.hist(x=data[1], bins='auto')
    print('============ histogram content ')
    print(h)
    print('===============================')
    plt.xlabel('Illuminance [{}]'.format(unit))
    mean = np.mean(data[1])
    median = np.median(data[1])
    stdev = np.std(data[1])
    print('Average = ' + str(mean))
    print('Average = ' + str(median))
    print('StDev   = ' + str(stdev))
    plt.show()
```

Modules can be given a nickname or an alias.

When importing modules we can give them a nickname (an alias) using import...as, as in

```
import pandas as pd
```

Aliases are useful for shortening names. We can even import only part of a module, called a submodule, using a syntax like

```
import matplotlib.pyplot as plt
```

where we ask the script to load the submodule pyplot of module matplotlib, identified as plt in the rest of the script.

sys gives access to system-wide functions and variables.

Module sys provides access to system-specific variables and functions. We use it at the beginning of the script to check whether the script is called with arguments. An array of strings called sys.argv is defined in the module, containing the name of the actual script in the first component sys.argv[0], and any other parameter on the subsequent ones.

When writing sys.argv, we mean the attribute argv of an object of class sys. In some cases, there is no need to explicitly instantiate an object of a class, as in unit = sys.argv[2], where unit is an object of the class String. The name of the class refers to a default object created *ad hoc*.

Running the script without any parameter on the command line results in a sys.argv array with just one component. We check the length of the array with the function len(). If it is less than or equal to one, the script prints an error message and exits.

The if statement works as follows: it checks the value of a Boolean expression; if it turns to be true, it executes the subsequent indented lines, otherwise, it optionally executes a different set of lines, if the else clause is present.

Boolean expressions (named after George Boole 1815–1864) are logical statements whose value can be either true or false, such as "it rains". Statements to which we cannot assign a truth value are not Boolean expression (e.g., "Hey teacher, leave those kids alone!").

The lines to be executed when the expression is true are written indented after a colon. Those to be executed alternatively, if any, are written, still indented, after the else: clause, which can be omitted. In the example we are analysing, if there is at least one parameter after the script name, the length of sys.argv is ≥ 2 and the string filename is given the name of the file, contained in sys.argv[1]. For example, if the script, whose name is analyse.py, is referred to as

```
analyse.py lightIntensity.csv
```

where the parameter lightIntensity.csv is the name of the file containing the data and the length of sys.argv is equal to 2, sys.argv[0] contains analyse.py, while sys.argv[1] contains lightIntensity.csv.

Once opened in read mode, data can be read using the native Python read() function. Here, we show a different method for exploiting the power of the pandas (alias pd) module. A statement like

```
f = pd.read_csv(filename)
```

opens and reads the content of a CSV file whose name is filename and puts its entire content in a data structure,

Sidebar notes:

The components of sys.argv contain all the parameters in the command line. It contains at least one component containing the script name.

The if statement allows for making choices. Statements to be executed when the Boolean expression is true are written indented, as are to be executed if the expression is false, after the else clause.

Modules make life easier. With pandas, we can open and read a whole CSV file in one line of code.

4

pandas organises data in Data Frames, which can be imagined as a matrix, whose transpose is accessed using the attribute T of the class DataFrame.

returned by the module and assigned to the object f. All in just one line of code!

Data are arranged in a structure called DataFrame, from which we want to extract the second column as a list. DataFrame.T is the transposed data frame, i.e., assuming the filename content is

```
"Time (s)","Illuminance (lx)"
0.00E0,1.39E2
1.10E-1,1.40E2
2.21E-1,1.38E2
3.33E-1,1.39E2
4.43E-1,1.38E2
...
```

the resulting data frame can be imagined as a matrix with two rows like

```
"Time (s)"  0.00E0 1.10E-1 2.21E-1 3.33E-1...
"Illuminance (lx)" 1.39E2 1.40E2 1.38E2 1.39E2...
```

In a computer's notation, numbers expressed in scientific notation $m \times 10^n$ are written as mEn.

It is worth noting that numbers are written in scientific notation as mEn, meaning $m \times 10^n$, such that 2.21E-1 corresponds to $2.21 \times 10^{-1} = 0.221$ and 1.40E2 to $1.40 \times 10^2 = 140$.

In fact, the data frame contains more information than just numbers. We can access the values with values and transform them into a **list** with tolist(). This way, data is a list of two lists: the first, whose index is zero, contains times, while the second, with index 1, contains the illuminance. Note that f.T still contains the information about the caption "Time (s)" and "Illuminance (lx)", while f.T.values contains just the numerical values.

Histograms can be done by passing individual measurements as an array to the method hist() of matplotlib. pyplot. They are graphically represented as in Fig. 4.1 with show().

The hist() method of matplotlib.pyplot (alias plt) takes the data passed as x and computes the number of events per bin. Each bin is a class of data that, in the example, are computed automatically by the method, based on the data found. It returns an array containing the number of events per bin, the edges of each corresponding bin and a data structure (rect) representing the graphical counterpart of the histogram, usually represented as a series of rectangles.

We assign a label to the horizontal axis of the histogram using the method xlabel(), to which we pass a string formatted such that it contains the proper unit. The latter is passed as the second parameter of the script. It defaults to a.u., but if the script is called with a second parameter, the latter is used to assign the unit object, from which we build the label.

The mean, the median and the standard deviations are computed on the second row of the transposed data frame exploiting `mean()`, `median()` and `stdev()` methods defined in `numpy` (for which we chose the alias np). With `plt.show()`, we ask Python to show a graphical representation of the histogram, which appears as in Fig. 4.1.

numpy allows us to easily compute the mean, the median and the standard deviation.

Summary

Data collected by an instrument fluctuate because of random effects. The result is that a measurement consists of a distribution rather than a number, unless the resolution of the instrument is poor.

A histogram is a plot showing how experimental data distribute.

Statistics

The value of the measurement is taken to be the mean of the data, defined as

$$\langle x \rangle = \frac{1}{N} \sum_{i=1}^{N} x_i \,.$$

The width of the distribution is taken as its standard deviation, defined as

$$\sigma = \sqrt{\frac{1}{N-1} \sum_{i=1}^{N} (x_i - \langle x \rangle)^2} \,.$$

Its square σ^2 is called the variance. Measurements are quoted as $\langle x \rangle \pm \sigma$, followed by their units, as in

$$\mathcal{L} = 35.9 \pm 0.5 \, \text{a.u.}$$

When the reading uncertainty δ dominates and $\sigma = 0$, the uncertainty is taken as $\simeq \frac{\delta}{3}$.

There is a 68% probability of finding a measurement within an interval of $\pm 1-\sigma$ around the mean. The percentage is 95% within $\pm 2-\sigma$ and 99.7% within $\pm 3-\sigma$.

We quote uncertainties with, at most, one significant digit and write the mean accordingly.

4

Python

`pandas`, numpy and `matplotlib` are useful modules for scientific applications.

We can choose among two or more options, checking the truth value of a Boolean expression using the `if` statement.

Data in a CSV file can be read as a data frame via `pandas`. From a data frame, we can extract columns to be used to compute the mean and the standard deviation thanks to numpy functions and build histograms with `matplotlib.pyplot.hist()`.

Establishing Physical Laws

Contents

© The Author(s), under exclusive license to Springer Nature
Switzerland AG 2021
G. Organtini, *Physics Experiments with Arduino and
Smartphones*, Undergraduate Texts in Physics,
https://doi.org/10.1007/978-3-030-65140-4_5

5

This chapter is devoted to illustrating the way in which physicists establish phenomenological physical laws, i.e., purely experimentally-based relations between physical quantities. In doing so, we will not make any attempt to understand the underlying physics, apart from a few very basic properties.

5.1 Light Transmission

In this chapter, we want to study light transmission by a transparent material. It is well known that light can be partly absorbed and reflected by a transparent filter, depending on the thickness and nature of the material of which is made of. Here, we want to quantitatively study this phenomenon. As already stated, conducting an experiment means finding numerical values, and not just observing something qualitatively. In this case, we need to find the precise relationship that exists between the intensity of the transmitted light and the properties of the filter, such as its thickness.

To set up such an experiment, we need a sufficiently intense and stable source of light, an instrument to measure its intensity and a series of transparent filters. As usual, there may very well be plenty of professional instrumentation at your university or college, however, we will try to perform the experiment using readily available material, such that you can do so at home, too.

Experiments must look good to be effective. Always pay attention to the realisation of the setup.

Irrespective of the place where you set up your experiment, it is of capital importance that the setup be clean, stable and safe. Take your time to fix everything in the proper way such that you do not risk introducing biases into your experiment due to bad alignments, parts moving, etc. Usually, a good criterion for determining whether an experiment has been well designed is aesthetic. A nice setup is certainly better than an ugly one.

Many experiments can be realised using readily available materials.

The light source can be either a tabletop lamp or the flash light of a smartphone. The instrument for measuring the light intensity can be either another smartphone equipped with PHYPHOX or an Arduino with a light sensor. Filters can be realised by cutting out 5-cm wide strips from an A4-crystal-clear plastic punched pocket wallet, stapled together on one of the short sides.

We need to measure the light intensity that emerges from a set of n filters, as a function of n.

The experiment consists in measuring the light intensity after it has traversed n strips, with n variable. Using PHYPHOX, we put the smartphone under the light, then cover the light sensor with a strip and begin data acquisition. Once enough data have been collected, we put another strip on top of the latter and restart data acquisition, and so on, for a number of strips.

To locate the light sensor on your mobile device, just start data acquisition, then move a finger along the surface of the phone until you see a drop in the intensity. When you start data acquisition, your hand must pass in front of the phone and can partially shield the light entering into the sensor. To avoid this effect, it can be useful to pause data acquisition for few seconds. Collecting data for 30−40 s is more than enough.

Save the data collected for each number n of strips in different files, named according to n (e.g., `lightIntensity-3.csv`, `lightIntensity-7.csv`, and so on). Using the techniques described above, build a table with n and the corresponding mean and standard deviation of the corresponding intensity of the light.

Take many measurements of the light intensity in timed runs of a few tens of seconds, avoiding influencing the measurement with your fingers. For each set of measurements, compute the mean and the standard deviation. Record the values in a text file as a function of n.

5.2 Taking the Average with Arduino

Arduino, in contrast to PHYPHOX, is programmable, and we can do many things with software, including computing the mean and the standard deviation directly during data acquisition (online).

Let's consider the following `loop()` function in the Arduino sketch:

```
void loop() {
   float S = 0.;
   float S2 = 0.;

   for (int i = 0; i < 1000; i++) {
      int k = analogRead(A0);
      int k2 = k*k;
      S += k;
      S2 += k2;
   }

   Serial.print(S/1000);
   Serial.print(", ");
   Serial.println(sqrt(S2/1000-S*S*1e-6));

   while (1) {
      // do nothing
   }
}
```

5

Data are stored in variables in a computer's memory, with their type determining how to interpret them.

A byte is a group of eight bits.

As soon as the function begins, two variables, S and $S2$, are declared as float and assigned the value 0. A variable, in programming, is a container for data. Data are represented in a computer's memory as sequences of bits, and their interpretation depends on their type. The same sequence, in fact, can be interpreted differently. Consider a 32-bit long variable whose content is the following:

01000101 11010010 01100000 00000000

where we separated groups of eight bits (a **byte**) for better readability. Its hexadecimal representation is 0x45D26000 (the prefix 0x identifies numbers in base-16). If interpreted as an integer number, its value corresponds to

$$n = \sum_{i=0}^{32} b_i \times 2^i , \qquad (5.1)$$

where $b_i = 0, 1$ is the ith bit and bits are counted from right to left. The calculation leads to $n = 1\,171\,415\,040$ (work it out by yourself). If interpreted as a string, it represents a three-character string. Each character in a string is, in fact, represented as a sequence of eight bits. The correspondence between a given character and the integer number represented by each sequence has been internationally established as the ASCII table (ASCII stands for American Standard Code for Information Interchange). Searching the Internet for that table, we find that the sequence 01000101, corresponding to 69, represents the character E. The next one, 210 if interpreted as an integer, represents the character Ê in the Latin-1 encoding. The third integer 01100000 corresponds to the ASCII code 96, representing the single quote ', while the latter is the so-called NULL character whose ASCII code is zero.

Characters in a string are represented as bytes. The correspondence between characters and integer numbers is given in the ASCII table.

In the ASCII table, numbers from 0 to 127 are fixed and internationally valid characters. Characters whose ASCII code is 128 or more depend on the encoding. ISO Latin-1 is one of the possible encodings and is frequently used in western countries. Changing the encoding, the same integer may represent a different character.

Strings are terminated by a NULL character, whose ASCII code is 0.

The NULL character marks the end of a string. A string made of three characters needs four bytes to be represented: the three characters plus the NULL one.

If interpreted as a floating point number, the 32 bits represent the number 6 732. Non-integer numbers are represented following the IEEE-754 standard. In this format, numbers are represented in a sort of scientific notation in base 2, as $m \times 2^e$, where m is written in the *normal* form and e in the excess-127 notation. To understand the form, imagine writing 6 732 as above with $1 < m < 2$, i.e., as $1.6435546875 \times 2^{12}$ corresponding to the binary code $1.1010010011 \times 2^{1100}$. Since, following this convention, the first digit is always 1, it is not worth representing it in the memory, and the so-called mantissa in normal form is 1010010011. The excess-127 notation is obtained adding 127 to the number, such that the exponent e is represented as $127 + 12 = 139$. The 32-bit number is composed as follows: the first bit represents the sign (0 for + and 1 for −), the following eight bits represent the exponent e in excess-127 notation (10001011 in binary), while the remaining 24 bits represent the mantissa in the normal form.

Floating point numbers are represented following the IEEE-754 rules: a sort of scientific notation in base 2.

Variables of the `float` type represent floating point numbers, and hence are stored in the memory following the latter convention.

Declaration and initialisation can be done at the same time. When declaring a variable specifying its name and type, we can assign it an initial value with the = operator. Each statement, as usual, ends with a semicolon.

= is the assignment operator.

The `for` loop is used to iterate over a given number of times. In this case, the integer variable `i` is initially set to 0 (the first expression in parenthesis, where `i` is declared as `int`, too). Before entering the loop, Arduino evaluates the middle expression in the `for`, `i < 1000`. If it is true, it executes the statements within the pair of curly brackets, otherwise, it abandons the loop.

Loops can be done utilising the `for` statement, used when you want to *count* the number of iterations.

This way, we repeat the following statements 1 000 times.

```
int k = analogRead(A0);
int k2 = k*k;
S += k;
S2 += k2;
```

5

The operator ++ is used to increment the variable to which it is applied by one. It can be put before a variable (the variable is incremented, then used) or after it (the variable is incremented after its usage).

After each execution, the third expression of the for, i++, is evaluated. ++ is a post-increment operator: it returns the value of i, then increases it by one.

In the loop, we put the value returned by analogRead (A0) into an integer variable k, and store its square in k2 (* being the multiplication operator). Then, we iteratively sum k, the last reading, to S and k2 to S2. The sum is obtained by the auto-increment operator +=, which adds the operand on its right to the variable on its left.

When we exit from the loop, S contains the sum of 1 000 measurements of the light intensity in arbitrary units and S2 the sum of their squares.

We then send over the serial line the value of S/1000, corresponding to the mean of the data, followed by

$$
\sigma = \sqrt{\frac{1}{N}\sum_{i=1}^{N} x_i^2 - \left(\frac{1}{N}\sum_{i=1}^{N} x_i\right)^2} = \sqrt{\langle x^2\rangle - \langle x\rangle^2},
$$

(5.2)

with $N = 1\,000$, separated by a comma from the mean. Remembering the definition of the variance of the population, we find that

The variance of a population can be evaluated as $\sigma^2 = \langle x^2\rangle - \langle x\rangle^2$.

$$
\sigma^2 = \frac{1}{N}\sum_{i=1}^{N}(x_i - \langle x\rangle)^2 = \frac{1}{N}\sum_{i=1}^{N}\left(x_i^2 + \langle x\rangle^2 - 2x_i\langle x\rangle\right)
$$
$$
= \langle x^2\rangle + \langle x\rangle^2 - 2\langle x\rangle^2 = \langle x^2\rangle - \langle x\rangle^2 .
$$

(5.3)

For $N = 1\,000$, there is not much difference between the variance of the population and that of the sample, so we take the standard deviation as the square root of the expression above.

Some may find the formula difficult to remember. In particular, it is difficult to remember whether σ^2 is equal to $\langle x^2\rangle - \langle x\rangle^2$ or to $\langle x\rangle^2 - \langle x^2\rangle$. Actually, it is quite easy to choose between the two options: just note that, in the case of two equal and opposite measures, $\langle x^2\rangle$ is certainly positive, while $\langle x\rangle^2$ may be worth 0. In this case, $\langle x\rangle^2 - \langle x^2\rangle$ would lead to a negative value for σ^2, which is impossible.

As in Python, in the Arduino language, while realises an iterative structure. The expression in parentheses is evaluated as a Boolean expression. On Arduino, true and false are represented as integer numbers that are different from and equal to zero, respectively. Since $1 \neq 0$, the expression is always true and the while loop never finishes. There are no statements in the loop. The double slash // represents a *comment*, intended to provide some information to the reader. Any character written after that sign is ignored by the Arduino compiler. In fact, the execution of the program stops here.

while realises an iteration structure, like for, but it is used when the number of cycles cannot be predicted.

We can read the mean and the standard deviation from the serial monitor. To restart data acquisition after adding a strip on top of the sensor, we just need to press the RESET button on one of Arduino's corners. This way, we can simply manually record the mean and the standard deviation for various n.

5.3 A First Look at Data

To get an idea of the way in which illuminance I depends on the thickness of the filter, we can produce an initial plot of the values of I as a function of n, as shown in Fig. 5.1. The vertical bars on each point represent the amplitude of the $1-\sigma$ interval.

A plot of the data is always useful.

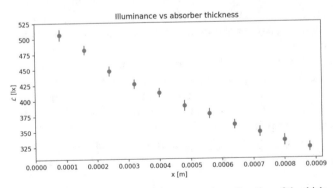

■ **Fig. 5.1**   The intensity of the light measured as a function of the thickness of the filter. Uncertainties are shown as vertical bars on the data

5

Conversion factors between measurements made with different instruments are obtained through data. They are affected by uncertainties, too, but they can be considered as constants as long as we are not interested in absolute values.

To make the plot, we used the techniques illustrated in the previous chapters. The measurements have been done using Arduino. Comparing the illuminance measured for $n = 0$ with PHYPHOX, $\mathcal{L}_p(0) = 532.0$ lx, with the one obtained with Arduino in arbitrary units, $\mathcal{L}_A(0) = 31.06$, we can easily obtain a calibration factor. If we assume that

$$\mathcal{L}_p(x) = C\mathcal{L}_A(x), \tag{5.4}$$

the calibration factor C is

$$C = \frac{\mathcal{L}_p(x)}{\mathcal{L}_A(x)}. \tag{5.5}$$

Since $\mathcal{L}_A(x)$ is given in arbitrary units (thus, dimensionless) and $\mathcal{L}_p(x)$ is in lux, the calibration factor is measured in lux, too, and a measurement taken with Arduino in arbitrary units can be cast into a measurement in lux by multiplying it for

$$C = \frac{532.0}{31.06} \simeq 17.13. \tag{5.6}$$

Clearly, both $\mathcal{L}_p(0)$ and $\mathcal{L}_A(0)$ are affected by uncertainties, hence C, too, cannot be known with infinite precision. For a first interpretation of the data, in which we want to study the relative change in illuminance, this problem can be neglected for the moment, so we will postpone its discussion.

The uncertainty about the thickness is dominated by the finite resolution of the instrument of $10\,\mu$m. The corresponding uncertainty is $\frac{10}{3} \simeq 3\,\mu$m.

The thickness of the filter is obtained by multiplying the number of strips by the thickness of the plastic sheet $dx = 80 \pm 3\,\mu$m, measured using a caliper. For the moment, we ignore the effects caused by the possible inhomogeneities in the strips' thickness.

At a very first glance, illuminance seems to decrease almost linearly. We might then try to model the data as

A phenomenological physical law is a relation between physical quantities conjectured on the basis of experimental data.

$$\mathcal{L} = \alpha x + \beta. \tag{5.7}$$

The equation above is what we call a physical law, i.e., a relationship between two or more physical quantities. The law is valid as long as we cannot spot any significant deviation from it in our experiments. For the time being, *significant* is only an adjective meaning "taking into account the experimental uncertainties". We give it a more quantitative meaning in the following.

Since a physical law is an equality between physical quantities, the equation must hold for measurement results. Thus, since the left side of the equation is measured in lux, so should the right one be. Moreover, the right hand side of the equation is made up of two terms added together. Manifestly, each term must exhibit the same properties: if the sum has to be measured in lux, β must be expressed in this same unit, as the product αx. On the other hand, x is a length and is measured in m, and to make αx an illuminance, α must be measured in units of illuminance divided by units of length, e.g., lx/m.

The above considerations are part of what physicists call dimensional analysis. Dimensional analysis is very useful, and not only for determining the units in which things are measured. It also helps in spotting mistakes in equations' manipulations. Since every physics equation must be dimensionally consistent, such a consistency must be manifest at each stage of computation. If, at a certain point, you find that the equation you have written is not dimensionally consistent, for sure, you made a mistake earlier. To make dimensional analysis easy, we use symbols enclosed in square brackets [...], where the symbols, represented by capital letters, indicate the *nature* of the quantity. For example, if we use I to indicate illuminance and L for length, we write

> An essential requirement for any physical law is to be dimensionally consistent. Both sides of the equation that represents it must be measurable in the same units, as well as all the terms in a sum or a difference. The arguments of trigonometric functions, logarithms and exponentials must be dimensionless.

$$[I] = [\alpha][L] + [\beta] , \tag{5.8}$$

such that

$$[\alpha] = \frac{[I]}{[L]} \quad \text{and} \quad [\beta] = [I] . \tag{5.9}$$

We can then assign the proper units by substituting the chosen one for each quantity, e.g., lux for $[I]$ and m for $[L]$.

5

Dimensional analysis

It is worth noting that dimensional analysis is also great in helping memory or guessing the right formula for something. For example, in special relativity, it is known that time appears dilated for systems moving with respect to an observer, i.e., $t \to t' = \gamma t$, where $\gamma > 1$ because $t' > t$. Since $[t] = [t']$, γ must be dimensionless and can always be written as a ratio between two numbers. It must also depend on the velocity $\mathbf{v}$ of the moving system. The higher the speed, the bigger γ, such that $\mathbf{v}$ must appear in the denominator. However, $\mathbf{v}$ is a vector and γ is a scalar. The only way to obtain a scalar from just one vector is to multiply the vector by itself, as in $v^2 = \mathbf{v} \cdot \mathbf{v}$. v^2, on the other hand, is not dimensionless, but $[v^2] = [L^2 T^{-2}]$, T being the symbol for time. In order to make a dimensionless combination, we need to divide v^2 for the square of another velocity. Since the only other velocity involved is the one of light $c = 3 \times 10^8$ m/s and is universal, manifestly, γ must depend on the ratio $\frac{v^2}{c^2}$. When $v = 0$, $\gamma = 1$, and we can then ultimately write

$$\gamma = \left(\frac{1}{1 - \frac{v^2}{c^2}} \right)^n .$$

(5.10)

It turns out that $n = \frac{1}{2}$. Besides dimensionless numerical factors, this is the only other information that dimensional analysis cannot provide.

Each time we have no reason to prefer one value among others, we average over them.

In order to determine the values of α and β, we can keep any pair of data points and evaluate the slope as

$$\alpha_{ij} = \frac{I_i - I_j}{x_i - x_j} ,$$

(5.11)

while the intercept can be obtained by imposing

$$I_i = \alpha_{ij} x_i + \beta .$$

(5.12)

How do we choose the pair to compute α_{ij} and which point (x_i, I_i) do we choose to find β? In principle, all the possible pairs should lead to the same value for α_{ij}, however, uncertainties lead to a different value for each pair. Choosing $i = 1$ and $j = N$, with $N = \max(n)$ could be a choice, giving us the largest leverage, but the extreme points can be problematic, leading to incorrect values.

As we learned in Chap. 4, averaging over many measurements gives better results. Thus, averaging over all the possible α_{ij} guarantees the best possible result. The number of possible pairs is given by combinatorics, according to which the number of combinations of k elements in a set of n without repetitions is given by

Combinatorics allows for the computation of **permutations** of n elements $n!$, dispositions $\frac{n!}{(n-k)!}$ and combinations $\frac{n!}{k!(n-k)!}$ of n elements in groups of k.

$$C_{nk} = \binom{n}{k} = \frac{n!}{k!\,(n-k)!}\,, \qquad (5.13)$$

where the symbol in parentheses is called the binomial coefficient and ! denotes the factorial $n! = \prod_{i=1}^{n} i$. In this case, $n = 11$ and $k = 2$, such that

Combinations are dispositions in which the order of the elements does not matter. Formulas are valid when elements do not repeat in the list.

$$C_{11,2} = \frac{11!}{2!9!} = 55\,. \qquad (5.14)$$

Given the 55 possible slopes, we can average them and evaluate the corresponding standard deviation. The latter is $\sigma_\alpha \simeq 114\,632$ lx/m. Taking into account our rule according to which we only keep one significant digit, $\sigma_\alpha \simeq 100\,000$ lx/m. It might appear to be huge, but α is, on average, $\alpha \simeq -450\,473$ lx/m, such that our result should be written as

$$\alpha = (-5 \pm 1) \times 10^5\,\text{lx/m}\,. \qquad (5.15)$$

Written this way, the uncertainty does not seem so large, being $\frac{1}{5} = 0.2$, i.e., 20%. The ratio between the uncertainty and the central value of a measure is called the relative uncertainty and gives a more objective feeling about how large the uncertainty is.

If you are uncertain about how many digits you should quote in your result, write the standard deviation using the scientific notation, keeping just one digit in the mantissa, then write the central value using the same exponent, keeping only the digits before the decimal point.

We can do a similar thing for β. We can average over all the possible values of i, assuming $\alpha = -5 \times 10^5$ lx/m. The result is

$$\beta = 520 \pm 10\,\text{lx}\,. \qquad (5.16)$$

Figure 5.2 shows the resulting line superimposed onto the data. The agreement between the data and the model is not bad, but certainly not very satisfactory.

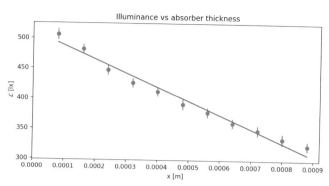

■ **Fig. 5.2**   The intensity of the light measured as a function of the thickness of the filter with a linear model superimposed

5.4 Plotting Graphs and Interpolating Data

Suppose that our CSV file contains three columns: the number of plastic strips covering the sensor, the measured illuminance in arbitrary units, and its uncertainty. We can read the file using Python and convert the number of strips into a thickness into units of length and the intensity in units of luminous flux per unit area, as follows:

```
f = pd.read_csv(sys.argv[1])
data = f.T.values.tolist()
x = [x*dx for x in data[0]]
y = [x*C for x in data[1]]
dy = [x*C for x in data[2]]
```

The first statement opens the file whose name is passed as the first argument to the script and reads it. It transposes the data, removes the headers, if any, and returns a list of three lists: data[0], containing the number of plastic strips, data[1], representing the illuminance in arbitrary units, and data[2], which stores the uncertainties about the latter.

We generate three new lists called x, y and dy in the last three lines, using a technique called "list comprehension", consisting in an implicit iteration over all the elements of a list. On each line, we iterate over the components of each list using for x in data[0] and store, in the corresponding component of the new array, the result of the transformation applied to x, representing the actual value. In the case of data[0], the number of strips, represented by x, is multiplied by dx, defined as

To iterate over a list to transform its elements, use [...] with a for *loop inside.*

```
dx = 80e-6
```

and representing the thickness of a single strip in m (80 μm). It is useful to see how the same effect can be obtained using a more *traditional* form of programming:

append() *adds an element to the end of a list. A list with given length N can be created with* x = [0]*N.

```
x = []
for i in range(len(data[0])):
    x.append(data[0][i]*dx)
```

The first statement creates an empty list whose name is x. Then, we iterate over data[0] components (remember that data[0] is itself a list) counting with i. len() returns the length of a list, while range(n) returns an integer range from 0 to n−1. Multiplying the i-th component of the list data[0], data[0][i], by dx, we obtain the thickness in units of length and store them in subsequent components of x via append().

The plot is produced with the following code:

```
plt.errorbar(x, y, yerr=dy, fmt='o')
plt.title('Illuminance vs absorber
    thickness')
plt.xlabel('x [m]')
plt.ylabel('${\cal L}$ [lx]')
plt.xticks(xticks)
plt.show()
```

The errorbar() method gets the two arrays x and y and plots the second as a function of the first. With yerr=dy, we inform it that the uncertainties about y are stored in dy, while fmt='o' tells the method to represent data with circles.

The name of the method is derived from the term by which measurement uncertainties are sometimes referred to, "errors". The name is a bit misleading, because it suggests a mistake. Though the term "uncertainty" is preferable, in what follows, we sometimes use the term "error" as a synonym.

The word "error" is used as a synonym for "uncertainty".

5

The name of the game

LaTeX can be considered an evolution of TeX, a document preparation system that has been widely adopted, especially in the scientific domain. LaTeX is pronounced "latek", because TeX was pronounced "tek".

The reason for that is explained by TeX's creator Donald Knuth (also known as the author of the monumental "The Art of Computer Programming"):

> English words like "technology" stem from a Greek root beginning with the letters $\tau\epsilon\chi$...; and this same Greek word means art as well as technology. Hence the name TeX, which is an uppercase form of $\tau\epsilon\chi$.

The "La" prefix that turned TeX into LaTeX refers to the name of its author, Leslie Lamport, who started writing *macros* to simplify the use of Knuth's invention to write scientific documents.

The reference to art in Knuth's words is not accidental. You will find a number of references to art scattered throughout this textbook. Science and coding are, in fact, arts, as long as art implies the "creation" of something new. The root for the word "artisan" is the same for the same reason. Moreover, art and technology could be considered as synonyms in ancient Greek, given that the latter, too, implies creation.

LaTeX is a document preparation system that is worth learning.

We decorate the plot with a title and labels, one of which is rendered using LaTeX (pronounced "latek"), a high quality document preparation system that is worth learning for scientific typesetting. We also force the tick marks on the horizontal axis to appear at the coordinates given in the xticks list, defined as

```
xticks = np.arange(0, 0.50e-3, 1e-4)
```

Defining functions is a way to refer to a set of instructions using a name. Their run may depend on parameters. Functions improve readability, reduce the chance of errors and make scripts easily maintainable.

To compute the average slope and its error, we define a function, i.e., a set of operations identified by a single name, whose behaviour depends on zero or more parameters, as in

```
m, sigma = averageSlope(x, y)
```

The function averageSlope() is defined as

```
from itertools import combinations as
    comb

def averageSlope(x, y):
    ijpairs = list(it.comb(np.arange(
        len(y)), 2))
    m = 0
    m2 = 0
    for ij in ijpairs:
```

```
      i = ij[0]
      j = ij[1]
      s = (y[i]-y[j])/(x[i]-x[j])
      m += s
      m2 += s*s
   N = len(ijpairs)
   m /= N
   sigma = np.sqrt(m2/N-m*m)
   return m, sigma
```

The set of instructions to be executed when calling the function are written indented with respect to the function definition, after the colon (:). The module `itertools` is useful for iterating over lists. In particular, `combinations(n,k)` returns a sequence of all the possible combinations of k elements in a set of n. `ijpairs`, then, contains all the combinations of the first integers from 0 to the length of the list passed as a second argument minus one: (0, 1), (0, 2), ..., (0, $n-1$), (1, 2), (1, 3), etc., cast into a `list`. In fact, in Python, a sequence is an abstract type, i.e., a type that can be cast into another one for which an implementation exists. For the sake of brevity, we renamed the method as `comb(n,k)` when importing the module with

> `itertools` is a useful module for iterating over elements in a *sequence*: an abstract type. Abstract types cannot be instantiated, i.e., variables of that type do not exist. They are used to describe a common behaviour of different types.

```
from itertools import combinations as
   comb
```

The two variables m and m2 are intended to represent, respectively, the mean of the slopes and the mean of their squares. The slope is computed in the `for` loop that, iterating over all the `ijpairs`, computes all the possible values of the slope s and sums them up in m and their squares in m2. The average of the slopes squared is used to compute the standard deviation.

> Python methods and functions can return zero or more objects.

Both the mean and the standard deviation are returned by this function.

5.5 An Approximated Model

It is not difficult to realise that the model described above is not a good one. The reason is very simple, indeed. The illuminance $\mathcal{L}$ is, by definition, positive, i.e., $\mathcal{L} > 0$. If $\mathcal{L}(x) = \alpha x + \beta$, there is always a value $x = x_0$ such that $\mathcal{L}(x > x_0) < 0$. On the other hand, a linear model seems to be quite good in intervals of thickness that are short enough. For example, for $x < 0.00015$ m, it is easy to find a line that describes the data well.

5

Taylor expansion is a useful technique for approximating a function in a small interval.

The reason for this is that any function $f(x)$ infinitely differentiable at a point a can be expanded in a Taylor series as

$$f(x) = \sum_{n=0}^{\infty} \frac{f^{(n)}(a)}{n!} (x-a)^n , \qquad (5.17)$$

$f^{(n)}(a)$ denoting the nth derivative of $f(x)$ at $x = a$. For small enough $(x-a)$, i.e., in the vicinity of the point $x = a$, the higher order terms become smaller and smaller and the series can be truncated.

Truncating the Taylor series at the first order transforms any function into a line. Functions that exhibit a minimum or a maximum are better described by a second order truncation near the extreme points.

For small enough $(x-a)$, any physical model $\mathcal{L}(x) = f(x)$ can be approximated by a linear one

$$f(x) \simeq f(a) + f'(x)(x-a) \qquad (5.18)$$

irrespective of how complicated $f(x)$ is. The approximated model looks better with increasing n. In general, any function (at least, those of interest to a physicist) can be approximated by a sufficiently high degree polynomial.

We use this feature when we study systems like springs and oscillators. Such systems are complex, indeed, but if the external forces to which they are subject are low, their state changes slightly, and whatever the correct model is, it can always be approximated by linear or quadratic ones.

5.6 Non-polynomial Models

Looking carefully at the plots suggests that the change $d\mathcal{L}$ of $\mathcal{L}$ diminishes with x. Since $\mathcal{L}$ also diminishes with x, we can write that

$$d\mathcal{L} = -\alpha\mathcal{L} . \qquad (5.19)$$

Often, in physics, we observe a change in a quantity that is proportional to the quantity itself.

Such an equation means that the change in illuminance is proportional to the illuminance itself. If the illuminance is high, the filter can absorb a lot of light and the luminous flux per unit area decreases by a relatively large quantity. On the other hand, if the light impinging on the filter is low, the filter cannot absorb more light than it receives, and the change in the flux is lower. The change is negative, because the illuminance after the filter is lower than that before it, and if $\mathcal{L} = 0$, $d\mathcal{L} = 0$, as expected.

This model already seems more physically plausible than the simple linear one of the previous section.

Moreover, it is clear that $d\mathcal{L}$ must be proportional to the thickness dx of the filter: the thicker the filter, the higher, in absolute value, the change in illuminance. This leads us to modify the model such that

$$d\mathcal{L} = -\beta \mathcal{L} dx, \tag{5.20}$$

with $\alpha = \beta dx$. Such an observation is also suggested by the fact that the left hand side of Eq. (5.19) contains an infinitesimal, while the right hand side is written as if it were a finite quantity. In order to make it explicit that even the right hand side of the equation is an infinitesimal quantity, we rewrite it as above.

Equation (5.20) is what mathematicians call a separable differential equation whose solution is relatively simple. Dividing both members by $\mathcal{L}$ and integrating, we find

$$\int_{\mathcal{L}(0)}^{\mathcal{L}(x)} \frac{d\mathcal{L}}{\mathcal{L}} = -\beta \int_0^x dx, \tag{5.21}$$

such that

$$\log \frac{\mathcal{L}(x)}{\mathcal{L}(0)} = -\beta x, \tag{5.22}$$

which can be rewritten by taking the exponential of both members as

$$\mathcal{L}(x) = \mathcal{L}(0) \exp(-\beta x). \tag{5.23}$$

The parameter β has to be determined from data and includes the modelling of various effects depending on the nature of the absorber. For example, a crystal clear filter absorbs less light than an embossed one.

The argument of the exponential must be dimensionless, thus $[\beta] = [L^{-1}]$. Sometimes, to make this explicit, we prefer to define

$$\lambda = \frac{1}{\beta} \tag{5.24}$$

Equation $d\mathcal{L} = -\beta \mathcal{L} dx$ is important, because it represents a prototype for many physical phenomena.

A differential equation is an equation relating functions and their derivatives.

The solution to the above model is always an exponential function.

Dimensional analysis requires that the argument of functions like log or exp be dimensionless. We often redefine constants such that their dimensions become apparent.

such that $[\lambda] = [L]$ and λ, measured in m, is called the mean free path. Finding the value for λ is easy if we use the linearised version of the law linking $\mathcal{L}$ to x:

$$\log \frac{\mathcal{L}(x)}{\mathcal{L}(0)} = -\frac{1}{\lambda}x. \tag{5.25}$$

Linearising an equation $y = f(x)$ means transforming it such that it appears as $w = az + b$, where $F(f(x)) = az + b$ and $w = F(y)$.

Linearisation consists in finding a transformation of an equation of the form $y = f(x)$, such that it appears as $w = az + b$, where $w = F(y)$ and $az + b = F(f(x))$. In the case of an equation such as (5.23), the transformation consists in taking the logarithm of both sides.

Figure 5.3 (left) shows the linearised plot, i.e., a graph in which we show the logarithm of the ratio $\frac{\mathcal{L}(x)}{\mathcal{L}(0)}$ versus the absorber thickness x. The average interpolating line, obtained as above, has been superimposed onto the plot. The model appears quite good, indeed. Now, data are found along almost the entire line, but not very many of them, because of statistical fluctuations (so, this is expected). Indeed, knowing that a $1 - \sigma$ interval contains more or less 70% of data, we expect about $\frac{1}{3}$ of the points not to lie on the interpolating curve. On the right side of the figure, the two models are compared. The linear model is represented in red, while the exponential one is in green. Manifestly, its agreement with the data is better.

To obtain these plots, we follow the same procedure as above, finding that the slope of the interpolating line in the left one is

$$\frac{1}{\lambda} = -1100 \pm 200 \, \mathrm{m}^{-1}, \tag{5.26}$$

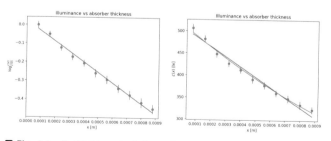

◘ Fig. 5.3 (Left) Linearised data of the illuminance versus the filter thickness with the average line superimposed. (Right) comparison of the two models: the linear model is in red, the exponential in green

while the intercept, which is dimensionless, is

$$q = 0.02 \pm 0.01 . \qquad (5.27)$$

From the equation, we could predict that $q = 0$. In this case, $q \neq 0$, but it is *only* $2-\sigma$ different from zero. Remember that there is a chance of 95% that the measured value lies within an interval whose amplitude is $2-\sigma$. Data, then, are still compatible with our model, if we accept that a fluctuation occurring with a probability of 5% is not so unlikely.

It is worth noting that such a model is better than the previous one, not only because the data follows it better, but, even more so, because it is grounded in reasonable considerations about the mechanisms involved in light absorption. Just finding the curve that best fits your data is not enough. You need to justify the assumptions with reasonable arguments that are not in contradiction with the rest of your knowledge. If this is already so, to be acceptable, the model must explain, possibly in an alternative way, the facts that are already known.

> To be acceptable, an alternative model must fit the data and be grounded in experimental knowledge and reasonable arguments.

Of course, intuition about how to describe data correctly is supported by looking at the ways in which data distribute, and that is why plotting is important.

The result on the slope tells us that the mean free path of the light inside the plastic of which the sleeves are made is about 0.001 m = 1 mm. Using the form

$$\mathcal{L}(x) = \mathcal{L}(0) \exp\left(-\frac{x}{\lambda}\right) \qquad (5.28)$$

for the model, we can see very clearly that, when $x \simeq \lambda$, the illuminance is reduced by a factor of $\frac{1}{e} \simeq \frac{1}{3}$. After three mean free paths, $\mathcal{L}(x) = \mathcal{L}(0)e^{-3} \simeq 0.05\mathcal{L}(0)$, i.e., the illuminance is reduced to 5%.

> Writing equations in the proper way makes their interpretation easier: $\beta \simeq 1\,100\,\mathrm{m}^{-1}$ is not as informative as $\lambda \simeq 1\,\mathrm{mm}$.

For completeness, you should know that only part of the observed effect can be ascribed to light absorption in the filter. Fresnel losses and interference may play a role.

5.7 The Exponential Model

As a matter of fact, the behaviour of most things in the Universe is either linear, exponential or periodic. Once you dominate the math of these models, you can model almost any physical phenomena.

The behaviour seen for light intensity is very common in physics. An exponential growth or decrease is observed any time some quantity y changes proportionally to itself and to another quantity x, as in

$$dy = \alpha \, y \, dx, \qquad (5.29)$$

where α is a constant having the dimensions of $[x^{-1}]$. In many cases, you will be able to draw results in regard to topics with which you are still not very familiar. Usually, it is enough to know a few basic laws, even without a deep understanding of them, to obtain important results. Examples of such behaviour are as follows.

Exponential models describe the change in amplitude of periodic phenomena due to dissipative forces.

When a pendulum oscillates, its amplitude A decreases with time because of friction. Friction can subtract a fraction of the pendulum's energy from it, leading the amplitude to reduce to a fraction of its value: $dA \propto -A$. Moreover, the longer the oscillation, the higher the decrease, so $dA = -\alpha A dt$ and $A(t)$ behaves as above, decreasing exponentially.

In many cases, the equation describing the emptying of a container is an exponential one. Electrical capacitors store energy accumulating charges on their plates.

In electrical circuits, capacitors store energy by placing charges on conductors. Charges moving in a circuit generate a current, whose intensity depends on the resistance of the circuit and the applied voltage. To predict their behaviour, knowing the above principles and two basic formulas (the Ohm Law and the definition of capacitance) is enough.

In an RC circuit, in which a capacitor discharges by injecting a current through a resistor R, the amount of electric charge dQ subtracted from the capacitor cannot be larger than the one stored in it, Q, and is larger when Q is larger, such that $dQ \propto -\alpha Q$. Again, the amount of charge subtracted from the capacitor is proportional to the elapsed time dt and $dQ = -\alpha Q dt$. As a result, the charge in the capacitor decreases exponentially and, given that

Capacitors are characterised by their capacitance given by the ratio between the charge Q on their plates and the voltage across them.

$$\Delta V = \frac{Q}{C}, \qquad (5.30)$$

ΔV being the voltage across the capacitor and C its capacitance, ΔV decreases exponentially, too. According to the Ohm Law, $I = \frac{\Delta V}{R}$, the current I flowing in the circuit drops exponentially, too.

If we connect a voltage source V_0 to a capacitor through a resistor R, we see similar behaviour. The current flowing into the circuit is again given by the Ohm Law, where the capacitor acts as a voltage source, such that

Ohm Law predicts that the current flowing in a circuit is proportional to the applied voltage.

$$V_0 - \Delta V = RI = R\frac{dQ}{dt}. \tag{5.31}$$

Substituting the expression of ΔV, we obtain

$$dQ = \frac{1}{R}\left(V_0 - \frac{Q}{C}\right)dt. \tag{5.32}$$

We integrate this differential equation taking $y = \left(V_0 - \frac{Q}{C}\right)$, such that $dy = -\frac{dQ}{C}$, and substituting so as to rewrite it as

$$-C\,dy = \frac{1}{R}y\,dt, \tag{5.33}$$

the equation can be rewritten in the usual form

$$\frac{dy}{y} = -\frac{1}{RC}dt, \tag{5.34}$$

leading to an exponential decrease of y that, observing that $Q(0) = 0$ and remembering that $\frac{Q}{C} = \Delta V$, can be cast into a law for ΔV as

$$V_0 - \Delta V(t) = V_0 \exp\left(-\frac{t}{RC}\right). \tag{5.35}$$

Measuring ΔV, we then expect it to rise indefinitely, tending to V_0 as $t \to \infty$:

$$\Delta V(t) = V_0\left(1 - \exp\left(-\frac{t}{RC}\right)\right). \tag{5.36}$$

Our very simple experiment allows us to develop a model that describes a variety of phenomena, from the simplest to the most intriguing, like the behaviour of subatomic particles.

The same behaviour is exhibited by any body reaching thermal equilibrium, whose temperature changes exponentially towards that of the equilibrium. Also, the radioactive decay rate follows an exponential model, as well as the absorption of subatomic particles in a beam impinging on a target. Thanks to our understanding of this model, Viktor Hess

(1883–1964) was able to discover cosmic rays in 1912 and we can treat cancer in radiotherapy facilities.

It is thus very important that you realise that, starting from a model for which $dy = -\alpha y \, dy$, you can predict that $y(t)$ behaves exponentially and, conversely, that if you observe one quantity changing exponentially when another one increases linearly, an appropriate model can be formulated as above.

Summary

Dimensional analysis consists in identifying the nature of a physical quantity in an equation and imposing or verifying that both sides of the equation have the same dimensions. Two quantities of the same kind are said to be **commensurable** (i.e., can be measured with the same tool).

Graphs showing the width of uncertainty intervals are useful. The amplitude of the uncertainty interval is shown as an error bar extending from $\mu - \sigma$ to $\mu + \sigma$, where μ is the central value.

Almost any function $f(x)$ in which physicists can be interested, can be rewritten as an infinite series of powers of x, i.e., like a polynomial. Often, the series can be truncated at very small orders, if we can neglect higher order terms.

When a physical quantity changes proportionally to itself and to another quantity like $dy = \alpha y \, dx$, the resulting model is always exponential, i.e., $y(x) = y(0) \exp(\alpha x)$. This model describes a long list of physical phenomena and it is very important that you master it.

Linearisation consists in transforming an equation such that it appears as linear in the transformed independent variable. It usually helps in finding the parameters of a model from data.

Arduino

A variable, in a programming language, is a place where a value is stored.

Data are represented in a computer's memory as sequences of bits. To interpret them correctly, we need to know their type.

A `for` loop is used to repeat one or more statements.

The content of a numeric variable can be altered using post-, pre- and auto-increment or decrement operators. Post- (e.g., `i++`) and pre- (e.g., `++i`) increment operators increase the content of a variable by one. The increment is done after or prior to using it in the current state-ment, depending on the position of the operator. The auto-increment operator increases the variable on the left by the amount on the right (e.g., S += k).

Iterative structures can be realised using while, too. The latter is preferred when we cannot predict the number of times the loop is run.

In programming languages, we can define our own functions: collections of statements whose behaviour can depend on parameters.

Statistics
The variance of a population can be estimated as $\sigma^2 = \langle x^2 \rangle - \langle x \rangle^2$.

Free-Fall and Accelerations

Contents

The original version of this chapter was revised: Typographical errors have been corrected. The correction to this chapter is available at ▶ https://doi.org/10.1007/978-3-030-65140-4_16

Electronic supplementary material The online version of this chapter (doi:▶ 10.1007/978-3-030-65140-4_6) contains sup-. plementary material, which is available to authorized users

© The Author(s), under exclusive license to Springer Nature Switzerland AG 2021, corrected publication 2022
G. Organtini, *Physics Experiments with Arduino and Smartphones*, Undergraduate Texts in Physics,
https://doi.org/10.1007/978-3-030-65140-4_6

In this chapter, we take our first measurement, the result of which can be compared to another experiment. Comparing results with those of other experiments is of paramount importance in physics. As we believe that physics is universal, i.e., its laws are valid at any time and at any place, we expect that results obtained by different people in different experiments should be consistent with each other. That does not mean that they must be the same in the mathematical sense.

6.1 Setting Up the Experiment

6

Bodies subject only to gravity are said to fall freely.

Free fall consists in the motion of a body subject solely to gravity. In this experiment, we want to study how objects fall and characterise their motion, i.e., determine their **state**. The latter consists in finding all the information needed to predict its state at time t, knowing the state at a different time t_0.

An experiment on free-fall consists in dropping a body from a certain height and measuring how much time it takes it to reach the floor.

To conduct the experiment, we must drop a body, and we should characterise its motion, measuring all the possible physical quantities affecting it. The shape and length of the path followed during the fall, for example, can be interesting quantities to measure, as is the time needed to follow the path. The shape of the object may be of some interest, however, it is not a good idea to start experimenting with objects whose shape is complicated enough to influence the motion. It is well known, for example, that the path followed by a sheet of paper freely falling is rather complicated, and it is thus difficult to study it. We should get rid of such effects. One possibility is to choose an object that has *no shape*, i.e., a point-like object. Manifestly, such an object does not exist. However, we can easily find objects whose shape and size are such that the details do not affect their free-fall.

When designing an experiment, many aspects can influence the result. We need to keep them negligible or, at least, constant.

The mass of the object may have an influence, too. Light objects may fall differently than heavy ones. A simple qualitative observation shows that heavy enough objects fall in almost the same way.

As a result, to conduct a meaningful experiment, we must choose a sufficiently heavy body whose shape and size do not affect the motion. This way, we get rid of quantities on which the state of the object may depend and that are difficult to control. In every experiment, we should always try to make the results depend on one or, at most, very few variables.

With these choices, we describe, in fact, the motion of a pointlike particle of mass m, subject only to gravity. For simplicity, it is better to start it from a stationary position. It is easy to show that the trajectory described by such a particle can be represented by a vertical segment of length h.

In principle, there are other characteristics identifying the state of the falling object: its temperature, its color, the material of which is made, etc. However, many of these characteristics either do not affect the motion or are meaningless in the context. For example, a pointlike particle cannot have a color, since it has no surface. On the other hand, the color of the real object that we are going to throw is not found to influence its motion. As a result, color is not part of the state of the object, and it is useless to collect its value, as well as its temperature, its nature, etc.

> In characterising the state of the object, many factors are negligible because they do not affect the motion or are meaningless in the chosen model.

In summary, an experiment consists in letting the object fall freely from a given height h and observing that the time needed to fall t depends on h, i.e., $t = f(h)$. Our goal is to find f.

6.2 Measuring Times

To measure times, we need a stopwatch. Every smartphone comes with a stopwatch application, so we can use that. The application gives the time elapsed between two successive taps on the display.

> A smartphone can be used as a stopwatch.

A first attempt could be the following: holding an object in one hand at a certain predefined height h, we tap on the smartphone when we release the object and tap it again when we hear the object hit the floor. Though it may work, there are a number of inconveniences in conducting the experiment this way.

First of all, it is not entirely obvious that the moment at which the object starts falling will be the exact same moment at which you tap the display on the smartphone to start measuring time. Moreover, your reaction time when you hear the noise of the object hitting the floor is not negligible and will affect the measurement. Moreover, the time needed for the object to reach the floor is short indeed.

6

Triggering an instrument by hands often results in large statistical fluctuations, due to the enormous variability in our response times to external stimuli.

As a result, the measurement is affected by large statistical fluctuations. Repeating the measurement several times will make it much clearer.

Comparing the results of the same measurement taken by different experimenters, you can easily see that there are sometimes significant differences in the measurements. The difference may result from the smartphone being used or from the person taking the measurement, whose reaction time could be longer or shorter.

We can easily get rid of the latter effect by assigning the task of activating the stopwatch to some automatic device. Here, we describe two possible solutions.

6.3 Photogates

Photogates use a light sensor to provide start and stop signals to a stopwatch. It is easy to make a photogate with Arduino.

A **photogate** consists of a light detector illuminated by some source of light. Precisely identifying the time at which something breaks the light beam passing between the source and the detector provides a measurement of the time at which the object's coordinate along its trajectory coincides with that of the gate.

It is easy to make a photogate with Arduino. We need a light intensity detector and a source of light. By continuously monitoring the illuminance, we can tell when it drops below a certain threshold, as in

```
#define THRESHOLD 15

void setup() {
   Serial.begin(9600);
}

void loop() {
   while (analogRead(A0) >
      THRESHOLD) {
      // do nothing
   }

   unsigned long t = micros();
   Serial.println(t);
}
```

Here, THRESHOLD is a **constant** whose value is 15. Entering the loop(), nothing happens until analog Read(A0) returns a number lower than 15. When this happens, we assume that it is because something interrupted the light beam, and we measure the current time using micros(). The latter returns the time that has elapsed since the beginning of the execution of the sketch in microseconds. The number is represented as un unsigned long, i.e., a positive integer of 32 bits.

In Arduino programming language, constants are symbols used to represent values.

We can measure the time needed to pass by points *A* and *B* with two photogates in *A* and *B* that measure the difference in their concealment times. For example, connecting the output leads of two analog light sensors to A0 and A1 pins, respectively, the time needed to go from sensor 1 to sensor 2 can be measured with the following loop() function:

```
void loop() {
  while (analogRead(A0) >
    THRESHOLD) {
    // do nothing
  }

  unsigned long t1 = micros();
  while (analogRead(A1) >
    THRESHOLD) {
    // do nothing
  }

  unsigned long t2 = micros();
  Serial.println(t2-t1);
}
```

Here, we first wait until the first sensor detects the passage of something between itself and a light source, and then put the time at which it happens in the variable called t1. Then, we wait for the same event to happen with sensor 2. The time elapsed between the two events is computed as the difference t2-t1 and is shown on the screen when the serial monitor is open.

6.4 Measuring Time with Arduino

Microprocessors operate at a fixed rate, the pace of which is given by a clock. Arduino's clock runs at 16 MHz. `delay()` and `delayMicroseconds()` pause the execution of the program for the given time.

Microprocessors' operations happen at a fixed rate regulated by a clock: a device producing pulses at regular intervals. The Arduino UNO clock runs at 16 MHz. It can then measure time simply by counting the number of clock pulses.

There are few functions that are dedicated to time management in the Arduino programming language.

The execution of a sketch can be paused using `delay()` or `delayMicroseconds()` functions. The first pauses the program for a number of milliseconds passed as an argument, e.g., `delay(1500)` suspends the execution of the sketch for 1.5 s. The latter works exactly the same way, but it suspends the execution for the number of microseconds passed as its argument: `delayMicroseconds(10)` waits for 10 μs.

Using its clock, Arduino can measure the time elapsed since the beginning of the sketch.

The time elapsed since the beginning of the sketch can be obtained in milliseconds using `millis()` and in microseconds with `micros()`. Both return an `unsigned long` number. `long` is a type representing a 32-bit-long integer. In the memory of computers, negative numbers are represented in the so-called *two's complement*, exploiting the fact that numbers are always represented using a fixed number of digits. The idea is as follows. Imagine a tailor's measuring tape arranged into a ring, with its end corresponding to 0 that coincides with the 100 mark and suppose you want to write all the numbers with, at most, two digits. If you need only positive numbers, you can clearly represent those from 0 to 99. However, if you want to work with negative numbers, too, you can imagine the numbers to the right of 0 as being positive and those at its left (99, 98, 97, ...) as being negative, such that 99 corresponds to -1, 98 to -2, and so on.

Negative numbers are represented in the two's complement.

To write the opposite of a number, write the complement of each digit, then add 1.

To write a negative number this way is very simple: find the **complement** of each digit, i.e., the digit that summed to the previous one gives the last digit in the chosen base, then add 1 to the result. For example, -3, with two digits, is -03 and can be written finding the complement of 3 in base 10, i.e., 6, and that of 0, i.e., 9. The result is 96. Adding 1 to the latter give us 97, which, in fact, is the third number on the left with respect to 0. Similarly, -23 can be represented as $76 + 1 = 77$.

The rule comes from the observation that, with n digits, you can represent numbers from 0 to $b^n - 1$, where b is the base of the numeral system, and that b^2 has $n + 1$ digits. Given a number m, we can always find a number k such that

$$m + k = b^n. \tag{6.1}$$

If we cannot represent more than n digits, the extra digit in b^n drops and the above equation is equivalent to

$$m + k = 0, \tag{6.2}$$

being that b^n is always written as 1 followed by n zeros, whatever the base. Finding k is easy:

$$k = b^n - m = (b^n - 1) + 1 - m. \tag{6.3}$$

In the above example, $b = 10$ and $n = 2$, such that $b^n - 1 = 10^2 - 1 = 99$. Subtracting the number of which we want to know the complement ($m = 3$) leads to 96. Finally, adding 1, we obtain 97.

In a two's complement, the base of the numeral system is 2. Finding the complement of a number is super easy: just invert all the digits, then add 1 to the result. For example, in a 4-bit representation, 6 is represented as 0110. The complement of each digit gives 1001, while adding 1 gives 1010, corresponding to $8 + 2 = 10$ if interpreted as a positive number. Interpreted as a two's complement one, it has the property such that

$$0110 + 1010 = 10000, \tag{6.4}$$

where the leading 1 does not fit in the 4 bits and is ignored, making the sum of 0110 and 1010 equal to 0000, i.e., 0110 is the opposite of 1010.

Using this technique, all negative numbers begin with 1 and there is only one representation for zero (0000). The highest positive number is 0111, corresponding to 7, while the lowest possible number is 1000, corresponding to -8. In general, using the b's complement technique with n digits, we can represent numbers ranging from $-\frac{b^n}{2}$ to $+\frac{b^n}{2} - 1$.

Variables declared to be `long` can then contain values between $\frac{-2^{32}}{2} = -2\,147\,483\,648$ and $\frac{2^{32}}{2} = +2\,147\,483\,647$.

Using the `unsigned` **qualifier**, we get rid of the negative numbers, and the variable can represent numbers from 0 to $2^{32} - 1 = 4\,294\,967\,295$. For the elapsed time,

Whatever the base b of the numeral system, b^n is always written as 1, the first non-zero digit, followed by n zeros.

In base 2, flipping the bits and adding 1 to the result is enough to find the two's complement of a number.

Variables of type `long` can represent integer numbers with 32 bits. All integer types can be qualified as `unsigned` to represent only non-negative values.

this type is appropriate and allows us to represent it for up to more than four billion microseconds, corresponding to almost 72 hours.

Using this number to represent milliseconds, the maximum representable time can be extended up to 50 days.

6.5 An Acoustic Stopwatch

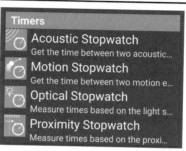

PHYPHOX, provides an interesting acoustic stopwatch, whose start and stop are triggered by sound.

Using PHYPHOX we have access to several types of stopwatch. One of them, the **acoustic stopwatch**, utilises the microphone of the smartphone as a sound detector. The stopwatch starts when a sound of sufficient intensity is detected and stops when a second sound exceeding the given threshold is recorded after a minimum delay τ, returning the time $t > \tau$ between the two events.

We start counting time when we pop a balloon sustaining a weight that falls to the floor. The noise produced by the impact stops the counter.

We can use the acoustic stopwatch as follows. We tie an inflated balloon to a relatively heavy object, such as a small bag of bolts, and suspend the system to a height h using some support, for example, a sieve from which the mesh has been removed. The smartphone is placed on the floor, closely adjacent to the spot where the bolts are expected to land (but far enough away not to risk the bolts falling on it). Popping the balloon with a needle starts the stopwatch, which, however, often stops before the falling object reaches the floor. The reason for this is that echoes in the room will reach the microphone before the bolts finish falling off. In order to get rid of echos we must measure the echo time as above and use it as τ. For example, if the time we read on the smartphone is 0.086 s, we use $\tau = 0.1$ s

We need to get rid of the echoes by properly setting the time during which the stopwatch is insensitive.

as a minimum delay and put the bolts at a height such that they take more than τ to fall. A rough estimation of the minimum height can be done by multiplying τ by five. In our example, $h_{min} \simeq 5 \times 0.1 = 0.5$ m.

Repeating the experiment with this setup gives us the right falling time, usually with a resolution of 1 ms. As usual, we should repeat the measurement several times to

6.5   An Acoustic Stopwatch

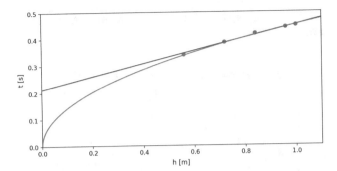

■ **Fig. 6.1** The time t needed for a body to fall from a height h. In red a linear interpolation is shown. The model for constant acceleration is in green

estimate the uncertainty. However, in this case, we can avoid repeating the measurements under the same conditions, instead taking them at different heights h_i for which we obtain a measurement of t_i. The height can be measured with a self-retracting metal tape measure with an uncertainty of ±0.3 mm.

Our data looks as it does in Fig. 6.1. The data seems to be aligned, but drawing the interpolating line, we clearly see that $t > 0$ when $h = 0$. There is no doubt that the time needed to fall from $h = 0$ must be null, hence this model is clearly wrong. The correct model can be recovered from the definitions of kinematic variables. Velocity, for example, is defined as

> To get a glimpse of the data, it is useful to make a plot of them.

$$v = \lim_{\Delta t \to 0} \frac{\Delta h}{\Delta t} = \frac{dh}{dt}, \tag{6.5}$$

thus, $dt = \frac{dh}{v}$, and integrating both members,

$$t = \frac{h}{v} \tag{6.6}$$

if v is constant. In this case, t grows linearly with h, and this is manifestly not the case: any line passing by $O = (0, 0)$ cannot pass through (in the vicinity of) all the data points. This means that v is not constant, and the acceleration

> It is easy to spot that v is not constant from the data. If $v \neq const$, the acceleration is not null.

$$a = \lim_{\Delta t \to 0} \frac{\Delta v}{\Delta t} = \frac{dv}{dt} \neq 0. \tag{6.7}$$

6

If a is constant, $v(t) = v(0) + at$. Since $dh = v\, dt$, integrating both members, we obtain

$$h = \int_0^t v(t)dt = v(0)\int_0^t dt + a\int_0^t t\, dt = v(0)t + \frac{1}{2}at^2 \,.$$

$$(6.8)$$

In our case, $v(0) = 0$, thus

$$t = \sqrt{\frac{2h}{a}} \,.$$

$$(6.9)$$

Data are well described by the hypothesis $a = const.$ There is no reason, then, to suppose that $a \neq const.$

The latter is shown in Fig 6.1 in green. In fact, the curve well interpolates the data, and for $h = 0$, $t = 0$. It is worth noting that, expanding the square root in a truncated Taylor series around $= h_0$ and writing $h = h_0 + y$,

$$\sqrt{2\frac{h_0 + y}{a}} \simeq \sqrt{\frac{2h_0}{a}} + \frac{y}{\sqrt{2ah_0}} \,,$$

$$(6.10)$$

i.e., the falling time grows almost linearly with y. To estimate a, we can linearise the relationship between t and h by squaring the equation to obtain

$$t^2 = \frac{2h}{a} \,.$$

$$(6.11)$$

Linearisation is always a good technique for obtaining the parameters of a curve. Slopes and intercepts are easy to determine.

The slope of the graph of t^2 as a function of h is $m = \frac{2}{a}$ and can be obtained computing all the possible ratios $\frac{\Delta t^2}{\Delta h}$. The result is

$$m = 0.19 \pm 0.02\, \frac{\mathrm{s}^2}{\mathrm{m}} \,.$$

$$(6.12)$$

From this number, we can obtain the acceleration as

$$a = \frac{2}{m} = \frac{2}{0.19} \simeq 10.5\, \frac{\mathrm{m}}{\mathrm{s}^2} \,.$$

$$(6.13)$$

The uncertainty σ_x on x propagates to $f(x)$ such that the uncertainty σ_f on the latter is given by $f'(x)\sigma_x$.

What about its uncertainty? To estimate it, let's analyse what happens to a when m fluctuates by σ_m:

$$a' = \frac{2}{m'} = \frac{2}{m + \sigma_m} \simeq \frac{2}{m} - \frac{2}{m^2}\sigma_m \,.$$

$$(6.14)$$

The variance of a is the square of its deviation, i.e.,

$$\sigma_a^2 = \left(a' - a\right)^2 = \frac{4}{m^4}\sigma_m^2, \qquad (6.15)$$

thus

$$\sigma_a = \frac{2\sigma_m}{m^2}. \qquad (6.16)$$

In this expression, we recognise the derivative of the expression for a. This is a general rule, in fact, as you can easily verify in numerous cases. The uncertainty about x **propagates** to a function $f(x)$ of it as $\sigma_f \simeq \frac{df}{dx}\sigma_x$. In our example, $\sigma_a = 0.2 \frac{m}{s^2}$, and our final result is

We always write our result keeping one significant digit in the uncertainty and writing the central value accordingly.

$$a = 10.5 \pm 0.2 \, \frac{m}{s^2}. \qquad (6.17)$$

According to the *Bureau Gravimetrique International* (BGI, ▶ http://bgi.obs-mip.fr/), the gravity acceleration in Rome, where we took our measurements, is $g = 9.803 \frac{m}{s^2}$. Our data differ from this measurement by

In many cases, we can compare our results with those of others, often available on the internet.

$$\Delta = \frac{10.5 - 9.803}{9.803} \simeq 0.07, \qquad (6.18)$$

i.e., by 7 %. Not bad considering the quality of the tools chosen. Despite the apparent agreement, considering the uncertainties, the data do not agree. In fact, the difference between the two numbers $10.5 - 9.803 \simeq 0.7 \frac{m}{s}$ is more than three standard deviations and there is little probability that this is due to some statistical fluctuation.

A discrepancy of three standard deviations is large, but not so large as to be ascribed to new phenomena. A conventional threshold for claiming a discovery is five sigmas. Assuming, then, that the BGI measurement is known with far greater precision than ours, we need to find the reason that leads to the **systematic** shift in our data.

The existence of some systematic effect is signalled by the fact that the intercept of the line interpolating data of t^2 vs. h is not null, even though it is expected to be so. In fact, it is $q = 0.013 \pm 0.002 \, s^2$. The consequence of $q \neq 0$ is that $t(0) = \sqrt{q} \simeq 0.1$ s. A large effect, indeed, if compared to the 0.3–0.4 s from our data. Moreover, there are other effects we have not considered. The sound from the explosion of the balloon takes $t_s = \frac{h}{c}$, where $c \simeq 340$ m/s is the speed of sound, to reach the phone. As a consequence, the effective travel time of the weight is longer than that measured by t_s, even if $t_s \simeq 0.002 - 0.003$ is small compared

Conventionally, a discovery is claimed when the discrepancy between a measurement and the theoretical prediction is larger than five standard deviations. This is a general rule that must be taken cum grano salis. *It depends on how solid the theory is and how credible the measurements are.*

6

to t_i. To correct for this effect, we can find the slope of

$$t^2_{corr} = \left(t + \frac{x}{c} \right)^2 \qquad (6.19)$$

as a function of x, finding $m = 0.20 \pm 0.02 \, \frac{s^2}{m}$ and $q = 0.007 \pm 0.002 \, s^2$, which is still significantly different from zero. The corresponding value for a is

$$a = 10 \pm 1 \, \frac{m}{s^2} . \qquad (6.20)$$

The source of any significant discrepancy in an experiment must be carefully analysed and may lead to the design of new experiments. Never simply discard values that are not fully understood.

Now, a is compatible with $g = 9.803 \, \frac{m}{s^2}$, because $a - g \simeq 0.2 \, \frac{m}{s^2}$, quite less than the uncertainty with which a is known. The fact that $q \neq 0$ can be ascribed to various causes. The smartphone takes a certain amount of time to start and stop counting. If this time is not taken into proper account, it adds to the measured time and results in $q > 0$.

In these cases, it is worth designing new experiments, in which systematic uncertainties are missing or, at least, different. New experiments may imply a completely different technology and methods, or can be conducted under different conditions. For example, repeating the above experiment for $h \gg 1$ m makes the possible offset t_s negligible.

Once we have established that a given object falls with an acceleration equal to **g**, we can start experimenting with objects that have different masses, to investigate whether acceleration depends on the mass and, if so, how. We can also investigate whether the acceleration depends on the shape or on the density of the falling object, keeping the mass fixed. In these cases, we may have to take into account that the time needed to reach the floor may depend on the object's orientation, so the experiment can be quite difficult, indeed. Eventually, we find that, at least for large enough masses, objects fall with the same acceleration, irrespective of mass, shape, density, etc. Very light objects (small leaves) or objects that are not as light but are large in size (a large piece of cardboard), as well as low density ones (inflated balloons) fall differently. Further studies reveal that such behaviour is due to the fact that these objects are not, in fact, freely falling: they are immersed in a fluid (the air) that imposes forces upon it that temper the pull of gravity. Removing the fluid (e.g., repeating the experiments in vacuum) makes it clear that even those objects fall with the same acceleration as heavy ones. The latter are apparently not subject to the effects of air: in fact, they are, but the effects on them are negligible and cannot be spotted by our measurements, due to limited precision.

On YouTube, it is possible to view interesting free-falling experiments, like the one conducted at the NASA Space Power Facility in Ohio, keepers of the world's biggest vacuum chamber, or the one conducted by Commander David Scott during the Apollo 15 mission, in which he let a hammer and a feather fall to the surface of the moon.

6.6 Uncertainty Propagation

Any uncertainty in the knowledge of the value of a quantity x affects, manifestly, each function $f(x)$ of it. The way in which the error σ_x propagates to $f(x)$ follows general rules.

In fact, if $f(x)$ is a function of x, any change from x to $x + \sigma_x$ results in a change of $f(x)$ to $f(x + \sigma_x)$. For small enough σ_x,

$$f(x + \sigma_x) \simeq f(x) + f'(x)\sigma_x . \tag{6.21}$$

The uncertainty about $f(x)$ is then

$$\sigma_f = |f(x + \sigma_x) - f(x)| \simeq |f'(x)\sigma_x| . \tag{6.22}$$

The absolute value operator is uncomfortable, and we prefer to write the variance of $f(x)$,

$$\sigma_f^2 \simeq f'^2(x)\sigma_x^2 . \tag{6.23}$$

If f is a function of more than one variable, e.g., $f = f(x, y, z)$, its uncertainty depends on the uncertainty about each quantity. Supposing that only x can fluctuate, we may write

$$\sigma_f^2 \simeq \left(\frac{\partial f}{\partial x}\right)^2 \sigma_x^2 , \tag{6.24}$$

where $\frac{\partial f}{\partial x}$ represents the partial derivative of f with respect to x, i.e., considering y and z as constants. Of course, other variables behave the same way, and since all the effects sum up, we can write

$$\sigma_f^2 \simeq \sum_{i=1}^{n} \left(\frac{\partial f}{\partial x_i}\right)^2 \sigma_i^2 . \tag{6.25}$$

For example, if the measurements of the three components of a vector are $x \pm \sigma_x$, $y \pm \sigma_y$ and $z \pm \sigma_z$, the magnitude of the

To find general rules about the propagation of the uncertainty about data to a function of them, we expand such a function as a Taylor series around the central value of the data, keeping only the first order term.

The variance of a function of many variables is the sum of partial variances. As a result, we say that uncertainties sum in quadrature.

vector is given by $v = \sqrt{x^2 + y^2 + z^2}$ and its uncertainty is computed as follows. First, compute the partial derivatives of v:

$$\frac{\partial v}{\partial x} = \frac{x}{v} \qquad \frac{\partial v}{\partial y} = \frac{y}{v} \qquad \frac{\partial v}{\partial z} = \frac{z}{v}. \tag{6.26}$$

Then, multiply each derivative by the corresponding uncertainty, square the result and sum all the terms together to obtain

$$\sigma_f^2 = \left(\frac{x}{v}\sigma_x\right)^2 + \left(\frac{y}{v}\sigma_y\right)^2 + \left(\frac{z}{v}\sigma_z\right)^2. \tag{6.27}$$

6.7 Measuring Accelerations

Collecting data from the accelerometer of a smartphone using PHYPHOX is a good exercise for illustrating uncertainty propagation.

It is instructive to work out a few examples using functions of real data. We can, for example, collect 30 s of accelerometer data from PHYPHOX. If the smartphone is at rest, the acceleration components must be those of the gravitational acceleration, whose magnitude is 9.8 m/s^2.

Smartphones use accelerometers to determine their orientation. They measure the three components of the acceleration in a reference frame centred on the phone, where the x axis is oriented along the width of the device, the y axis along its height and the z axis is perpendicular to the screen, as in the figure.

Smartphones use accelerometers to determine their orientation.

Comparing the size of the x and y components, one can tell whether the phone is kept in portrait or in landscape mode. Raw data from the accelerometer are collected by PHYPHOX, allowing us to use these data to engage in some physics.

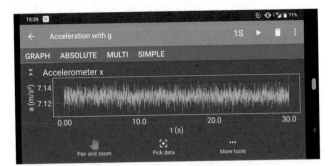

◘ Fig. 6.2 Data collected using the accelerometer of a smartphone, when it was kept almost in something close to landscape orientation

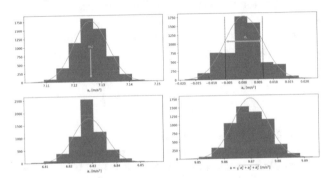

◘ Fig. 6.3 Histograms of the values of the three components of the acceleration vector and of its magnitude with the phone at rest. The mean value and the standard deviation are labeled, respectively, as a_x and a_y

To take the measurements, we kept the smartphone in landscape mode and started a timed run with a delay of 1 s in order to avoid registering vibrations due to the tap on the start button. Figure 6.2 shows a screenshot of the phone at the end of the data-taking.

During the measurements, the smartphone was at rest, thus the measured acceleration is $\mathbf{g} = (g_x, g_y, g_z) = const.$ Figure 6.3 shows the histogram of the values collected for each component and for the vector magnitude.

We can notice that data distribute in such a way that the histograms resemble a bell. The distribution is more or less symmetric around the mean value, to which data tend to accumulate. The width of the distribution represents the uncertainty: the larger the latter, the wider the distribution. Computing averages and standard deviations of the components of $\mathbf{g}$ results in the following:

Data distribute such that they appear denser around certain values.

$$g_x = 7.126 \pm 0.006 \, \text{ms}^{-2}$$
$$g_y = 0.001 \pm 0.006 \, \text{ms}^{-2} \qquad (6.28)$$
$$g_z = 6.829 \pm 0.007 \, \text{ms}^{-2} .$$

The acceleration measured by a phone at rest is the opposite of the gravitational acceleration.

It is worth noting that, while $g_y \simeq 0$, g_x and g_z are non-null and positive. The reason for that is that, during the measurements, we kept the phone in landscape orientation leaning against our laptop display. The positive direction of the x and z axis is, respectively, to the right side of the phone and exiting from its display. The acceleration measured by the device thus has the opposite direction with respect to the gravitational acceleration. The next section explains why.

A Gaussian well describes the shape of data distribution.

The orange curve superimposed onto the histograms is the so-called **Gaussian** curve, whose expression is

$$G(x) = \frac{C}{\sqrt{2\pi}\sigma} \exp\left(-\frac{1}{2}\left(\frac{x-\mu}{\sigma}\right)^2\right). \qquad (6.29)$$

Gaussians describing data have their maximum close to the average value and their width is proportional to the standard deviation.

A Gaussian has its maximum at $x = \mu$ and its width is determined by the σ parameter. It can be shown that the curve width at half maximum (FWHM: Full Width at Half Maximum) is about $2\sqrt{2\ln 2}\sigma \simeq 2.4\sigma$. C is a **normalization** coefficient, representing the height of the curve at its maximum. In fact, for $x = \mu$, the exponential is 1 and $G(\mu) = \frac{C}{\sqrt{2\pi}\sigma}$, hence

$$C = \sqrt{2\pi}\sigma \, G(\mu) . \qquad (6.30)$$

The factor $\sqrt{2\pi}\sigma$ makes the area under the curve equal to

$$\int_{-\infty}^{+\infty} G(x)dx = C . \qquad (6.31)$$

The magnitude of the average acceleration vector is thus $\sqrt{g_x^2 + g_y^2 + g_z^2} \simeq 9.870 \, \text{ms}^{-1}$. The predicted error for this value is given by the above formula (6.27) and is $\sigma_g = 0.006 \, \text{ms}^{-1}$. From the data histogrammed in Fig. 6.3, we can easily see that the standard deviation is, in fact, $\sigma_v^{exp} = 0.006 \, \text{ms}^{-1}$.

The above considerations are very general, indeed. Histograms of statistically independent measurements often distribute as Gaussians, whose mean coincides with the average of the measurements and whose width depends on their standard deviation. The uncertainty of data propagates in their functions such that their variances sum according to the uncertainty propagation formula (6.25).

6.8 MEMS Accelerometers

Knowing your instruments is among the most important rules in experimental physics. Never use an instrument without understanding its working principle. Smartphones are no exception.

A physicist must always know how his/her instruments work.

It is well known that, being in a non-inertial, i.e., accelerated, reference frame, Newton's second Law is not valid unless we add **fictitious forces** to those acting as a result of some interaction:

Smartphone accelerometers exploit the Newton's second Law in non-inertial reference frames.

$$\mathbf{a} = \frac{\sum_i \mathbf{F}_i - \mathbf{F}_{fict}}{m}. \tag{6.32}$$

In a system where $\sum_i \mathbf{F}_i = 0$, then $\mathbf{F}_{fict} = -m\mathbf{a}$. For example, consider a smartphone at rest on the seat of a merry-go-round that is also at rest. The smartphone is subject to the gravitational force $\mathbf{F}_g$, directed towards the earth's center, and to the normal force $\mathbf{F}_N$, applied by the seat, oriented in the opposite direction such that they cancel each other out: $\mathbf{F}_g + \mathbf{F}_N = 0$. Its acceleration is thus zero.

Letting the system rotate, the smartphone remains at rest only until there is sufficient friction. At a certain point, we see the smartphone moving radially, and it eventually falls. What happens is the following.

When the merry-go-round starts rotating, thanks to static friction acting on the phone, it applies a centripetal force to the latter that, as a consequence, moves along a circular path. At a certain point, the static friction eventually vanishes and the phone is no longer subject to any force. According to Newton's second Law, it moves with constant velocity and, while the seat keeps rotating, it slides along a line tangent to its original path until it reaches the border of the seat and falls. The above description is fully coherent with Newtonian mechanics, as it is done within an inertial reference frame.

A smartphone on a merry-go-round follows a circular path because of the **centripetal** *acceleration provided by static friction. When the friction vanishes, the smartphone starts moving according to the Newton's first Law.*

Seen from a reference frame attached to the rotating system, the smartphone's motion can be described differently. Initially, the gravitational and normal forces cancel each other out and the smartphone is at rest on the seat. At a certain point, it starts sliding along a radius until it falls from the merry-go-round.

6

Manifestly, for the non-inertial observer, an acceleration had to develop and, according to Newton's Law, there must be a force to produce it. However, there is no such force, or, more precisely, there are no interactions between the phone and other systems such that the phone is *attracted* towards the border of the carousel. In fact, if friction was acting at the beginning of the rotation while the system was being observed from the outside, it would have to keep acting even for anyone who was observing the phone from inside the rotating frame. Contrastingly, while there are no forces acting on the phone when it is sliding along the tangent, there cannot be any force for anyone who is observing it while standing on the rotating frame.

Fictitious forces are introduced to make predictions by different observers consistent.

The only way to make the two descriptions coherent is to add fictitious forces in non-inertial reference frames. Fictitious forces are just a mathematical trick for making Newton's laws valid in non-inertial frames. Since the systems experience an acceleration in those systems, we can pretend there are forces producing them, according to Newton's Law.

When a physicist says that forces measured in non-inertial frames are fictitious, he/she means that they are not the result of some interaction. There are no sources of fictitious forces, while there are sources for gravitational, electrostatic and magnetic forces.

The term "fictitious" is somewhat misleading because, in fact, according to our experience, these forces are not at all fictitious. They are among the most real and frequently experienced forces. Even more so than gravity. In fact, we do not *feel* gravity as a force, since our senses are immersed in the gravitational field, too. The feeling of falling does not coincide with the feeling of being drawn down. On the contrary, when we execute a turn in a car, we clearly feel a force pushing us towards the outside of the road.

In fact, fictitious forces are as real as gravity, but they are not the result of an interaction. In other words, there is no source for the **centrifugal** force, as it is called, while there are sources for gravitational, elastic, electrostatic and magnetic forces. We should differentiate between a force and an interaction that did not exist in Newton's time. Interactions produce forces, and forces develop in non-inertial frames as well. It is worth noting that the (local) equivalence between an accelerated frame and an inertial frame immersed in a gravitational field is the basis for one of the most advanced physical theories: **general relativity**.

Measuring accelerations is equivalent to measuring forces.

According to the above considerations, fictitious forces can be written as

$$\mathbf{F}_{fict} = -m\mathbf{a} \,, \tag{6.33}$$

where **a** is the acceleration of the reference frame. Note, too, that, in this case, Newton's second law takes the form $\mathbf{F}_{fict} = m\mathbf{a}$ and not $\mathbf{a} = -\frac{\mathbf{F}_{fict}}{m}$. In the above formula, in fact, $\mathbf{F}_{fict}$ is *defined* as such, and we state that the force is, in fact, equal to the product of m and **a**.

Such a long preamble is meant to introduce the way in which accelerations are measured by an instrument called an **accelerometer**. Suppose we attach a mass m to a spring and suspend this system such that the spring stretches because the mass pulls it towards the floor with a force $F = mg$. If we move the suspension point abruptly upwards, the spring initially stretches even more, as if it were subject to a force whose intensity is $F_{fict} = ma$. Moreover, if we accelerate the suspension point downwards, the spring shortens as if a force of intensity $F_{fict} = ma$ was acting upon it. The force can easily be measured: a spring is, in fact, a dynamometer, and $\mathbf{F} = -k\Delta\mathbf{x}$, where k is the elastic constant of the spring and $\Delta\mathbf{x}$ the variation of the position of one of its ends with respect to the other. A measurement of the length Δx can then be turned into a measurement of the intensity $F = k\Delta x$ that, in turn, can be cast into an acceleration as $a = k\frac{\Delta x}{m}$.

A system like this measures the acceleration in the vertical direction. A horizontal spring measures the acceleration along the direction of its axis. With three orthogonal springs, we can then measure the three components of the acceleration vector.

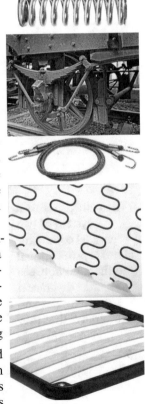

Manifestly, when this system is at rest, the springs are stretched in the vertical direction by $\Delta\mathbf{x} = \frac{m\mathbf{a}}{k}$ and, in this condition, we measure $\mathbf{a} = \mathbf{g}$. On the other hand, if the system is falling, it finds itself in a non-inertial reference frame moving with acceleration **g**, hence the experienced acceleration is $\mathbf{a} - \mathbf{g} = 0$.

This is exactly the way in which smartphones' accelerometers work. Remember that the system described above is a **model**: it represents a mass attached to something exhibiting an elastic behaviour. The fact that the mass is represented as being pointlike is an oversimplification of the mathematical model. It just means that the details of the shape of the mass are irrelevant to its dynamics. The spring is often represented as a zig-zag segment $\mathcal{WV}$, inspired by the shape of a kind of spring. However, as shown on the side of the page, springs exist in a variety of shapes. Its shape is irrelevant to the dynamics. What is important is that its length changes proportionally to the applied force.

6

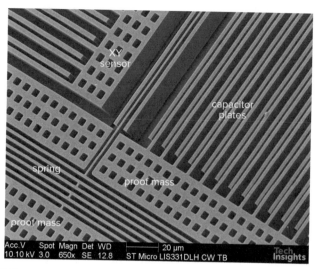

☐ **Fig. 6.4** Microphotograph of a MEMS (left). The spring is made of long structures interconnected by tiny joints, resembling the springs of a bed (visible in the lower left corner of the figure). The mass resembles a brick and is connected to a set of parallel plates that move together with the mass with respect to other plates fixed on the substrate. Courtesy of TechInsights Inc.

Smartphones' accelerometers are **microelectromechanical systems** (MEMS): microscopic mechanical devices obtained by assembling tiny components whose size can be as short as a few tens of micrometers. In these systems, three microscopic masses are suspended to three microscopic springs such that they can freely move along as many mutually orthogonal directions parallel to the sides of the phone.

The voltage across a system of two parallel conductor plates is proportional to the ratio between the distance d between plates and the area S that they share: $V \propto \frac{d}{S}$.

The elongation $\Delta\mathbf{x}$ is measured by exploiting the variation of some electrical property of the device as it changes its shape. Often, for example, the MEMS consists of a pair of metallic plates, one of which is attached to a spring. Such a system of conductors works as a **capacitor**: a system for storing electrical energy. As soon as the distance between the plates or the area of a surface exposed to the other changes, an electrical current flows from or to the capacitor. Measurement of such a current can be cast into a measurement of the acceleration. Because of the above considerations, smartphones' accelerometers do not return 0 when they are at rest, but rather **g**.

Figure 6.4 shows a microphotograph of such a device (namely, the LIS331DLH by ST-microelectronics).

6.9 Annotating Graphs

It can be useful, when presenting experimental data, to decorate graphs with graphical elements like arrows or text, as shown in the above examples.

annotate() decorates graphs with text and graphical elements.

This is done, in Python, using the annotate() method of pyplot. The annotate() method takes at least two parameters: a string and a pair of coordinates. A statement like

```
plt.annotate('$\\langle a_{x}\\rangle$',
          xy=(mu, sigma))
```

prints the string $\langle a_x \rangle$ at coordinates mu and sigma. Coordinates refer to those used in the plot, by default. The meaning of the coordinates can be changed with the optional parameter xycoords='...'.

In the above example, the text to be printed is given in LaTeX and the string \langle a_x \rangle is rendered as $\langle a_x \rangle$. \langle and \rangle represent, respectively, the left and right acute parentheses, while the characters inside braces after the underscore are rendered as a subscript. The double backslash present in Python code is needed, since the backslash \ is a meta-character representing an *escape* sequence. A character following an escape character is not interpreted as such, but rather as a non-printable character. For example, to represent the newline character, we use \n, where n is interpreted as a newline character, rather than a plain n character, being preceded by the backslash. Since the latter is a meta-character, in order to represent it, we need a meta-character, too. This is done by writing \\. A full example follows.

Acute parentheses are represented in LaTeX as \langle and \rangle. Subscripts follow an underscore.

Meta-characters, also known as escape characters, are used for different interpretations of the characters that follow them.

```
plt.annotate('$\\sigma_y$',
      xy=(muy, C*0.65),
      color='white')
plt.annotate('',xy=(muy-sigmay, C*0.6),
      xytext=(muy+sigmay, C*0.6),
      arrowprops=dict(ec='orange',
                arrowstyle='<|-|>',
                lw=4))
```

This code writes a white LaTeX symbol σ at coordinates represented by the variables muy and C multiplied by 0.65. Then, it makes a horizontal orange arrow with heads at

The appearance of the decorations is highly customisable.

both extremes (set by the `arrowstyle` option) and a line width of 4 points. `xytext` and `xy` are the starting and ending coordinates of the arrow, while `ec` stands for *edge color*, and is used to assign a color to the arrow. Note that the corresponding string in `annotate()` is empty, such that only the arrow is shown.

In such a command, `arrowprops` is called a *dictionary* (hence, `dict(...)`), i.e., a list of key-value pairs.

6.10 Instrument Characteristics

To quantify the discrepancy between two numbers, we compute their difference in units of uncertainty.

With the experiment conducted with the smartphone at rest, we clearly see that taking just one measurement is not enough and that we must repeat the measurement several times, then compute the average to obtain a reasonably reliable value. We found that the gravitational acceleration is

$$g = 9.870 \pm 0.006 \, \frac{m}{s^2} . \tag{6.34}$$

In order to compare this value to the one obtained from BGI, we need to take the difference and evaluate their *distance* in units of uncertainty. For the time being, we assume that the BGI value $g = 9.803 \, \frac{m}{s^2}$ is known with infinite precision (i.e., it has negligible uncertainty). We find

$$\Delta = \frac{9.870 - 9.803}{0.006} \simeq 11 . \tag{6.35}$$

The probability of finding a given value can be approximated by the relative frequency with which we have found that particular value. Modelling the distribution of the data as a Gaussian, we can use its values as the observed frequency.

This is a huge discrepancy, a fortiori if we consider that the uncertainty of $0.006 \frac{m}{s^2}$ is overestimated, as we will see in the next chapter. The probability that two values differ by more than ten standard deviations just because of statistical fluctuations, in fact, is extremely small. Indeed, the probability for a Gaussian distributed variable to lie between $-x$ and x is

$$P(x) = \frac{1}{\sqrt{2\pi}\sigma} \int_{-x}^{+x} \exp\left(-\frac{1}{2}\left(\frac{x-\mu}{\sigma}\right)^2\right) dx . \tag{6.36}$$

Putting $t^2 = \frac{1}{2}\left(\frac{x-\mu}{\sigma}\right)^2$, the integral becomes

$$P(x) = \frac{1}{\sqrt{\pi}} \int_{-t}^{+t} \exp\left(-t^2\right) dt = \frac{2}{\sqrt{\pi}} \int_0^{+t} \exp\left(-t^2\right) dt ,$$

$$(6.37)$$

where the latter equivalence comes from the fact that the function being integrated is symmetric around $t = 0$. $P(x)$ is thus a universal function, since it does not depend on particular values of μ and σ and is called the **error function**,

$$\text{erf}(t) = \frac{2}{\sqrt{\pi}} \int_0^{+t} \exp\left(-t^2\right) dt ,$$

$$(6.38)$$

whose values can be computed numerically. Clearly, substituting the expression of t in $\text{erf}(t)$ gives

$$\text{erf}\left(\left|\frac{x - \mu}{\sqrt{2}\sigma}\right|\right) = P(x) ,$$

$$(6.39)$$

hence $\text{erf}\left(2^{-\frac{1}{2}}\right) \simeq 0.68$ corresponds to the fractional area below a Gaussian in a $\pm\sigma$ interval. Similarly, $\text{erf}\left(\frac{2}{\sqrt{2}}\right) \simeq 0.95$ and $\text{erf}\left(\frac{3}{\sqrt{2}}\right) \simeq 0.997$.

An analytic expression of the integral of a Gaussian does not exist. The integral can, however, be computed numerically as $\sum P(x_i)\Delta x$, for small enough Δx.

The **complementary error function** $\text{erfc}(x)$ is defined as

$$\text{erfc}(x) = 1 - \text{erf}(x) = \frac{2}{\sqrt{\pi}} \int_x^{\infty} \exp\left(-t^2\right) dt .$$

$$(6.40)$$

If x is distributed as a Gaussian, $y = \frac{x-\mu}{\sigma}$ is said to be **normally distributed**, where a **normal distribution** is a Gaussian with $\mu = 0$ and $\sigma = 1$, i.e.,

The normal function is a Gaussian with null mean and unit width.

$$P_N(y) = \frac{1}{\sqrt{2\pi}} \exp\left(-\frac{y^2}{2}\right).$$

$$(6.41)$$

Given the above definitions, the probability of a normally distributed variable x being larger than $x_0 > 0$ is

$\text{erf}\left(\frac{1}{\sqrt{2}}\right) = 0.68,$

$\text{erf}\left(\frac{2}{\sqrt{2}}\right) = 0.95$ and

$\text{erf}\left(\frac{3}{\sqrt{2}}\right) = 0.997.$

$$P_N(x > x_0) = \frac{1}{\sqrt{2\pi}} \int_{x_0}^{\infty} \exp\left(-\frac{x^2}{2}\right) dx = 1 - \frac{1}{\sqrt{2\pi}} \int_{-\infty}^{x_0} \exp\left(-\frac{x^2}{2}\right) dx =$$

$$1 - \left(\frac{1}{\sqrt{2\pi}} \int_{-\infty}^{0} \exp\left(-\frac{x^2}{2}\right) dx + \frac{1}{\sqrt{2\pi}} \int_0^{x_0} \exp\left(-\frac{x^2}{2}\right) dx\right) =$$

$$\frac{1}{2} - \frac{1}{2}\text{erf}\left(\frac{x_0}{\sqrt{2}}\right) .$$

$$(6.42)$$

Known values for erf(x) are erf$(0) = 0$, erf$\left(\frac{1}{\sqrt{2}}\right) = 0.68$, erf$\left(\frac{2}{\sqrt{2}}\right) = 0.95$ and erf$\left(\frac{3}{\sqrt{2}}\right) = 0.997$ (you can recognise the values of the fractions of data found within one, two and three standard deviation intervals described in Sect. 7.4).

We found a value that is 11 standard deviations greater than expected. Such an event can occur with a probability of

$$P_N(x > 11) = \frac{1}{2} - \frac{1}{2}\text{erf}\left(\frac{11}{\sqrt{2}}\right) \simeq 2 \times 10^{-28}, \quad (6.43)$$

if the discrepancy is ascribed only to random fluctuations. We believe that an event with such a probability is extremely unlikely to happen, and we must thus ascribe the discrepancy to something else.

Instruments returning correct values for measured quantities are said to be accurate. Smartphone accelerometers are not very accurate, although they are rather sensitive and precise.

A possible explanation can be that the tool we are using to take this measurement is not very **accurate**. Indeed, smartphone MEMS are designed to be quite sensitive, i.e., to respond to very tiny signals, but not to be accurate, i.e., to provide an exact value. A smartphone accelerometer is designed to provide a fast non-null response, even for very small accelerations, since they are usually impressed by handling it. However, the absolute value of the acceleration is irrelevant to the device knowing whether it is in landscape or portrait orientation, these devices are simply not very carefully calibrated, to keep their cost affordable.

Instruments can be classified according to different characteristics.

The **resolution** of an instrument is the smallest measurable unit.

The **sensitivity** is the ratio between the instrument's smallest response and the corresponding measured quantity. In an accelerometer working as above, the sensitivity is given by $\frac{\Delta x}{\Delta a}$, where Δx is the variation of the length of its spring and Δa the corresponding acceleration.

Precision is the measure of the ability of an instrument to provide the same result for the same quantity. It can be defined as the closeness between different measurements of the same object. This characteristic is often misunderstood. Sometimes, it is confused with sensitivity. More often, it is believed that an instrument that always gives the same result in response to the same stimulus is necessarily very precise. In fact, a very precise instrument is often an instrument that provides a very approximate result.

Accuracy, finally, is the degree of conformity of the result of the measurement to a correct value.

Consider, for example, an analog bathroom scale and a kitchen scale, such as those shown in the figure. The resolution of the bathroom scale is 1 kg, while that of the kitchen scale is 20 g. The sensitivity of the latter can be expressed as

$$s_k = \frac{\Delta\theta}{\Delta m} = \frac{2\pi}{5\,\mathrm{kg}} = 0.4\,\mathrm{kg}^{-1}, \tag{6.44}$$

while that of the bathroom scale is

$$s_b = \frac{2\pi}{150\,\mathrm{kg}} \simeq 0.013\,\mathrm{kg}^{-1}. \tag{6.45}$$

The first is more sensitive because its response to the same stimulus is higher than that of the second one. Putting a weight of 3 kg on the kitchen scale, the needle rotates $0.4 \times 3 = 1.2$ radians (69°), while the needle of the bathroom scale rotates $0.013 \times 3 = 0.039$ radians (2.2°).

In digital scales, the sensitivity can be expressed as the inverse of the resolution. For example, if the scale has a resolution of 1 g, it responds with one digit per gram, thus the sensitivity is $\frac{1}{1\,\mathrm{g}} = 1\,\mathrm{g}^{-1}$.

Both scales are quite precise. They always give the same response to the same stimulus. However, weighing several boxes of pasta with both scales will probably result in the same reading with the bathroom scale and slightly different readings with the kitchen scale. Contrary to common beliefs, then, the former can be considered more precise with respect to the latter, as an instrument. Note, however, that the uncertainty of the result is expected to be better in the case of the kitchen scale, which, in fact, provides a more precise result. In fact, even if we always obtain the same result repeating the measurement in both cases, the closeness of the measurements among them can be represented by the corresponding resolution. When talking about instruments, being more precise does not always imply being *better*, however, when talking about measurements properly quoted, the word "precision" takes on its commonly agreed upon meaning.

6

The accuracy of an instrument may not be constant with time. When taking measurements, we should always check the accuracy of the instruments against known values.

The accuracy of the instruments depends on the quality of the calibration. It may also depend on environmental conditions and it may change with time, due to wear. The kitchen scale used at home is probably less accurate than the one at your preferred deli counter, which is subject to periodical revisions by the metrological authorities.

The accelerometer of a smartphone is quite sensitive and has relatively good precision, but is quite inaccurate. Though it is not a good idea in using it to measure absolute accelerations, we can be confident to use it to measure acceleration differences. In fact, the inaccuracy can be thought of as an offset to the correct value that is canceled out when two values are subtracted.

Summary

An experiment is aimed at measuring a physical quantity y as a function of another quantity x. It is possible that y depends on other physical quantities u, v, z, etc. Typically, we try to let only one of them vary, either by making the effects of a variable negligible or by keeping it constant. For example, the time needed for a body to fall may depend, in principle, on its height, its shape, its mass, etc. To understand the laws of motion, we study how bodies fall by changing their height, keeping their shape and mass constant. Once we have fully understood the motion in this case, we can study how shape and mass influence the results.

To measure times, we need a stopwatch. Triggering the watch by hand introduces large fluctuations and biases. It is much better to use an automatic trigger.

Photogates are systems that monitor the intensity of a light source. If something passes between the light source and the sensor, the luminous flux is interrupted for a while. The abrupt change in intensity can be used to trigger a device.

Free-fall happens at constant acceleration. We realise this from the fact that the falling time t as a function of the height h does not go to zero for $h \to 0$, while it does so if $a = const$.

Linearisation is a way to find the parameters of a curve that describes experimental data.

Instruments are characterised by their resolution (the smallest measurable unit), their sensitivity (the ratio between the instrument's response to the stimulus), their precision (the ability to give the same response to the same stimulus) and their accuracy (the closeness of the response to the correct value of the measured quantity).

Arduino

With Arduino, a photogate can be realised using an analog light sensor illuminated by an external source such as a lamp, a laser or an LED.

Arduino provides several functions for managing time: `delay()` and `delayMicroseconds()` cause a pause in the execution of the program, whose duration is expressed, respectively, in milliseconds and microseconds, as an argument. `millis()` and `micros()` return the time elapsed since the beginning of the execution of the sketch in milliseconds and microseconds, respectively.

Negative numbers are represented in the memory of a computer in a two's complement. Variables intended to store non-negative numbers can be declared as `unsigned`, extending the range of representable numbers by a factor two.

phyphox

PHYPHOX includes an acoustic stopwatch whose start and stop are triggered by sound. It starts when the phone detects a sound higher than the given threshold and stops when a second sound is detected after a user-defined delay. During the delay, sounds do not stop the counter.

An experiment in free-fall can be realised by suspending a weight from an inflated balloon. Popping the balloon triggers the start of the stopwatch, which stops counting when the noise of the weight hitting the floor is detected. A proper delay must be set to avoid cessation of the measuring time because the echo of the balloon explosion is detected.

Smartphones use accelerometers to determine their orientation. Data from the accelerometer can be collected using PHYPHOX.

When at rest, an accelerometer measures the opposite of the gravitational acceleration.

Accelerometers exploit the Newton's second Law in non-inertial frames so as to measure their acceleration. They measure the force applied on microscopic springs attached to microscopic masses.

Statistics

The uncertainty σ of a quantity x propagates to a function $f(x)$ as $\sigma_f = f'(x)\sigma$.

Uncertainties about more than one variable sum in quadrature, i.e., the variance of a function of more than one variable is the sum of the variances of the function taken as a function of only one variable at the time.

Experimental data often distribute as Gaussians peaked at the mean value and with a width proportional to their standard deviation.

To compare two results, we measure their difference in units of σ.

The error function $\text{erf}(x)$ is a universal function that can be used to evaluate the probability of observing a given value in an experiment. If the average value of the data is μ, the probability of observing a number higher than $\mu + x_0$ can be estimated as $\frac{1}{2} - \frac{1}{2}\text{erf}\left(\frac{x_0}{\sigma}\right)$, where σ is the standard deviation of the measurements.

6

Understanding Data Distributions

Contents

The original version of this chapter was revised: Typographical errors have been corrected. The correction to this chapter is available at ▶ https://doi.org/10.1007/978-3-030-65140-4_16

In this chapter, we examine the reasons why data distribute as observed. One can formulate a rigorous, formal theory of data analysis based on statistics and probability. Here, we concentrate on an intuitive explanation, leaving a more formal treatment to the following chapters and, for some of the more advanced topics, to specific courses.

7.1 On the Values Returned by Instruments

Results of experiments often distribute over a range of values. The shape of the distribution can be different, depending on the experiment and on the instruments used to perform the measurements. Yet, data most often appear to be distributed as a bell. There must be a reason for this. Indeed, we can assume that data fluctuate around *true* values because of the occurrence of random effects that cause the instruments to respond differently with respect to the ideal case. For example, we believe that the measurement of the acceleration of a phone at rest should return a value for g, instead of a distribution. However, the measurement of the acceleration is affected by the fact that the phone has to be placed on some kind of support that, for sure, will vibrate, even if only subtly. Any attempt to isolate it from external vibrations will fail: one can reduce the amplitude of the vibrations, but it is impossible to completely eliminate them. Even if we succeeded, the thermal noise would always affect the accelerometer: its temperature depends, in fact, on the kinetic energy of the particles of which it is composed, and it cannot be zero. Despite the fact that random fluctuations are random, we can take them into account in a relatively rigorous way thanks to statistics.

It must be noted that we implicitly assumed that there is, indeed, a *true* value for g. This is somewhat prejudicial, justified by the fact that data distribute closely around 9.8 m/s^2, so we are induced to believe that there must be a constant value of g with arbitrary precision that is impossible to reach because of our limited technology. In fact, this is likely to be untrue. If we had a *perfect* instrument, we would soon realise that g varies from point to point because its value is affected by all the masses in the vicinity of the point at which it is measured, even that of the instrument used to obtain it. Moreover, since everything is moving in the Universe, g changes with time. Maybe it only changes in regard to the tenth or the hundredth digit after the decimal point, but we are still unable to define a

Due to uncertainties, the result of an experiment is often a set of data, distributed over a range of values.

From the shape of the most common distributions, we can infer that the result of measuring a physical quantity fluctuates around a value, which we define as a *true value*, due to stochastic effects out of our control.

The concept of a *true value* is convenient, but not necessarily correct from an epistemological point of view.

constant true value of it. On the other hand, there is no need for that. We need to know the value of g with sufficient precision to make predictions using a model that, in any case, will be approximated, neglecting some effect that happens in reality or oversimplifying something else. We agree that, by *true* value, we mean the value that we would have measured if all the unwanted perturbing effects were removed. It is useless try to know it, and would also be useless from a practical standpoint even if you did, but it is useful to assume it exists.

7.2 Probability

Given that we are going to talk about random variables, it is worth expending a few words on the concept of **probability**. Regardless of any individual intuitive notions of what probability is, it took centuries for an internationally accepted definition to emerge.

An initial attempt to define probability was based on combinatorics and is credited to Pierre-Simon Laplace (1749–1827) [2]. In his essay, Laplace defined the probability P of the occurrence of a given **event** as the ratio between the number of cases in which the event can occur and the total number of possible events.

Laplace defined the probability P of the occurrence of a given event as the ratio between the number of cases in which the event can occur and the total number of possible events.

Here, an event is defined as something to which we can assign a truth value, i.e., something that can be true or false.

A very straightforward example of such a definition can be made using dice. The probability of obtain 3 in the throw of a single die is $\frac{1}{6} \simeq 0.17$, because there is one *favourable* case (the throw resulted in 3) out of six possible cases (the numbers 1 through 6). Similarly, the probability of extracting an ace from a deck of 54 cards (52 face cards and 2 Jokers) is $\frac{4}{54} \simeq 0.074$, because there are four aces in a deck.

Such a definition is unsatisfactory, because, in order to compute the probability of an event, we must assume that the probability of each event is the same, i.e., that each event is equiprobable. From a formal point of view, it sounds odd to define the concept of probability by assuming that the probability of each event is the same. We need to know what it means to be equiprobable to define it.

A second attempt to define probability is the so-called frequentist approach, which can be ascribed to many authors (see the corresponding Wikipedia page for a comprehensive account of contributions). In this approach, the probability is defined after observed frequencies of event occurrence.

The need for this new definition comes from the fact that it is often difficult, if not impossible, to assign an *a priori* probability to some event based on symmetry principles, as in the case of the dice or the cards in a deck. Clearly, the probability of obtaining 3 as a result of a throw is $\frac{1}{6}$ if we believe that the die is symmetric. If it is not so (and this is most likely the case, due to imperfections in its construction), we cannot assign the same probability to all possible events. However, we can estimate it as the frequency f_3, defined as

$$f_3 = \frac{n_3}{N}, \tag{7.1}$$

where n_3 is the number of times a throw resulted in 3 in a set of N. Clearly, this is only an approximation of the probability. In fact, if we make a new set of N throws, it is very unlikely to again result in the exact same number of 3, n_3. However, we expect the new number n_3' to be close to it, i.e., $n_3' \simeq n_3$, if N is *large enough*. The higher N, the better the estimation. In principle, one can say that the probability of obtaining 3, $P(3)$, is

$$P(3) = \lim_{N \to \infty} \frac{n_3}{N}. \tag{7.2}$$

However, in this case, the limit operation cannot be taken as being very well defined as in mathematics. This is an *experimental* limit and it is manifestly impossible to execute an infinite number of trials. Even in this case, then, the definition of probability is not so obvious and depends on the events that happened in the past, which must be assumed to be representative of the whole set of possible events, i.e., they must have the same probability as similar events in the future.

In both cases, the following rules hold. If an event is impossible, its probability is null (either $n_i = 0$ or the number of possible cases is zero). If an event is certain, its probability is 1 (either $n_i = N$ or there is only one favourable case in a set of only one possible case).

The frequentist definition of probability is based on the empirical observation according to which the frequency of events tends toward their predicted combinatorial probability for a large number of trials.

For frequentists, the limit for an infinite number of trials of the frequency is equal to the probability, but the limit computation cannot be taken using ordinary methods.

According to Bayes, probability is a subjective measure of the confidence in the occurrence of an event. It can be evaluated using the whole set of available information.

7

A more modern approach consists in the so-called Bayesian probability, after Thomas Bayes (1701–1761). In this approach, the probability is taken as a subjective evaluation about the degree of belief that something is going to happen. It is often expressed in terms of a bet.

Suppose, for example, that you are going to bet an amount A that a given event E is going to happen. Your opponent bets B on the opposite event $\bar{E}$, such that the winner will get $S = A + B$. Manifestly, the amount you are willing to risk is proportional to the probability $P(E)$ that you can estimate on the occurrence of that event. Formally speaking, we can say that

$$A \propto P(E), \tag{7.3}$$

even if we don't know $P(E)$. The amount A you are willing to bet is clearly proportional to the total amount S that you are going to get if you win, i.e.,

$$A \propto S. \tag{7.4}$$

Since A must be proportional to both $P(E)$ and S, we have

$$A \propto P(E)S, \tag{7.5}$$

from which one can derive that

$$P(E) = \alpha \frac{A}{S} = \alpha \frac{A}{A+B}, \tag{7.6}$$

where α is a constant. It follows that

$$P(\bar{E}) = \alpha \frac{B}{S} = \alpha \frac{B}{A+B}. \tag{7.7}$$

1 represents an event happening with certainty.

From the last two equations, we obtain $A = \frac{P(E)}{\alpha} S$, $B = \frac{P(\bar{E})}{\alpha} S$ and $S = A + B = \frac{P(E)+P(\bar{E})}{\alpha} S$. The actual value of α is practically irrelevant, thus it is arbitrary, and choosing $\alpha = 1$, $P(E) + P(\bar{E}) = 1$, i.e., the probability of obtaining any of the possible events is 1. Moreover,

$$P(E) = \frac{A}{A+B}. \tag{7.8}$$

The value of $P(E)$ determines our propensity to bet. As a concrete example, suppose that you bet on the possibility that you will obtain a 3 by throwing a die. Is it worth betting 1 € on that? And what about 1 000 € or more?

1 € could be worth it, but it depends on how much you are going to win. If the amount you can win is 1 €, it is probably not worth it. If it is 10 €, maybe yes. However, risking 1 000 € to win 10 000 is perhaps not.

If you believe that $P(E) = 0$, then $A = 0$, i.e., you are not going to bet in any case, irrespective of S. On the other hand, if you believe that $P(E) = 1$, you are going to bet up to $A = S$.

The probability of impossible events is 0.

It is worth noting that $P(E)$ is subjective, in the sense that it includes, often implicitly, your prior knowledge about the conditions that may influence the result. If we believe that a die is perfectly realised, we can attach a probability of $\frac{1}{6}$ to the event $E = 3$, but if we found, in a set of throws, that 3 appears more often than expected, we may assign a higher probability to it. Though the higher observed frequency may be due to statistical fluctuations, it may also come from manufacturing imperfections within the instrument that favour the event $E = 3$. If, according to your previous experience, the die appears to be sound, you may be reluctant to bet $A = 1$ € against $B = 1$ € on $E = 3$, while you may be tempted to do that if $B = 9$ €. The reason is that betting 1 € against the same amount on $E = 3$ corresponds to assigning a probability of

$$P(3) = \frac{1}{1+1} = 0.5, \tag{7.9}$$

to $E = 3$, i.e., to overestimating the probability of observing the event with respect to the frequentist or combinatorial approach that anyone would judge as being reasonable if the die is close to perfect. On the other hand, betting 1 € against 9 may be worth it, since the assigned probability of the event is

$$P(3) = \frac{1}{1+9} = 0.1. \tag{7.10}$$

The *objective* probability is higher, so we can be confident that we have more chances to win with respect to those that are assumed subjectively, while the subjective probability is still relatively high. For the same reason, the number of people willing to participate in a lottery increases when the jackpot increases. The amount we are willing to lose increases with $S = A + B$ and with $P(E)$. Even if $P(E)$ does not change, the increase in S makes the bet more interesting.

The subjective probability depends on our knowledge about the system, as well as our expectations. Our propensity in betting depends on the objective probability compared to the subjective one.

On the other hand, should we bet $A = 1\,000\,€$ for $E = 3$ against $B = 10\,000\,€$, it would imply that

$$P(3) = \frac{1\,000}{1\,000 + 10\,000} \simeq 0.09 \,, \tag{7.11}$$

a much lower probability, indeed.

The subjective probability of an event occurring is called the marginal or prior probability.

$P(E)$, the probability of the event E, is called the "marginal" or "prior" probability (see below for an explanation of the terminology). For a symmetric die, it is reasonable to assume that $P(3) = \frac{1}{6}$. As a general rule, we usually assume that there is no reason to believe that systems should not behave according to principles of symmetry. As a consequence, we will estimate probabilities using either combinatorics or frequencies (in this case, they are equivalent to the Bayesian one), unless we have reasons to act differently.

7

Random number generators

In order to experimentally test our mathematically-derived results, we need some source of random events. In fact, it is hard, if not impossible, to find such a source.

Any device inevitably has imperfections that somewhat alter the theoretical distribution of the purely random output of that device. Most natural sources of random events are systems whose physics can be described by quantum mechanics. For example, nuclear decay is a random process whose characteristics are well described by theoretical predictions, even if each single event happens at random times. The problem is that, in order to detect these events, we need detectors whose behaviour is subject to a number of conditions and that do operate under 100% efficiency. As a result, no perfect source of random events can be found, either natural or artificial.

Computers can be used to simulate random events. There are, indeed, functions that return random numbers with the specified distribution and within a given interval. However, this source of random events is not at all random. Computers are fully deterministic, and there is nothing inside a computer able to generate random numbers. Those generated by a computer are, in fact, called pseudorandom, because they come from an algorithm, even if they appear to be random. A sequence of pseudorandom numbers is perfectly predictable. Once you know the last number generated by the algorithm, you can actually compute the next number to be generated.

Sequences of pseudorandom numbers simply exhibit statistical properties similar to those expected for the distribution of the random numbers
for which they were designed. Usually, pseudorandom generators provide
numbers uniformly distributed between 0 and 1. Other distributions can
be generated from them using proper techniques.

Most pseudorandom generators use the linear congruential method to
yield a sequence of uniformly distributed numbers. The linear congruential
method exploits the following recurrence relation:

$$x_{i+1} = (ax_i + b) \bmod c, \tag{7.12}$$

where x_i is the pseudorandom numbers to be generated (x_0 is said to be
the seed of the generator), a, b and c are integer constants and mod represents the modulo operation that returns the remainder after the division
of the operands. The quality of the generator depends on the choice of a,
b and c. Its period, i.e., the number of extractions after which the sequence
will repeat, is an important parameter.

A common choice, used in the GNU C-compiler, is $a = 1\,103\,515\,245$,
$b = 12\,345$ and $c = 2^{31}$.

The importance of the subjective definition of probability is made manifest in certain situations, discussed below
in the text.

For the time being, we can adopt a combinatorics
approach to estimate the probabilities under a few important conditions. Bearing that in mind, we can start computing the probabilities of certain events, aiming for the
discovering of general rules. Let $P(E_i)$ be the probability that event E_i occurs. Dealing with a die, for example,
$P(E_i)$ is constant and equal to $\frac{1}{6}$, and thus the set contains
six events: $\{1, 2, 3, 4, 5, 6\}$ and $E_i = i$. To examine a more
interesting case of events with a different probability, imagine we have a box with a total of six balls: three of them are
red (R), two are green (G) and one is blue (B). The probability of extracting a blue ball is $P(B) = \frac{1}{6}$, that of extracting
a green ball is $P(G) = \frac{2}{6} = \frac{1}{3}$, while $P(R) = \frac{3}{6} = \frac{1}{2}$. Let's
also suppose that we have two such boxes, with identical
contents. We define the following concepts:

- The "joint probability" is the probability of the occurrence of two simultaneous events. It is important to
  realise that, here, "simultaneous" means that the events
  are part of the same event, even if the two events happen
  at different times. For example, we may be interested
  in the joint probability of extracting a red and a green
  ball.

In the following, we
often estimating the
prior probability
using combinatorics
or frequentist
approaches, yet we
know that these are
just convenient ways
of doing that.

— The "marginal probability" is defined as the probability of a single event, irrespective of all other events. For example, we can compute the marginal probability of extracting a green ball from box 1.
— The conditional probability is the probability of an event given the occurrence of another event. If, for example, we extract a green ball from box 1, we can ask ourselves how large the probability is of extracting a second green ball from the same box.

Consider the event (G, R) consisting in the extraction of a green ball and a red ball. In order to compute the probability of occurrence of this event, we need to take into account:
— the independence of the two events and
— the ordering of the two events.

Two events are independent if the occurrence of one does not change the probability for the other occurring. The joint probability is the probability of the occurrence of two or more simultaneous events.

Two events are said to be independent if the occurrence of one of them does not change the probability for the other occurring. This is the case when we extract the two balls from different boxes (note that this is equivalent to extracting a ball from one box, putting it back in and then extracting a second ball from the same box).

The joint probability of such an event is the product of the probability $P(G)$ of extracting a green ball from box 1 and the probability $P(R)$ of extracting a red ball from box 2, i.e.,

$$P\,(G\,\text{and}\,R) = P(G)P(R) = \frac{1}{3} \times \frac{1}{2} = \frac{1}{6}. \qquad (7.13)$$

The joint probability of two events is the product of their prior probabilities, if they are independent.

It is easy to check that the result is correct. There are 36 possible results in this experiment, because, for each ball in box 1, we can extract 6 balls from box 2. The event (G, R) can be any among (G_1, R_1), (G_2, R_1), (G_1, R_2), (G_2, R_2), (G_1, R_3), and (G_2, R_3), where (G_i, R_j) represents the event in which the i-th green ball in box 1 is extracted together with red ball j from box 2. There are then six favourable events out of 36, and the probability is

$$P\,(G,\,R) = \frac{6}{36} = \frac{1}{6}. \qquad (7.14)$$

In the case of dependent events, the probability of extracting the green ball from box 1 is $P(G) = \frac{1}{3}$. If this event occurs, it modifies the probability of extracting a second red ball from $P(R) = \frac{1}{2}$ to $P'(R) = \frac{3}{5}$, because there are now three red balls in a box with a total of five balls.

In general, we can say that the joint probability $P(G, R)$ is equal to the product of the probability of one of the events occurring, times the probability of the second to occur given that the first happened. The latter is the conditional probability. Indicating, as $P(R|G)$, the probability of event R given G, we can write

$$P(G, R) = P(G)P(R|G). \tag{7.15}$$

In the example, such a probability is $P(G, R) = \frac{1}{3} \times \frac{3}{5} = \frac{1}{5}$. The latter is a general rule. In fact, when the events are independent, $P(R|G) = P(R)$, because the extraction of a green ball has no effect on the probability of extracting the red ball.

Till now, we have considered the events as if the order in which they occur matters. If we do not care about the order, the event (G, R) is equivalent to (R, G). Repeating the steps above, we arrive at the conclusion that

$$P(G, R) = P(R, G), \tag{7.16}$$

i.e., the joint probabilities are symmetric.

If two events are mutually exclusive, i.e., if the occurrence of one event forbids the occurrence of the other, the probability of observing either one or the other is the sum of the probabilities of each event. To check the result, let's start from a simple case. The extraction of a green ball from box 1 in an experiment in which one ball is extracted from one box prevents the extraction of a red ball, and vice versa. Hence, the probability of extracting either a green or a red ball is

$$P(G \text{ or } R) = P(G) + P(R) = \frac{1}{3} + \frac{1}{2} = \frac{5}{6}. \tag{7.17}$$

There are, in fact, five favourable events out of six. The extraction of a green ball from the first box and a red ball from the second makes it impossible to observe the event consisting in the extraction of a red ball from the first box and a green ball from the second. In this case, too, the probability of extracting two balls, of which one is red and the other is green, irrespective of the order, is

> The conditional probability is the probability of an event given the occurrence of another event. The joint probability of two dependent events is the conditional probability of the first, given the second, times the prior probability of the latter.

> The probability of any of two events occurring is the sum of the marginal probability for each, if the events are mutually exclusive.

$$P((G, R) \text{ or } (R, G)) = P(G, R) + P(R, G) = \frac{1}{6} + \frac{1}{6} = \frac{1}{3}.$$

(7.18)

We can compute this probability by considering all the possible outcomes of the experiment. To make the counting easy, we can associate each ball with a number from 1 to 6. Balls 1 to 3 are red, 4 and 5 are green and 6 is blue. There are 36 possible events that can be identified as (i, j), $i, j = 1, \ldots, 6$. The favourable events are $(1, 4)$, $(1, 5)$, $(2, 4)$, $(2, 5)$, $(3, 4)$, $(3, 5)$ and those in which i and j are swapped. There are, then, 12 favourable events out of 36 and the probability is $P((G, R) \text{ or } (R, G)) = \frac{12}{36} = \frac{1}{3}$.

Consider now the probability of extracting two green balls from the two boxes. In this case, following the above prescriptions, the favourable events are $(4, 4)$, $(4, 5)$ and those in which i and j are swapped. The probability is thus

$$P((G, G) \text{ or } (G, G)) = \frac{4}{36} = \frac{1}{9} \neq P(G, G) + P(G, G) = \frac{2}{9}.$$

(7.19)

If events are not mutually exclusive, the probability of the occurrence of any two events is the sum of the marginal probabilities of each minus the one both occur.

This is because the events are not mutually exclusive. In this case, the total probability is the probability computed as if the events were mutually exclusive minus the one in which both occur, i.e.,

$$P((G, G) \text{ or } (G, G)) = P(G, G) + P(G, G) - P((G, G) \text{ and } (G, G)) =$$
$$P(G, G) + P(G, G) - P(G, G)P((G, G)|(G, G)) =$$
$$\frac{1}{9} + \frac{1}{9} - \frac{1}{9} \times 1 = \frac{1}{9}.$$

(7.20)

In general,

$$P(A \text{ or } B) = P(A) + P(B) - P(A)P(B|A).$$ (7.21)

The marginal probability is the probability of an event, irrespective of all other events.

The marginal probability of the event G in which a green ball is extracted from box 1 is the probability of such an event occurring irrespective of any other event. This is the probability that we computed initially, $P(G) = \frac{1}{3}$. The reason why it is called "marginal" is explained by considering the marginal probability $P(E)$ as the joint probability $P(E, X)$, where X represents any possible event. Let's build a table in which we fill each cell with the joint probability of the events obtained by joining the event on a row with the event in a column for all the possible events.

	R	G	B
R	$\frac{1}{4}$	$\frac{1}{6}$	$\frac{1}{12}$
G	$\frac{1}{6}$	$\frac{1}{9}$	$\frac{1}{18}$
B	$\frac{1}{12}$	$\frac{1}{18}$	$\frac{1}{36}$

The probability $P(G, R)$ is found by looking for the cell that corresponds to row G and column R, or vice versa (note that the joint probabilities are symmetric, i.e., $P(G, R) = P(R, G)$). Then, let's sum the rows and the columns.

Joint probabilities are symmetric: $P(A, B) = P(B, A)$.

	R	G	B	
R	$\frac{1}{4}$	$\frac{1}{6}$	$\frac{1}{12}$	$\frac{1}{2}$
G	$\frac{1}{6}$	$\frac{1}{9}$	$\frac{1}{18}$	$\frac{1}{3}$
B	$\frac{1}{12}$	$\frac{1}{18}$	$\frac{1}{36}$	$\frac{1}{6}$
	$\frac{1}{2}$	$\frac{1}{3}$	$\frac{1}{6}$	

The probabilities on the right and bottom *margins* of the table are the marginal probabilities of the event indicated, respectively, in the row and in the column header.

In fact, in order to compute the probability $P(G)$, we can consider the following joint probabilities:

$$P(G, R) = \frac{1}{6} \qquad P(G, G) = \frac{1}{9} \qquad P(G, B) = \frac{1}{18}.$$
$$(7.22)$$

The probability of any of these three events happening is the sum of the probabilities. Remember, in fact, that $P(G, R) = \frac{1}{6}$ because there are six out of 36 possible events giving the desired result (G, R). Similarly, there are four out of 36 events leading to the result (G, G) and there are only two events whose result is (G, B). There is thus a total of $6 + 4 + 2 = 12$ events out of 36 for which a green ball factors into the event, thus the probability $P(G) = \frac{12}{36} = \frac{1}{3}$.

The conditional probability $P(G|R)$ can be computed starting from the joint probability, as

$$P(G|R) = \frac{P(G, R)}{P(R)}. \qquad (7.23)$$

The formula is only valid if $P(R) \neq 0$. Indeed, if $P(R) = 0$, it is impossible that G can occur if R occurred and $P(G|R) = 0$, as well as $P(G, R)$.

Since the joint probability is symmetric, we have

$$\begin{cases} P(G, R) = P(G)P(R|G) \\ P(R, G) = P(R)P(G|R) \end{cases}, \qquad (7.24)$$

and since the LHS of the equations are equal,

$$P(G)P(R|G) = P(R)P(G|R),\qquad(7.25)$$

The Bayes Theorem is a cornerstone in probability theory and states that $P(A|B) = \frac{P(A)P(B|A)}{P(B)}$.

from which we find the "Bayes Theorem":

$$P(R|G) = \frac{P(R)P(G|R)}{P(G)}.\qquad(7.26)$$

An alternative formulation of the theorem can be made by observing that

$$P(G) = P(G|R)P(R) + P(G|G)P(G) + P(G|B)P(B).\qquad(7.27)$$

Substituting into the expression of the Bayes Theorem above,

$$P(R|G) = \frac{P(R)P(G|R)}{P(G|R)P(R) + P(G|G)P(G) + P(G|B)P(B)}.\qquad(7.28)$$

The posterior probability, the LHS of the Bayes Theorem, is computed after the prior probability.

In Bayesian terminology, the LHS of this equation is called the posterior probability, while $P(R)$ is called the "prior probability". $P(G|R)$ is the likelihood, while the denominator is called the evidence.

Let's check the result in the case of two extractions from the same box, bearing in mind that $P(R|G) = \frac{3}{5}$, as we already computed. We know that $P(R) = \frac{1}{2}$, $P(G) = \frac{1}{3}$, while $P(G|R)$ is the probability of G given R. When R has been extracted from the box, there are two green balls out of five and $P(G|R) = \frac{2}{5}$. Using the first formulation of the theorem, we find

$$P(R|G) = \frac{P(R)P(G|R)}{P(G)} = \frac{\frac{1}{2} \times \frac{2}{5}}{\frac{1}{3}} = \frac{3}{5},\qquad(7.29)$$

as expected. For the alternative formulation, we have

$$P(R|G) = \frac{\frac{1}{2} \times \frac{2}{5}}{\frac{2}{5} \times \frac{1}{2} + \frac{2}{5} \times \frac{1}{6} + \frac{1}{5} \times \frac{1}{3}} = \frac{3}{5}.\qquad(7.30)$$

The most general form of the Bayes Theorem can be written as

$$P(R|G) = \frac{P(R)P(G|R)}{\sum_i P(G|E_i)P(E_i)},\qquad(7.31)$$

where the sum is extended to all events. We can easily remember it as

$$P(R|G) \propto P(R)P(G|R), \qquad (7.32)$$

where marginal and conditional probabilities alternate, as do the events indicated as the probability arguments: $RGRGR$, the denominator working as a normalisation factor.

This textbook was prepared during the COVID-19 pandemic. Naso/oropharyngeal swabs are performed to diagnose the disease. This kind of test produces a positive or negative response, respectively on whether the virus is detected or not.

Often, their reliability is not perfect: sometimes, the swabs return a negative result for a sick person (false negative) or a positive result for a healthy one (false positive). Doctors define the PPV (positive predictive value) and the NPV (negative predictive value) as the rate of true positive and true negative responses. The table below reports the PPV and the NPV of a sample of swabs, according to [1].

PPV	95	89	86	95	89	68	85	85	97	94	97	94
NPV	100	94	95	100	94	96	100	96	99	97	99	97

The average PPV is 89.5%, while the average NPV is 97.25%. Suppose that we were subject to a swab, having been randomly chosen within the population (i.e., we do not exhibit symptoms and we have not engaged in any risky behaviour), and that it resulted in a positive response. The probability of being affected by COVID-19 could naively be estimated as $P(\text{COVID}) \simeq 89.5\%$ or, more precisely, considering the possibility of false negatives, as $P(\text{COVID}) \simeq 89.5 + (100 - 92.25) = 92.25\%$. However, when we update our knowledge, the posterior probability is updated, too, and the difference could be dramatic.

According to the Bayes Theorem,

$$P(\text{COVID}|+) = \frac{P(\text{COVID})\, P(+|\text{COVID})}{P(+)}, \qquad (7.33)$$

where $P(\text{COVID}|+)$ represents the probability of actually having the disease after a diagnosis from a positive swab, $P(+|\text{COVID})$ the probability that people affected by COVID-19 have tested positive with a swab (PPV), while $P(+)$ is the probability of receiving a positive diagnosis when being tested with a swab:

$$P(+) = \text{PPV} \times \text{COVID} + (1 - \text{NPV}) \times \text{healthy} = 3.1\%. \qquad (7.34)$$

P (COVID) is the probability of being infected, which can be estimated from the fraction of the population that has been diagnosed as such. In Italy, at the time of this writing, this fraction is 0.35%.

It is useful to build a table as follows, in which we arrange the probabilities of receiving a positive/negative response, depending on your status:

	COVID (0.35%)	healthy (99.65%)
+	89.5	2.75
−	10.5	97.25

Persons affected by COVID-19 belong in the first column. They have a chance PPV of receiving a positive answer from a swab and a chance $1 - \text{PPV} = 10.5\%$ of receiving a negative answer. Healthy people belong in the second column: they receive a negative diagnosis in NPV= 97.25% of the possible cases, however, there is a non-null probability, equal to $1 - \text{NPV} = 2.75\%$, of their diagnosis being positive. The sum of all the values in each column is 100.

Using the values in the Bayes formula, we get

$$P\,(\text{COVID}|+) = \frac{0.35 \times 89.5}{3.1} = 10\%. \tag{7.35}$$

It has been argued that the actual fraction of infected people could be as large as ten times those declared as such by authorities. Even in this case, the chance of being affected by COVID-19 having received a positive response from a swab is about 54%.

It must be noted that the result of the last exercise is not in contradiction with the commonly used frequentist approach. In fact, even if the fraction of true positives is large, the number of people who are actually healthy is huge. A small fraction of them can thus represent a larger group than those who are truly positive and infected. In fact, using the frequentist approach, we can say that, out of N patients, those diagnosed with a positive swab who are actually infected are a fraction of $0.895 \times 0.0035 \simeq 0.0031$ of N, while those who simply received a positive diagnosis from a test are $0.895 \times 0.0035 + 0.9965 \times 0.0275 \simeq 0.031$ times N. The ratio between the two is

$$\frac{0.0031}{0.031} = 0.10, \tag{7.36}$$

corresponding to 10%. The Bayes Theorem, in essence, allows us to update our initial estimate concerning the probability of being ill, taking into account the increase in information represented by the knowledge of the fraction of people infected overall. The probability is subjective, in the sense that it is estimated based on limited information. Each time we obtain new information, we can update our estimation of the probability.

Despite the apparent impossibility of drawing scientifically sound conclusions from what appears to be a completely arbitrary opinion, the Bayes Theorem works perfectly and allows for very solid results.

7.3 Bayes Theorem and Physics

The content of the previous section seems to be quite far removed from the topic of this book. In fact, besides being the foundation of the study of probability distributions, its conclusion, Bayes' Theorem, has direct consequences on the way in which the results of an experiment should be interpreted.

Probability theory is important for understanding data distributions. Bayes Theorem is also important on matters of principles.

We can rewrite its last formulation as

$$P(E|H) \propto P(E)P(H|E), \qquad (7.37)$$

where E represents the outcome of an experiment and H the *prior* knowledge of the physics involved in that experiment. To remember this formula, think of the arguments in parentheses as a sequence of alternate symbols $EHEHE...$ signifying the arguments of alternate functions $P_C P P_C...$, where P_C represents the conditional probability and P the prior probability.

This formula states that the probability of a certain interpretation of experiment E being true, $P(E)$, is updated when an additional hypothesis H is known being true. $P(H|E)$ is the probability of H to be true in the context of the experiment E, while $P(E|H)$ is the updated probability of E being true, when we consider $P(H|E)$.

The Bayes Theorem has important implications when interpreting data, especially for rare processes or searches. From the technical point of view, its application is not always straightforward, and many of its interesting applications are beyond the scope of this book.

Nevertheless, even if it is sometimes not possible to compute the probabilities involved in the Bayes Theorem, its (subjective) approach to probability definition is of great importance in science and in physics, even when - indeed, particularly when -when we cannot (realistically) quantify them. Physics, in fact, is not just a matter of collect-

To remember the Bayes Theorem, read the events from left to right, top to bottom, alternating the events, and using them as the arguments of the probability $P()$ as $ABABAB \rightarrow$ $P(A|B) = \frac{P(A)P(B|A)}{P(B)}$. Alternatively, think of the sequence as $P(A)P(B|A) = P(B)P(A|B)$.

ing numbers during experiments and formulating physical laws in the form of equations. Physicists must interpret the results of an experiment and draw a coherent picture of the Universe. They cannot just write down the results of their experiments, ignoring the complete formal and informal picture, or formulate beautiful theories that ignore the results of experiments.

R. Feynman quote: "It doesn't make any difference how beautiful your guess is, it doesn't make any difference how smart you are who made the guess, or what his name is. If it disagrees with experiment... it's wrong.".

In a famous lecture given by Richard Feynman (1918–1988), whose recording is available on YOUTUBE, he explained the *essence of the scientific method* as follows:

> It doesn't make any difference how beautiful your guess is, it doesn't matter how smart you are who made the guess, or what his name is. If it disagrees with experiment... it's wrong.

To understand the very meaning of such a statement, we have to discuss the context in which it was formulated. Part of the scientific community believes in the *power of beauty*. In other words, they are convinced that the probability of a theory being true is higher if the resulting equations are *beautiful*. According to this point of view, a theory whose equations are ugly is likely to be wrong.

The Bayes Theorem tells us that the probability of a theory being true has nothing to do with beauty. Given an *a priori* probability that it is true, such a probability must be updated taking into account both implicit and explicit knowledge, acquired in the past or to be acquired in the future. The new posterior probability can become higher or lower, depending on the effect on the prior of $P(H|E)$.

A very instructive example of a non-quantitative application of the Bayes Theorem is the publication of what, at that time, seemed a thrilling discovery by the Opera collaboration [3], which claimed that neutrinos were found to move with a speed faster than light.

Opera ran at an underground laboratory at Gran Sasso to investigate the properties of neutrinos coming from CERN. Neutrinos are weakly interacting particles that can travel across the earth practically without interacting with it. After a careful investigation of many possible sources of systematic errors, the collaboration decided it was worth publishing such a result, whose statistical significance was quite high. The results obtained in two runs of the experiment actually differed from expectations by about six standard deviations, and the probability that they could be ascribed to random fluctuations was, in fact, extremely low.

Despite such striking evidence, the scientific community was very reluctant to accept the result as genuine. It is worth noting that, in certain cases, results with less evidence had been accepted seamlessly, as in the case of the discovery of the Higgs boson, which was claimed with five-sigma evidence, or the existence of dark matter: an unknown kind of matter whose presence can be inferred from cosmological observations.

However, there was, in fact, good reason to be skeptical: Einstein's relativity forbids the possibility of particles moving at a speed faster than light, and this theory is among the most successful in history. A discovery like the one claimed by the Opera collaboration, in fact, opened up a problem in the reliability of such a theory that had to be resolved in turn. The prior probability $P(v > c)$ that, indeed, neutrinos travel with a speed v greater than that of light c is, according to the frequentist evaluations of the physicists of the collaboration, very high. However,

$$P(v > c|\text{relativity}) \propto P(v > c)P(\text{relativity}|v > c), \quad (7.38)$$

i.e., the probability that $v > c$, given the theory of relativity, is much less, due to the fact that it is proportional to $P(\text{relativity}|v > c)$, the probability of the relativity being correct given the Opera result. Actually, it is zero.

From the story told in the box, it is clear that what makes a discovery acceptable or unacceptable to the scientific community is the increased probability that, overall, both prior and posterior knowledge are corroborated. Often, the decision whether or not to support new discoveries is driven by subjective evaluations. Contrary to common beliefs, this is not at all far removed from a scientific point of view (of course, provided the evaluations are honest enough), thanks to the Bayesian point of view. The fact that consolidated discoveries have finally reached a state of presumed beauty is not due to a mysterious preference of Nature for beautiful theories, but rather to the fact that the supporters of those theories have worked hard to make them look as good as possible.

The *beauty argument* probably had its origin in an interview [4] given by Paul Dirac (1902–1984) to Thomas Kuhn and Eugene Wigner in 1962. In that interview, Dirac said

The reluctance of scientists to abandon established theories is not driven by religious or political arguments, nor by convenience or faith. It is the natural consequence of a rational attitude formalised by Bayes Theorem

7

P.A.M. Dirac admitted that his arguments in favour of searching for beauty in physical equations was a peculiarity of his own, in a sense, confirming the non-existence of a "method" for reaching rational conclusions, as argued by Feyerabend.

[The idea of spin] came out just from playing with the equations rather than trying to introduce the right physical ideas. A great deal of my work is just playing with equations and see in what they give. Second quantization I know came out from playing with equations. I don't suppose that applies so much to other physicists; I think it's a peculiarity of myself that I like to play about with equations, just looking for beautiful mathematical relations which maybe don't have any physical meaning at all. Sometimes they do.

In a paper [5] written in 1982, Dirac stated

It seems to be one of the fundamental features of nature that fundamental physical laws are described in terms of a mathematical theory of great beauty and power [...and we...] could perhaps describe the situation by saying that God is a mathematician of a very high order, and He used very advanced mathematics in constructing the universe. It seems that if one is working from the point of view of getting beauty in one's equations, and if one has really a sound insight, one is on a sure line of progress.

Not all scientists share Dirac's view, and yet they still believe in the power of mathematics, to which many credit a somewhat magical quality. This is mostly due to Galileo Galilei (1564–1642), who wrote

A common interpretation of G. Galilei writings is that the Universe is necessarily designed as a mathematical entity. In a sense, mathematics has some special hidden role in the existence of the Universe.

the universe, [...] cannot be understood unless we first learn to understand its language and know the characters in which it is written. It is written in mathematical language, and the characters are triangles, circles, and other geometrical figures, without which it is impossible to understand it humanly; without them, it is vain wandering through a dark labyrinth.

Mathematics is a powerful language. This is the reason why it is so successful in describing Nature.

Indeed, our opinion is that there is nothing magic in mathematics, nor does it exist thanks to the power of beauty. Mathematics is, in fact, a language and, like all languages, it is rather flexible; actually, in the case of mathematics, extremely flexible, as it turns out. Its dictionary and its syntax can be redefined almost uninhibitedly and this accomodates the possibility of introducing new concepts and structures that lend themselves to better and more effective ways of correctly and unambiguously describing physical phenomena of interest.

Using the appropriate language, physicists can look deeper into the most profound meanings of equations and the derivation of new results is greatly simplified.

An astonishing example of this is represented by the equations of electromagnetism. Modern textbooks report four beautiful symmetric equations as the Maxwell Equations (although in slightly different forms). It turns out that these four equations do not look anything like those written by James Maxwell (1831–1879) in his Treatise on Electricity and Magnetism [6]. Maxwell's original formulation of the equations was quite cumbersome, while the version currently used, credited to Oliver Heaviside (1850–1925), is considered to be much more *beautiful*. Both, however, are, in fact, the same equations. They just look different, but their meaning is manifestly the same. The reason that the later version looks nicer is that, since Maxwell's time, the mathematics of vector fields has introduced and developed new operators for playing with vectors, greatly simplifying notation.

> The modern form of Maxwell Equations is credited to Oliver Heaviside.

A very similar argument can be made for what is probably the most famous equation in physics, $F = ma$, known as the Newton's second Law. Isaac Newton (1643–1727), in fact, never wrote it. He wrote a long paragraph, whose meaning can be summarised by the above equation, appropriately ascribed [7] to Leonhard Euler (1707–1783), that was not yet in its modern form as a vectorial equation.

> There is no $F = ma$ equation in Newton's *Principia*. It was Leonhard Euler who introduced it.

Physical laws look beautiful simply because physicists write them in the proper way. They are not beautiful *per se*. However, the need to reformulate equations involves a lot more than just making them look nicer and easier to remember. The proper look also makes some results manifest or easier to be derived. We can compare the effect of a clever reformulation of physical laws in terms of a different *language* to what happens in literature, in which long, yet descriptive text used to illustrate a scene in a novel can be surpassed, in clarity, by poetry. Let's examine an excerpt from *The Pickwick Papers* by Charles Dickens (1812–1870):

> Beauty is not useless: it makes it easier to find hidden information in physical laws.

7

> That punctual servant of all work, the sun, had just risen, and begun to strike a light on the morning of the thirteenth of May, one thousand eight hundred and twenty-seven, when Mr. Samuel Pickwick burst like another sun from his slumbers, threw open his chamber window, and looked out upon the world beneath

Reading this sentence, one can almost *see* the sun rising close to the horizon, light filtering through the window, Mr. Pickwick being suddenly awakened, rising from his bed, going to the window to open it completely and looking around.

Compare this rather long description to the following *haiku* (translated from the original Japanese) by Yosa Buson (1716–1784):

> such a moon
> the thief pauses to sing.

Science is not different from Art. Both are activities that lead to "creation". Collider physicists often say that they "create" particles in their laboratories, but, in fact, they only make (produce) them. However, they "create" the interpretation of the data they collect, as theories. Subatomic particles exist irrespective of scientists; theories do not.

In these few syllables, there is a full description of a scene, even more detailed than the one depicted by Dickens. Yet, each reader can enjoy the picture in a variety of ways, which, nevertheless, are almost the same. In those two short lines, the reader can not only *see* the landscape, but also *feel* the atmosphere, *hear* the sounds, *smell* the air.

Moving from the original, ugly formulation of the Maxwell Equations to Heaviside's version produces a similar effect. The description of the physics is not simply much shorter. While maintaining the same informational content, the latter allows more people to *see* the relevant characteristics of the electromagnetic fields and technically skilled people can derive new information from them with a lot less effort.

Indeed, science is a creative activity, like art. In art, language evolves, too. From the very ingenuous drawings of the past to the extraordinary techniques achieved by Raffaello Sanzio in painting, we can clearly recognise the evolution of physics: from the very raw and vague concepts of the ancient Greeks to the perfection of thermodynamics and electromagnetism. However, when the technical perfection reaches its peak, new demands arise. Modern painters no longer need to reproduce Nature as it appears through the senses, but rather feel compelled to reproduce things that are impossible to see or touch, such as emotions. They thus need to develop a completely new lan-

Fig. 7.1 Mural reproducing "Guernica", by Pablo Picasso, installed in the town of Guernica. Photograph by Papamanila

guage. Some paintings, like "Guernica" by Pablo Picasso (1881–1973), have the power to evoke a feeling of fear and terror, such as that felt by the inhabitants of that city during a terrible bombardment. Similarly, when new unexpected phenomena were discovered, like the photoelectric effect, a new mathematics had to be invented (created). Quantum mechanics allowed physicists to *see* a completely new world, impossible to describe with classical dynamics. That world, in fact, is no less real than the one to which we are used to living in, just as fear, pain, and love are no less real than a tree in a landscape, a fish in a still life or a woman in a portrait. They just require, and deserve, a different language to be expressed.

It is probably no coincidence that many religions attribute to humankind the characteristic of being made in God's image and likeness. For believers, indeed, God is the Creator: men and women, contrary to other living creatures, who are committed only to feeding and breeding, are, like God, able to "create" through art and science.

7.4 Statistical Distribution of Data

From the considerations made while collecting data and analysing them, it is clear that we ascribe the fluctuations observed in data collected by an instrument to random effects. We can thus understand their distribution by studying the distribution of random variables.

Since each instrument always returns a value with a limited number of digits, the variables in which we are interested are always **discrete**. For example, on a ruler, we can

> Fluctuations in the results of a measurement can be ascribed to random effects.

only read values that differ by 1 mm. The possible results of the measurement of the width of an A4 sheet of paper can be 239, 240 or 241 mm. Results like 239.3, 240.152 73, 239.187 328 460 134 958 61, and so on, are not included in the set of possible results, which always form sets of discrete, numerable elements, often finite.

It is worth starting the study of statistics by dealing with discrete random variables. Measurements, in fact, always return discrete values.

Let's then start by analysing the distribution of simple discrete random variables.

The simplest random variable is a "uniformly distributed" one. In this case, all events have the same probability $p_i = p$, $\forall i$. The normalisation condition implies that

$$\sum_{i=1}^{N} p_i = \sum_{i=1}^{N} p = Np = 1 \tag{7.39}$$

and $p = \frac{1}{N}$, where N is the number of possible events. The scores of the throw of a die are an example of such a distribution with $N = 6$.

We cannot get random numbers from a computer, yet we can generate sequences of pseudorandom numbers that appear as purely random.

In order to work with random variables, it is convenient to simulate random events using computers. As computers are deterministic machines, it is impossible to extract random numbers from them. However, techniques exist for generating lists of numbers that appear to be random sequences, called pseudorandom. In the following, when not explicitly stated otherwise, we use the words "random" and "pseudorandom" as synonyms, for the sake of simplicity.

7.5 Uniform Distribution

numpy.random. uniform() returns a pseudorandom number between 0 and 1.

It is rather easy to generate uniformly distributed random numbers with Python. We can fill a list of 10 000 events as follows:

```
import numpy as np

x = []

for i in range(10000):
    x.append(np.random.uniform())
```

The method `uniform()` from `numpy.random` generates a uniformly distributed random number between 0 (included) and 1 (excluded). In order to get an idea of the shape of the data distribution, we can produce a histogram: a plot of the number of occurrences of events against the events themselves. In other words, for each event E_i, we count how many times N_i occurs and produce a plot in which we list all the possible events on the abscissa and draw a point at height N_i for each one (or, more often, a rectangle of height N_i). A histogram of the values in the list can be produced and shown using

```python
import matplotlib.pyplot as plt
```

```python
plt.hist(x)
plt.show()
```

The result is

> Histograms are plots of the number of occurrences of events, represented as points on the *x*-axis.

> `matplotlib.pyplot.hist(x)` divides the interval of the numbers in x into classes or bins and counts how many events fall into each of them, producing a histogram.

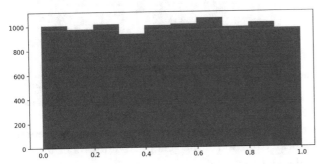

From the graphical point of view, histograms are often represented as bar charts in which each event (or group of events) is represented by a bar whose height is proportional to N_i. Note that the frequency f_i of an event E_i is proportional to N_i, too, so histograms also give information about the distribution of frequencies. Indeed,

$$f_i = \frac{N_i}{\sum_i N_i} . \tag{7.40}$$

The shape of the distribution clearly shows that the numbers are uniformly distributed, since all the $f_i \simeq p_i$ are almost the same, the differences being interpreted as statistical fluctuations (remember that, in the frequentists' view, $f_i = p_i$ only for $N = \sum_i N_i \to \infty$).

The optional parameter `size` tells `uniform()` to generate a list.

In fact, `uniform()` returns a list. By default, its length is one, however, you can change it with the `size` parameter:

```
x = np.random.uniform(
    size=10000)
```

The `rwidth` optional parameter changes the bars' relative widths in a histogram.

You can also change the appearance of the histogram by changing the relative width of each column, as in

```
plt.hist(x, rwidth=0.9)
plt.show()
```

The result is the histogram below, in which each column has a width equal to 0.9 times the width of each bin or class.

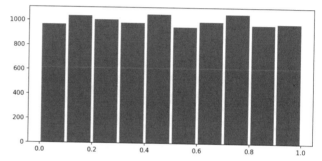

In order to generate numbers between a and b, it is enough to linearly transform the generated numbers in the interval $[0, 1)$, i.e.,

$$y_i = a + (b - a)x_i . \tag{7.41}$$

Linearly transforming uniformly distributed random numbers from 0 to 1 allows us to generate random numbers in any interval $[a, b)$. The same result is achieved using the `low` and `high` optional parameters.

This is done by Python when we pass the parameters `low` and `high` to `uniform()` as in

```
x = np.random.uniform(
    low=-1, high=1, size=10000)
```

The number of classes or bins in which the interval is divided to build the histogram can be arbitrarily chosen. Moreover, in order to plot the relative frequencies rather than the absolute number of events, we can ask the histogram to normalise data using the keyword `density`. With

```
plt.hist(x, rwidth=0.9,
         bins =50, density=True,
         color='green')
plt.show()
```

we plot a green-coloured histogram with 50 bins between $a = -1$ and $b = +1$, properly normalised such that the area under the histogram is 1.

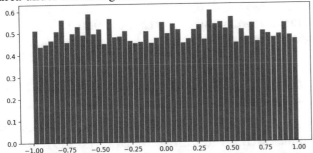

The three distributions appear very similar and reasonably uniform. Being (pseudo)random, fluctuations in each bin are observed. They appear to be larger for the last histogram, in which each bin contains fewer events. In all cases, we generated 10 000 events. While, in the first two, each bin contains, on average, $\frac{10\,000}{10} = 1\,000$ events, the bins of the last histogram contain, on average, $\frac{10\,000}{50} = 200$ events. The bin width is $\frac{b-a}{n} = 0.04$. The area under the histogram is

$$1 = \sum_{i=1}^{n} p_i \Delta y_i = \Delta y \sum_{i=1}^{n} p_i, \tag{7.42}$$

where p_i is the height of the i-th bin and $\Delta y_i = \Delta y = 0.04$ is the bin width and is the same for all i. If $p_i = p$ were constant, $1 = np\Delta y$ and $p = \frac{1}{n\Delta y} = 0.5$, as can be seen in the plots.

7.6 Expected Value, Variance and Moments

If we compute the average of x, we always obtain a value close to the center of the distribution (0.5 in the first two cases and 0 in the third one). The average of a sequence of random numbers is, in fact, an estimator of its expected value, defined as

$$E[x] = \mu = \sum_{i=1}^{N} x_i P_i, \tag{7.43}$$

where P_i is the probability of obtaining a value in the class x_i and N is the total number of classes. For example,

The expected value of a random variable is the sum of all the possible values times the corresponding probabilities. It can be estimated from the arithmetic mean (the average) of the values in a random sequence.

in the first set, we have ten classes corresponding to the following x_i: 0.05, 0.15, 0.25, 0.35, 0.45, 0.55, 0.65, 0.75, 0.85, 0.95 (we defined the classes according to the central value of the interval in which a number falls when counted in the histogram). Since, in this case, all the classes have the same probability $P_i = p = \frac{1}{10}$, the expected value is

$$\mu = \frac{1}{10} \sum_{i=0}^{9} 0.05 + 0.10 \times i. \tag{7.44}$$

The sum on the right can be easily computed using a formula ascribed to Carl Friedrich Gauss (1777–1855). The sum of the first and the last number in the sum is $0.05 + 0.95 = 1$, as the sum of the second and the penultimate is $0.15 + 0.85 = 1$. The same happens for all the possible sums taken between two extremes, once those just used have been removed. If there are N terms in the sum, there are $\frac{N}{2}$ such sums, each of which gives $x_1 + x_N$. The result is then

$$\mu = \frac{1}{10}\frac{10}{2}(0.05 + 0.95) = \frac{1}{2} = 0.5. \tag{7.45}$$

Note that the Gauss algorithm applies to any sum of a linear transformation of i, even if N is odd. In fact,

$$\sum_{i=0}^{N-1} i = \frac{N(N-1)}{2}. \tag{7.46}$$

If N is even, $N+1$ is odd and $\frac{N}{2}$ is integer; if N is odd, $N+1$ is even and $\frac{N+1}{2}$ is integer. Moreover,

$$\sum_i a + b \times i = Na + b\sum_i i = Na + b\frac{N(N-1)}{2}. \tag{7.47}$$

In our case, $a = 0.05$, $b = 0.10$, $N = 10$. Clearly, if the sum starts with 1, the Gauss formula becomes

$$\sum_{i=1}^{N} i = \frac{N(N+1)}{2}. \tag{7.48}$$

It should be noted that, while μ is perfectly defined and constant for a distribution, P_i and x_i being constant, this is not the case for the average of a sample of random variables. In fact, in this case, while P_i may be thought of as constant, the sequence of x_i is random and is different for each sequence. In the examples above, we found, respectively, 0.500 and 0.496 for the first two distributions and -0.010 for the third (for which the expected value is zero).

Even if we always deal with discrete random variables, it is useful to extend the reasoning to the case of continuous ones. In fact, if Δy is small enough, the distribution can be approximated by a continuous curve. In this case, $P(x)$ is called the "probability density function" (PDF). Even if the interval of possible values of x is finite, there are infinite possible values of x in this interval and, from one of the possible definitions of the probability, it follows that $P(x) = 0$, $\forall x$, if we interpret it as in the discrete case. However, from the discrete case, we also know that $\sum_i P_i \Delta y_i$ represents the probability of finding y in an interval whose width is Δy_i, thus $P(x)dx$ can be thought of as the probability of finding the random variable x in the interval $[x, x+dx)$. The normalisation condition becomes

> If the random variables are continuous, the probability density function $P(x)$ is introduced, such that $P(x)dx$ is the probability that the variable lies between x and $x + dx$.

$$\int_{-\infty}^{+\infty} P(x)dx = 1. \tag{7.49}$$

> PDF's obey the normalisation condition $\int_{-\infty}^{+\infty} P(x)dx = 1$.

If $P(x)$ is the PDF of a uniformly distributed variable within the range $a \le x < b$, $P(x) = 0$ for $x < a$ and $x \ge b$, and $P(x) = p = const$ in the interval, hence

$$\int_{-\infty}^{+\infty} P(x)dx = \int_{-\infty}^{a} 0 \cdot dx + \int_{a}^{b} p\,dx + \int_{b}^{+\infty} 0 \cdot dx = p\,(b-a) , \tag{7.50}$$

from which $p = \frac{1}{b-a}$. The expected value is

$$\mu = \int_{a}^{b} xp\,dx = p \int_{a}^{b} x\,dx = \frac{1}{b-a}\frac{x^2}{2}\Big|_{a}^{b} = \frac{1}{b-a}\frac{b^2-a^2}{2} = \frac{b+a}{2} . \tag{7.51}$$

It is useful to introduce the so-called "moments" of the distribution, defined as

> The moments of a distribution are defined as the expected value of powers of the differences between the variable and its expected value.

$$M_{\mu}^{(k)} = \int_{-\infty}^{+\infty} (x-\mu)^k P(x)\,dx . \tag{7.52}$$

It is easy to realise that $M_0^{(0)} = M_\mu^{(0)} = 1 \; \forall \mu$ and $M_0^{(1)} = \mu$. Moreover, $M_\mu^{(1)} = 0$. The smallest interesting moment with respect to μ is

$$M_\mu^{(2)} = \int_{-\infty}^{+\infty} (x - \mu)^2 \, P(x) \, dx \,, \tag{7.53}$$

i.e., it is the expected value of $(x - \mu)^2$, $E[(x - \mu)^2]$. Its discrete counterpart is

$$M_\mu^{(2)} = \sum_{i=1}^{N} (x_i - \mu)^2 \, P(x_i) = \frac{1}{N} \sum_{i=1}^{N} (x_i - \mu)^2 \,, \tag{7.54}$$

7

The second moment of a distribution can be estimated from the variance of a sample of it.

which, for $\mu = \langle x \rangle$, coincides with our definition of variance of the population. The variance of the sample, then, is an estimator of the second moment of the distribution with respect to its expected value and is a measure of its width. In fact, $(x - \mu)^2$ is the square of the distance of x from μ, and taking the integral weighted by $P(x)$ is equivalent to computing the expected value of this quantity. In other words, the variance of a distribution is the average distance squared of the values with respect to their expected value. Taking into account that $P(x) = 0$ for $x < a$ and for $x > b$, for a uniform distribution, we have

$$M_\mu^{(2)} = \int_a^b (x - \mu)^2 \, P(x) \, dx \,. \tag{7.55}$$

Setting $y = x - \mu$, $dx = dy$ and

$$M_\mu^{(2)} = \int_{y(a)}^{y(b)} y^2 \frac{1}{b - a} \, dy = \frac{y^3}{3 \, (b - a)} \bigg|_{a - \mu}^{b - \mu} \,. \tag{7.56}$$

The limits of integration are

$$a - \mu = \frac{a - b}{2} \quad \text{and} \quad b - \mu = \frac{b - a}{2} \tag{7.57}$$

such that

$$M_\mu^{(2)} = \frac{(b - a)^2}{12} \,. \tag{7.58}$$

It is worth noting that the standard deviation of the population $\sqrt{M_\mu^{(2)}} = \frac{b-a}{\sqrt{12}} \simeq \frac{b-a}{3}$, as we empirically obtained in Chap. 4. As it turns out, the uncertainty associated with the reading error comes from the fact that we have no reason to prefer one value or another within the interval between two consecutive ticks of the instrument scale. We are thus forced to assign to each possible value in that interval a uniform probability, the standard deviation of which is the amplitude of the interval divided by $\sqrt{12} \simeq 3$.

> The standard deviation of a uniformly distributed random number is given by the width of the interval of the possible values divided by the square root of 12.

7.7 Combining Errors, Revisited

In Sect. 6.6, we show that uncertainties combine in quadrature, i.e., if $c = a+b$, $\sigma_c^2 = \sigma_a^2 + \sigma_b^2$. From the definition of the second moment with respect to the mean, we can make a formal proof of such a result. In fact,

> Uncertainties sum in quadrature only if the measured quantities are uncorrelated.

$$\sigma_c^2 = E[(c - E[c])^2] = E[(a + b - E[a+b])^2]. \quad (7.59)$$

To simplify the notation, let's define $E[x] = \bar{x}$, such that $\sigma_c^2 = E[(c - \bar{c})^2]$. The expected value is a linear operator, thus $E[a + b] = E[a] + E[b]$ and

$$\begin{aligned}
\sigma_c^2 &= E[(a + b - E(a+b))^2] = E[((a - \bar{a}) + (b - \bar{b}))^2] \\
&= E[(a - \bar{a})^2] + E[(b - \bar{b})^2] + 2E[(a - \bar{a})(b - \bar{b})] \\
&= \sigma_a^2 + \sigma_b^2 + 2E[(a - \bar{a})(b - \bar{b})].
\end{aligned}$$

$$(7.60)$$

The last term in the sum $E[(a - \bar{a})(b - \bar{b})] = \text{Cov}(a, b)$ is called the "covariance" of a and b and

> The covariance, defined as $E[(a - \bar{a})(b - \bar{b})]$, is null for uncorrelated variables, but not for correlated ones.

$$\begin{aligned}
\text{Cov}(a, b) &= E[ab - a\bar{b} - b\bar{a} + \bar{a}\bar{b}] \\
&= E[ab] - E[a]E[b] - E[b]E[a] + E[a]E[b] \quad (7.61) \\
&= E[ab] - E[a]E[b].
\end{aligned}$$

By definition,

$$E[a]E[b] = \int aP(a)da \int bP(b)db \quad (7.62)$$

and

$$E[ab] = \int\int abP(a, b)da\,db, \quad (7.63)$$

where $P(a, b)$ is the joint probability of observing a and b. If a and b are independent (or uncorrelated), $P(a, b) = P(a)P(b)$ and the covariance is null. The result found in Sect. 6.6 is thus a direct consequence of the statistics of uncorrelated random variables. On the other hand, it is only valid when the variables are uncorrelated. If they are not, we need to correct the error estimation by including their covariance. Note that $\text{Cov}(a, a) = \sigma_a^2$. Moreover, the average being an estimator of the expected value,

The covariance between a and b can be estimated as $\text{Cov}(a, b) \simeq \langle ab \rangle - \langle a \rangle \langle b \rangle$. Because $\text{Cov}(a, a) = \sigma_a^2$, the variance of a random variable can be estimated as $\sigma_a^2 \simeq \langle a^2 \rangle - \langle a \rangle \langle a \rangle$.

$$\text{Cov}(a, b) \simeq \langle ab \rangle - \langle a \rangle \langle b \rangle , \tag{7.64}$$

which, for $b = a$, reads as

$$\text{Cov}(a, a) = \sigma_a^2 \simeq \langle a^2 \rangle - \langle a \rangle \langle a \rangle , \tag{7.65}$$

providing a formal justification for the rule to estimate σ_a^2. In general, then, if $c = a + b$,

When variables are correlated, the uncertainty in their combination must include the covariance.

$$\sigma_c^2 = \sigma_a^2 + \sigma_b^2 + 2\text{Cov}(a, b) . \tag{7.66}$$

This rule can be easily extended to any combination of two variables $f(a, b)$. Suppose that we measured N values a_i and b_i, $i = 1, \ldots, N$, from which we obtained their averages a and b. From these measurements, we can obtain another quantity $f(a, b)$. For example, having measured the height $h = \frac{1}{N} \sum_i h_i$ of a freely falling body in a time $t = \frac{1}{N} \sum_i t_i$, we can compute the gravitational acceleration

$$g = f(h, t) = \frac{2h}{t^2} . \tag{7.67}$$

In fact, if a and b are both affected by errors, then so is $f(a, b)$, and, indeed, we can compute N values of the quantity as $f(a_i, b_i)$ (in the example, we will have N values of g, $g_i = \frac{2h_i}{t_i^2}$).

The uncertainty about $f(a, b)$ can be estimated as the average of $f(a_i, b_i) - f(a, b)$. As usual, the uncertainty is defined as a positive quantity, thus we could take the average of $|f(a_i, b_i) - f(a, b)|$ as the uncertainty, but the absolute value operator $|\ldots|$ is difficult to manipulate. Better to evaluate σ_f^2 as the average of the squares of the differences, whose square root is positive by definition. For the sake of simplicity, let's define $f = f(a, b)$ and $f_i = f(a_i, b_i)$, then

$$\sigma_f^2 \simeq \frac{1}{N} \sum_{i=1}^{N} (f_i - f)^2, \tag{7.68}$$

expanding f_i around f, truncating the expansion at the first order,

$$f_i \simeq f + \frac{\partial f}{\partial a} (a_i - a) + \frac{\partial f}{\partial b} (b_i - b) \tag{7.69}$$

such that

$$\sigma_f^2 \simeq \frac{1}{N} \sum_{i=1}^{N} \left(\frac{\partial f}{\partial a} (a_i - a) + \frac{\partial f}{\partial b} (b_i - b) \right)^2 =$$
$$\left(\frac{\partial f}{\partial a} \right)^2 \sigma_a^2 + \left(\frac{\partial f}{\partial b} \right)^2 \sigma_b^2 + 2 \frac{\partial f}{\partial a} \frac{\partial f}{\partial b} \left(\frac{1}{N} \sum_{i=1}^{N} (a - a_i)(b - b_i) \right). \tag{7.70}$$

The last factor in parentheses is nothing but an estimator of the covariance, thus, in general,

$$\sigma_f^2 \simeq \left(\frac{\partial f}{\partial a} \right)^2 \sigma_a^2 + \left(\frac{\partial f}{\partial b} \right)^2 \sigma_b^2 + 2 \frac{\partial f}{\partial a} \frac{\partial f}{\partial b} \mathrm{Cov}(a, b), \tag{7.71}$$

Errors sum in quadrature, but if they are correlated, we must include the covariance in the error propagation.

which reduces to the simple sum in quadrature when the covariance is null, i.e., the measurements are uncorrelated. An important property of the covariance is that

$$\mathrm{Cov}(a, b) \leq \sigma_a \sigma_b. \tag{7.72}$$

In fact, given that $\mathrm{Cov}(a, b) = E[(a - \bar{a})(b - \bar{b})]$, its square is

$$\mathrm{Cov}^2(a, b) = E^2[(a - \bar{a})(b - \bar{b})]$$
$$\leq E[(a - \bar{a})^2] E[(b - \bar{b})^2] = \sigma_a^2 \sigma_b^2. \tag{7.73}$$

The covariance between two variables is always less than or equal to the product of their standard deviation: $\mathrm{Cov}(a, b) \leq \sigma_a \sigma_b$.

The last step in the demonstration above comes from the fact that

$$E^2(ab) \leq E(a^2) E(b^2). \tag{7.74}$$

To prove that $E^2(ab) \le E(a^2)E(b^2)$, we build a random variable z as a linear combination of a and b; observe that $E(z^2) \ge 0$ and find γ such that the latter condition is true.

The proof of this property is not straightforward. Consider the random variable

$$z = a - \gamma b. \tag{7.75}$$

Being that $z^2 \ge 0$, then $E(z^2) \ge 0$ and

$$E(z^2) = E\left[(a - \gamma b)^2\right] = E\left[a^2 + \gamma^2 b^2 - 2ab\gamma\right] = \\ E(a^2) + \gamma^2 E(b^2) - 2\gamma E(ab) \ge 0 \tag{7.76}$$

which is verified if

$$\gamma \ge \frac{E(ab)}{E(b^2)} + \Delta \quad \text{or} \quad \gamma \le \frac{E(ab)}{E(b^2)} - \Delta, \tag{7.77}$$

Δ being the absolute value of the square root of the discriminant of the associated equation divided by $E(b^2)$. Choosing $\gamma = \frac{E(ab)}{E(b^2)}$,

$$E(a^2) + \left(\frac{E(ab)}{E(b^2)}\right)^2 E(b^2) - 2\frac{E(ab)}{E(b^2)}E(ab) = E(a^2) + \frac{E^2(ab)}{E(b^2)} - 2\frac{E^2(ab)}{E(b^2)} \ge 0, \tag{7.78}$$

from which

$$E^2(ab) \le E(a^2)E(b^2), \tag{7.79}$$

leading to the result given above.

7.8 The Binomial Distribution

When events can be assigned to two classes, their distribution is binomial.

Let p be the probability that an event occurs. Because of the normalisation condition, the probability of that event not occurring is $q = 1 - p$. We want to compute the probability of observing n events in N trials. Since there are only two possible outcomes for this random variable (the event either occurs or not), the corresponding distribution is called binomial.

The simplest binomial experiment consists in tossing a coin, for which there can only be two results. The same happens with the decay of some subatomic particles. For example, the K particle sometimes decays (transforms) into $\pi\pi$, sometimes into $\pi\pi\pi$, while the τ lepton decays into *leptons* or *hadrons*. In general, any system in which one can divide the possible events into two classes follows binomial distribution.

Using binomial distribution, we can determine the probability of observing a given sequence of events. A very common example is the following: how large is the probability of observing n times "3" in a sequence of 10 throws of a die? A more physical example is the following. Consider a system of N electrons, whose spin, a quantum number, can only have two values: $s = \pm\frac{1}{2}$. Many measurable quantities of such a system depend on the number of electrons with *spin up*, i.e., $s = +\frac{1}{2}$. Being that it is impossible to measure all the spins for any individual electron, we can only compute such a number statistically. To estimate the probability of observing n spins up, we count how many states lead to each configuration and divide by the number of all possible configurations.

In one configuration, $n = 0$ and all spins are *down* or $s = -\frac{1}{2}$. There are, then, N configurations for which $n = 1$. The only electron with spin up can be the first, the second, the third, ..., the N-th. The number of configurations with $n = 2$ is $\frac{N(N-1)}{2}$. For each of the N configurations with $n = 1$, there are $(N-1)$ electrons that can have spin up, however, two configurations are equivalent (both have $n = 2$), so we have to divide by two. For $n = 3$, we have $\frac{N(N-1)(N-2)}{2\times3}$, and so on. In total, there are

> In a binomial system of N elements, the number of configurations with n in the "success" state is given by the binomial coefficient.

$$\binom{N}{n} = \frac{N!}{n!\,(N-n)!} \tag{7.80}$$

such combinations, where $N! = 1 \times 2 \times 3 \times \cdots N$ and $0! = 1$. If p is the probability for an electron to have $s = +\frac{1}{2}$, the joint probability of each single configuration is $p^n (1-p)^{N-n}$, thus the total probability is

$$P(n, p) = \binom{N}{n} p^n (1-p)^{N-n}. \tag{7.81}$$

The average number of spin up electrons is

$$\langle n \rangle = \sum_{n=0}^{N} n \binom{N}{n} p^n (1-p)^{N-n}. \tag{7.82}$$

To compute this, we observe that

$$(p + q)^N = \sum_{n=0}^{N} \binom{N}{n} p^n q^{N-n} , \tag{7.83}$$

and taking the derivative with respect to p, we obtain

$$N(p + q)^{N-1} = \sum_{n=0}^{N} n \binom{N}{n} p^{n-1} q^{N-n} . \tag{7.84}$$

If we multiply both members by p, the right-hand side of the equation reproduces our formula for the expected value:

$$Np(p + q)^{N-1} = \sum_{n=0}^{N} n \binom{N}{n} p^n q^{N-n} , \tag{7.85}$$

7

The mean of the binomial distribution is given by the product Np.

which, for $q = 1 - p$, becomes

$$Np = \sum_{n=0}^{N} n \binom{N}{n} p^n (1 - p)^{N-n} = \langle n \rangle . \tag{7.86}$$

The variance of the binomial distribution is $Np(1 - p)$.

Similarly, we can compute the variance that is

$$M_\mu^{(2)} = Np(1 - p) . \tag{7.87}$$

The relative width of the distribution decreases as $\frac{1}{\sqrt{N}}$.

Note that $\sigma = \sqrt{Np(1 - p)}$ and that the uncertainty about n increases with N, but only as $\sqrt{N}$. The relative uncertainty is

$$\frac{\sigma}{\mu} = \frac{\sqrt{Np(1 - p)}}{Np} = \sqrt{\frac{1 - p}{Np}} , \tag{7.88}$$

i.e., it decreases as $\frac{1}{\sqrt{N}}$.

In order to study the properties of fundamental interactions, physicists cause particles to collide head-on and look at the products of the collision. One can define the *asymmetry* as

$$A = \frac{N_+ - N_-}{N_+ + N_-} , \tag{7.89}$$

where N_+ and N_- are, respectively, the number of events with a given property (e.g., the electric charge) identified by the $+$ sign and those remaining, identified with the $-$ sign. Because of charge conservation, in $e^+ + e^-$ collisions (where the total charge is null), we expect the charge asymmetry to be zero.

Suppose that in an experiment, we counted $N_+ = 4728$ positively and $N_- = 5021$ negatively charged particles. The charge asymmetry is then

$$A = \frac{4728 - 5021}{4728 + 5021} \simeq -0.03. \tag{7.90}$$

In fact, the experiment is equivalent to drawing a random variable with two possible states $+1$ and -1, $N = N_+ + N_- = 9749$ times, when $p = q = 0.5$.

The expected number N_+ of positive particles is then $Np = 4875$, with a standard deviation of $\sigma_+ = \sqrt{Npq} = 49$. Of course, the same values apply to N_-, being that $q = p$. The uncertainty about the asymmetry is given by the uncertainty propagation:

$$\sigma_{A_{th}}^2 = \left(\frac{\sigma_{N_+}}{N}\right)^2 + \left(\frac{\sigma_{N_-}}{N}\right)^2 = 5 \times 10^{-5}, \tag{7.91}$$

hence $\sigma_{A_{th}} = 0.007$. In other words, the expected value for A is

$$A_{th} = 0.000 \pm 0.007. \tag{7.92}$$

Our results differ from this by

$$\frac{A_{th} - A}{\sigma_{A_{th}}} = \frac{0.03}{0.007} \simeq 4, \tag{7.93}$$

i.e., we found a value for A that is four standard deviations out of theoretical predictions (physicists say that there is a *tension* between data and theory). In a successive box, we find that such a tension, in fact, does not exist and is just the result of neglecting the fact that even the number of events counted in an experiment is affected by uncertainty.

We emphasise that there are no physicists who would trust you if you claimed a discovery with such a discrepancy. Even if we observed a discrepancy of a bit more than 5 standard deviations, conventionally established as a threshold for claiming a discovery, the discrepancy can be ascribed to many other effects that are not random at all and are called "systematics". The apparatus used to take the measurements, for example, may not be fully or consistently efficient, for example. You are thus required to prove that the discrepancy is genuine and that it does not come from systematic effects. This can be done repeating the experiments with different instruments and methods or studying the distribution of data as a function of the statistics, time, and other variables that, in principle, can influence the result.

Moreover, we did not consider N_+ and N_- to be affected by any uncertainty. Indeed, they are measurements, too, and, as such, have an associated uncertainty that must be evaluated. As a consequence, $N = N_+ + N_-$ has an uncertainty, as well as A. The uncertainties about this number are the topic of the section on the Poisson distribution below.

7.9 The Shape of the Bimonial Distribution

numpy.random. binomial() returns lists of random numbers distributed according to the binomial distribution.

In order to understand the shape of this distribution, we can exploit the ability of Python to generate (pseudo)random numbers with various distributions.

Consider, e.g., the following code:

```
import matplotlib.pyplot as plt
import numpy as np

i = 1
binwidth = 1
while i < 40:
    x = np.random.binomial(100, i*1e-2,
        100000)
    plt.hist(x, histtype = 'step', rwidth=1,
             bins=range(min(x),
             max(x)+binwidth, binwidth),
             label='p = {}'.format(i*1e-2)))
    i += 10

plt.xlim(0,50)
plt.legend()
plt.show()
```

The np.random.binomial(N, p, m) returns a list of m numbers according to a binomial distribution with parameters N and p. The expected value of these numbers is Np. Each number on the list is the number of times a binomial random variable assumes its "success" value in a series of N experiments.

At the first iteration, the script draws 100 000 numbers from a binomial distribution with $N = 100$ and $p = 0.01$, each representing the number of successful events occurring in 100 draws, when each event has a probability of 0.01. The expected number of those events is then $100 \times 0.01 = 1$. Manifestly, the possible values of the random number can be 0 (the event never occurs), 1, 2, ..., 100 (the event occurs in all draws). Since each event has a probability of 0.01, it is very unlikely that we can observe high values, while 1 being a relatively high probable value, observing 0 or 2 or some other value close to 1 should not be a too rare of an event.

Draws are repeated four times with different probabilities besides $p = 0.01$: $p = 0.1$, $p = 0.2$, $p = 0.4$. Correspondingly, we expect the drawn values to be distributed around $Np = 10$, $Np = 20$, and $Np = 40$. It is worth noting that the variance (hence, the standard deviation) grows with Np. The distribution of values is then expected to become larger.

The distribution can be observed by creating a histogram of the values, i.e., counting how many times a given number x is drawn from the distribution and plotting such a number as a function of x. This is done via

```
plt.hist(x, histtype = 'step',
         rwidth=1,
         bins=range(min(x),
         max(x)+binwidth,
         binwidth))
```

The only mandatory parameter is x, a list containing all the numbers drawn from the distribution. The histtype parameter tells us how to represent the histogram: data can be shown as bars, stacked bars, or filled or unfilled steps. rwidth determines the fraction of each bin to be occupied by the bar or the step, while bins defines the way in which the horizontal axis is divided into classes. In this case, we chose to have one bin per value: bins is thus a list of integer numbers from 1 to 101 (there are simpler ways to make such a list: we used this because it is very generic and can be used in many cases). Finally, label assigns a label to the data set. The labels are used to make the legend with plt.legend().

matplotlib is a very flexible library. The user can choose how to represent a histogram (histtype, rwidth) and how to divide the interval into classes (bins).

Plots of the histograms are shown in Fig. 7.2. For low Np, the distribution is asymmetric and narrow. It becomes wider and more symmetric as Np increases. It also moves to the right, showing a peak close to Np. The integral of the distribution is clearly equal to the number of values drawn, i.e., 100 000. Since the width of the distribution increases with Np as $\sigma = \sqrt{Np(1-p)}$, its height decreases.

☐ Fig. 7.2 Histograms of numbers drawn from binomial distributions with $N = 100$ and various probabilities p

7

The cumulative
distribution function
is the probability of
the random variable
being lower than x.

With Python, we can easily compute the probability of having observed a number $n_+ < n_b p$ defined in the previous section.

Given a probability distribution $P(x)$, its cumulative distribution is defined as

$$C(a) = \int_{-\infty}^{a} P(x)dx .$$ (7.94)

In the case of a discrete distribution like the binomial one, the integral becomes a sum, i.e.,

$$C(a) = \sum_{i=0}^{a} P(n) .$$ (7.95)

Clearly,

$$\lim_{a \to \infty} C(a) = 1$$ (7.96)

is a consequence of the normalisation condition and $C(a)$ is the area under the normalised distribution, i.e., the distribution divided by a constant such that its integral is one, for $x < a$. It represents the probability that $x < a$. For each random number generator defined in numpy, the corresponding theoretical distribution is defined in the scipy module. To compute the value of $P(n, p)$ for $n = 4728$ when $p = 0.5$ and $N = 10\,000$, we can use the following code:

```
from scipy.stats import binom

P = binom.pmf(4728, 10000, 0.5)
```

pmf stands for probability mass function or PMF. With this term, we identify the function that gives the probability of finding the value given as the first parameter in a set of random numbers that follow the given distribution with the given parameters (in this case, $N = 10\,000$ and $p = 0.5$). For continuous distributions, what makes sense is the probability distribution function $P(x)$ (PDF), given by the pdf() method. In a certain sense, PDF and PMF are synonyms, and we may confuse them below.

The cumulative $C(4728)$ can be obtained with the cdf() method (cumulative distribution function), as in

```
from scipy.stats import binom

P = binom.pmf(4728, 10000, 0.5)
C = binom.cdf(4728, 10000, 0.5)

print('P = binom(4728; N=1e4; p=0.5)')
print('P ={:.2e}'.format(P))
print('sum_0^4728[P] = ' + str(C))
```

The result is the following:

```
P = binom(4728; N=1e4; p=0.5)
P = 2.97e-09
sum_0^4728[P] = 2.9732458769875568e-09
```

There is thus a probability as low as $p \simeq 2.8 \times 10^{-8}$ of obtaining such a value from a binomially distributed dataset with $N = 10\,000$ and $p = 0.5$.

Here, we show how to format a string, too. The second print uses the + operator to join two strings: the first is a constant string, written within single quotes, while the second is the result of a conversion from a floating point number C into a string (i.e., its character representation) obtained with the casting operator str(). In general, you can convert a variable or a constant from one type to another using operators that identify the final type like int(), float() or str(). This operation is called type-casting.

scipy.stats.pmf() returns the probability that the random variable x takes the given value. For continuous random variables, it returns the PDF $P(x)$. The cumulative is given by the cdf() method.

Strings can be concatenated using the + operator. Numbers must be cast to strings before they are lexicographically added to other strings.

7

The output can be formatted according to a particular format, established by descriptors, represented as a semicolon followed by alphanumeric characters describing the output format.

When using the join operator +, the strings are represented as they appear by default using `print`. For example, if you call `print(C)`, you will see `2.7974014 420427173e-08`. The previous line of code renders the string according to a given format. Each pair of braces in the string is substituted with a value passed as an argument to `format()`: a method applied to the string. If no format descriptors are given in the braces, then the value is reproduced in the string as in its default representation, otherwise, it is formatted accordingly. In general, a format descriptor follows a colon (:) and is composed of an alphabetic character with optional preceding numbers. The alphabetic character tells the system how to interpret the value passed in the `format()` method. There are many descriptors, the most common being d for integers (digit), f for floating point numbers, and s for string, while e is used to represent numbers using the scientific notation.

Optionally, the descriptor character can be preceded by one or more numbers that depend on it. For d, the preceding number represents the number of characters with which the value is going to be written. For non-integer numerical descriptors, the descriptor can be preceded by a length specifier in the form n.m, where both n and m are integers. The first represents the total length of the field, and the second the number of digits after the decimal point to be shown. Consider the following examples:

```
print('n = {:d}'.format(10000))
print('n = {:8d}'.format(10000))
print('s = {:s}'.format('test'))
print('x = {:f}'.format(np.exp(1)))
print('x = {:.3f}'.format(np.exp(1)))
print('x = {:8.4f}'.format(np.exp(1)))
print('x = {:5.1e}'.format(np.exp(1)))
```

Executing a script containing these lines, we should see something like

```
n = 10000
n =    10000
s = test
x = 2.718282
x = 2.718
x =    2.7183
x = 2.7e+00
```

In the latter case, for example, we asked that the value of $e \simeq 2.7$ be printed using five characters, of which just one is after the decimal point and in the scientific notation. The result, being longer than five characters, is represented aligned to the left. The previous line requires writing the same number using eight characters, of which four are after the comma. Since the total number of characters needed to write this is six, two characters on the left are left blank.

7.10 Random Walk

Consider a particle moving uniformly along the x-axis of a reference frame, starting from its origin. If such a particle is subject to random symmetric effects, it deviates by ± 1 a.u. in the y direction, with equal probability. After N steps along x, it can be found at coordinates (N, y), where y is a random variable.

The average position along the y-axis after one step is

$$\langle y \rangle = \sum_{i=1}^{2} y_i P_i = +1 \times \frac{1}{2} - 1 \times \frac{1}{2} = 0 , \tag{7.97}$$

The mean coordinate in a random walk of length N is null, while its variance is N.

while, for the variance, we get

$$\sigma_y^2 = \sum_{i=1}^{2} (y_i - \langle y \rangle)^2 P_i = \sum_{i=1}^{2} y_i^2 P_i = 1 . \tag{7.98}$$

For N steps, then,

$$\langle y \rangle = 0 \tag{7.99}$$

and

$$\sigma_y^2 = N , \tag{7.100}$$

because the average of a sum is the sum of the averages and the variance of a sum is the sum of the variances (see Sect. 6.6). It is, in fact, straightforward to prove that the expected value is a linear operator, i.e., given a random variable x,

The expected value is a linear operator.

$$E(ax + b) = aE(x) + b , \tag{7.101}$$

where a and b are constants. After N steps in the x direction, then, our walker will still be on the x-axis, on average. However, in n experiments, we will find that $y(N)$ is distributed with zero mean and a standard deviation of $\sigma = \sqrt{N}$. Each step in the walk can be considered as a binomially distributed random variable, having just two possible values. Defining a step in the positive direction as a "success", the average number of successes after N steps is $Np = \frac{N}{2}$, with a standard deviation of $\sqrt{Npq} = \frac{\sqrt{N}}{2}$. The average coordinate will then be

On average, a random walker does not move.

$$\langle y \rangle = +1 \times P(y > 0) - 1 \times P(y < 0) = +1 \times \frac{N}{2} - 1 \times \frac{N}{2} = 0$$

$$(7.102)$$

The uncertainty in the position of the walker after N steps is $\sqrt{N}$.

while

$$\sigma_y^2 = \sigma_{y>0}^2 + \sigma_{y<0}^2 + 2\sigma_{y>0}\sigma_{y<0} = \frac{N}{4} + \frac{N}{4} = \frac{N}{2} + 2\frac{N}{4} = N.$$

$$(7.103)$$

Indeed, the steps towards positive values of y and those towards negative ones, in this case, are fully correlated. Consider, in fact, a random variable $N = n_+ + n_-$ (in the example, N is the number of steps of the random walk, n_+ the number of steps in the positive direction and $n_- = N - n_+$ those in the opposite direction). The variance of N can be written as

$$\sigma_N^2 = \sigma_{n_+}^2 + \sigma_{n_-}^2 + 2\mathrm{Cov}(n_+, n_-).$$

$$(7.104)$$

Given that the total number of steps is fixed to be N, if there are n_+ steps with $y > 0$, those with $y < 0$ are $n_- = N - n_+$, i.e., $n_- = f(n_+)$, and the covariance attains its maximum value.

In order to verify this *experimentally*, we can simulate a random walk using Python as follows. First of all, we define a function that, given N, simulates N steps of a random walk with $dy = \pm 1$ as

```python
def distance(N):
    d = 0
    for i in range(N):
        dy = 1
        if np.random.uniform() > 0.5:
            dy = -1
        d += dy
    return d
```

Functions, in Python, are defined using the keyword `def` followed by the name given to the function with the list of needed parameters in parentheses (if the function does not require parameters, a pair of parentheses is mandatory). The function declaration is terminated by a colon (`:`), while its body, representing the operations to be done on the parameters (if any), is written indented. The function returns zero or more values via the `return` keyword. It is worth noting that, in contrast to C or C++, variables do not need to be declared in Python.

Moreover, in C and C++, variables are always local to the scope in which they are declared, the scope being identified with the region of code comprised within a pair of braces. This means that a variable a defined in a function is different with respect to a variable with the same name defined in another function, irrespective of the fact that they share the name. Variables must be thought of, in fact, as containers whose name is arbitrary and irrelevant, except for the programmer. There are as many containers as declarations. Because of their *locality*, variables cannot be shared among scopes and it is forbidden to use a variable that has been declared in another scope. Consider, for example, the following excerpt:

```
void loop () {
    int x = analogRead (A0);
    float h = 1.;
    Serial.print(x)
    if (x < 30) {
        float V0 = 5.;
        float h = V0/1023;
        x *= h;
    }
    Serial.print("x");
    Serial.print(h)
    Serial.print("=");
    Serial.println(x);
}
```

Python functions are defined using the keyword `def` followed by the name given to the function with the list of needed parameters in parentheses. The body of the function is written indented on the following lines, after a colon (`:`).

Variables in C and in C++ must be declared and exist only within the scope in which they are declared and in those nested to it. Two variables with the same name declared in different scopes are different and can be used as if they had different names. Variables are containers and their names are only meaningful for the programmer.

It reads the value from the A0 analog input of Arduino and multiplies it by $\frac{5}{1023}$ if it is less than 30. Suppose, then, that x contains 51. Using the serial monitor, we should see

```
51x1=51
```

If, on the other hand, $x = 23$, we would expect to see

```
23x0.0049=0.1127
```

but, in fact, what we get is

```
23x1=0.1127
```

The reason being that the variable h is redeclared within the scope of the if. Its content is correctly set to $\frac{5}{1023} \simeq 0.0049$ and x is changed accordingly, however, exiting from the scope, in which h no longer exists, Serial.print(h) just prints the only accessible h whose value is 1.

Moreover, if we try to print V0, too, outside of the if scope, the compiler returns an error, being that it does not exist outside that scope.

Needless to say, this can cause a bit of confusion for beginners. Python overrides this behaviour and local variables follow rules that are somewhat more natural. Variables are always considered as local within a scope, but since it is not necessary to declare them, their value is taken from those used above, if needed. Function parameters always imply, as in C and C++, the declaration of local variables whose names coincide with those of the parameters. Consider the following example.

```
a = 3

def fun1():
    b = 2 * a
    return b

def fun2(a):
    a *= 2
    return a

def fun3(b):
    a = 7
    return a

print('{} {}'.format(a, fun1()))
print('{} {}'.format(a, fun2(a)))
print('{} {}'.format(a, fun3(a)))
print(a)
```

The first print() shows

```
3  6
```

In fact, `fun1()` multiplies a by 2 and returns it. Since a is not found on the left side of an operator in `fun1()`, it is taken from the *global* scope and its value is assumed to be 3. The second print produces the same output. Here, a is considered to be a parameter and it implies the creation of a new local variable a that may have anything to do with the global one. In the function, its value is altered, but when we use it in the third `print()`, its value is still 3, because the latter is a different container with respect to the one used in the function. In `fun3(b)`, the parameter is called b and is not used. So, when we pass the content of a to the function, it is ignored: the function just assigns the value 7 to a and returns it. What we see on the screen is, in fact,

3   7

Again, even if the content of a is assigned in `fun3()`, the change does not reflect upon the global a. In fact, being that a is on the left side of an operator in the function, it implies the creation of a new local variable and the last `print()` shows that, in fact, a is still equal to 3.

This behaviour may appear odd at first glance, but, in fact, it is very natural and does not require the programmer to consider the scope of a variable when using it. It is worth noting that Python allows for using the keyword `global` to turn a local variable into a global one, however, we suggest completely forgetting it, as in the case of the `goto` statement in C: it exists, but using it qualifies the programmer as an uneducated novice.

To simulate $m = 100$ walks with the above function and compute their average and their standard deviation, we can write another function as

```
def simulate(n):
    y = []
    for k in range(100):
        y.append(distance(n))
    return np.mean(y), np.std(y)
```

We are then ready to simulate random walks of different lengths, for example, as

```
nsteps = 1
n = []
t = []
st = []
for k in range(10):
    ymean, sigma = simulate(nsteps)
    n.append(nsteps)
```

In Python, new local variables are automatically allocated when they are found on the left side of an operator. In all other cases, the content of a variable is taken from the outer scope.

To generate a random event with a given probability p, we generate a random number $x \in [0, 1)$. The event occurs if $x < p$ or if $x > 1 - p$.

Through the use of functions, complex problems become easier and less error prone to implement, because we decompose them into smaller and simpler chunks.

```
t.append(ymean)
st.append(sigma)
nsteps *= 2
```

At the first iteration, we compute the average $\langle y \rangle$ and its σ after just $N = 1$ step, then after $N = 2$, $N = 4$, $N = 8$, ..., $N = 1024$ steps. For each simulation, we store N, $\langle y \rangle$ and σ in as many lists as we can use to make the following plot:

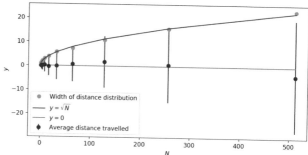

where blue dots represent $\langle y \rangle$. While $\langle y \rangle$ does not depend on N and is always close to zero, the standard deviation of the distribution (shown by the orange dots, as well as by vertical bars whose amplitude is σ) grows by $\sqrt{N}$, as expected.

The random walk is a relatively common phenomenon in physics. As an example, consider the multiple scattering of a particle A after a collision with other particles, where the x-axis is defined as the direction of the initial momentum of particle A.

After N collisions, on average, the particle continues moving in the same direction as its initial momentum. However, the distribution of momenta of m particles will have a width proportional to $\sqrt{N}$: the higher the number of collisions, the larger the width.

This result can be used to estimate the density of scattering centres in a material, for example. When a relativistic particle of momentum p collides with a particle at rest, its momentum direction changes, on average, by an angle of $\theta \propto \frac{1}{p}$ (the larger the momentum, the lower the scattering angle). After many collisions, then, the average angle between the initial momentum and the final one will be zero with a distribution whose width is proportional to

$$\sigma_\theta \propto \sqrt{\theta} \propto \sqrt{\frac{1}{p}}. \tag{7.105}$$

It is instructive to look at alternative ways of computing $\langle y \rangle$ and its standard deviation. The distance traveled in the $+y$ direction is a binomially distributed variable: the probability of going in the positive direction is $p = \frac{1}{2}$ and, after N trials, the average number of steps with $y > 0$ is $E[n_+] = Np = \frac{N}{2}$. The exact same applies to n_-, the number of steps with $y < 0$. On average, then, the sum of positive and negative steps gives

A random walk can be considered a sequence of binomially distributed random variables.

$$E[y] = E[n_+ - n_i] = E[n_+] - E[n_-] = \frac{N}{2} - \frac{N}{2} = 0.$$
$$(7.106)$$

The variance of positive steps is $\sigma_+^2 = Np(1-p) = \frac{N}{4}$, equal to that of the number of negative steps. Remembering that $\sigma_x^2 = E[x^2] - E[x]^2$, we can write

$$\sigma_+^2 = \frac{N}{4} = E[n_+^2] - E[n_+]^2 = E[n_+^2] - \frac{N^2}{4} \qquad (7.107)$$

and

$$E[n_+^2] = \frac{N}{4} + \frac{N^2}{4}. \qquad (7.108)$$

The same result applies to $E[n_-^2] = \frac{N}{4} + \frac{N^2}{4}$. Similarly, with $y = n_+ - n_-$,

$$\sigma_y^2 = E[y^2] - E[y] = E[y^2] = E[(n_+ - n_i)^2] =$$
$$E[n_+^2] + E[n_-^2] - 2E[n_+n_-] = \frac{N^2 + N}{2} - 2E[n_+n_-].$$
$$(7.109)$$

By definition,

$$E[n_+n_-] = \sum n_+n_- P_{n_+n_-} \qquad (7.110)$$

where the sum is extended to all the possible values of n_+n_- and $P_{n_+n_-}$ is the joint probability of observing n_+ and n_- counts at the same time. Since $N = n_+ + n_-$, if we observe n_+ events with $y > 0$, the probability of observing $n_- = N - n_+$ events with $y < 0$ is equal to 1 and

$$E[n_+n_-] = \sum n_+(N - n_+)P_{n_+} = N \sum n_+ P_{n_+} - \sum n_+^2 P_{n_+} =$$
$$NE[n_+] - E[N_+^2] = N\frac{N}{2} - \frac{N^2 + N}{4} = \frac{N^2 - N}{4}.$$
$$(7.111)$$

Substituting in the equation for σ_y^2,

$$\sigma_y^2 = \frac{N^2 + N}{2} - 2\frac{N^2 - N}{4} = N. \tag{7.112}$$

7.11 The Poisson Distribution

When $p \to 0$ and $N \to \infty$, such that Np is constant, the binomial distribution tends to the Poisson one.

In many cases, the probability of success p in a binomial experiment is extremely low. In those cases, in order to observe a statistically significant number of events, N has to be large, such that $M = Np$ is not negligible. In this case, the binomial distribution tends to a continuous distribution called the Poisson distribution, named after Siméon-Denis Poisson (1781–1840), who introduced it. To obtain the Poisson distribution, we need to let $N \to \infty$, keeping $p \to 0$ such that $Np = M = const$ and $p = \frac{M}{N}$.

The binomial distribution can be rewritten as

$$P(n, p) = \binom{N}{n} p^n (1 - p)^{N-n}. \tag{7.113}$$

Substituting p, it becomes

$$P(n, p) = \frac{N!}{n!\,(N-n)!} \left(\frac{M}{N}\right)^n \left(1 - \frac{M}{N}\right)^{N-n}, \tag{7.114}$$

and we take the limit to $n \to \infty$:

$$\lim_{N \to \infty} P(n, p) = \left(\frac{M^n}{n!}\right) \lim_{n \to \infty} \frac{N!}{(N-n)!} \left(\frac{1}{N^n}\right) \left(1 - \frac{M}{N}\right)^N \left(1 - \frac{M}{N}\right)^{-n}. \tag{7.115}$$

To evaluate the limit of a product, we compute the limit of each factor.

The first factor is

$$\frac{N!}{(N-n)!} = \frac{N(N-1)(N-2)\cdots(N-n)(N-n-1)(N-n-2)\cdots 2 \cdot 1}{(N-n)(N-n-1)(N-n-2)\cdots 2 \cdot 1} \tag{7.116}$$

and only the terms $N(N-1)(N-2)\cdots(N-n+1)$ survive. There are n factors in this product, so this number grows as N^n. The product of this number times $\frac{1}{N^n}$, then, tends to 1 as $N \to \infty$.

Defining $x = -\frac{N}{M}$, the third factor within the limit,

$$\lim_{N \to \infty} \left(1 - \frac{M}{N}\right)^N, \tag{7.117}$$

becomes

$$\lim_{N \to \infty} \left(1 + \frac{1}{x}\right)^{-xM}. \tag{7.118}$$

Letting N go to infinity is equivalent letting x go to infinity and

$$\lim_{x \to \infty} \left(\left(1 + \frac{1}{x}\right)^x\right)^{-M} = e^{-M}. \tag{7.119}$$

The latter term, when $N \to \infty$, approaches 1, hence

$$\lim_{N \to \infty} P(n, p) = \left(\frac{M^n}{n!}\right) \times 1 \times e^{-M} \times 1 = \frac{M^n}{n!} e^{-M}. \tag{7.120}$$

Such a probability no longer depends on p and is the probability density function for the Poisson distribution:

$$P(M, n) = \frac{M^n}{n!} e^{-M}. \tag{7.121}$$

The probability mass distribution is correctly normalised. In fact, given that the Taylor expansion of e^M is

$$e^M = \sum_{n=0}^{\infty} \frac{M^n}{n!}, \tag{7.122}$$

it is straightforward to check that

$$\sum_{n=0}^{\infty} \frac{M^n}{n!} e^{-M} = e^M e^{-M} = 1. \tag{7.123}$$

We can use the same result to compute the mean and the variance of the distribution. The mean is

$$\sum_{n=0}^{\infty} n \frac{M^n}{n!} e^{-M} = M \sum_{n=1}^{\infty} n \frac{M^{n-1}}{n(n-1)!} e^{-M} = M, \tag{7.124}$$

while the variance can be obtained as $\sigma^2 = \langle x^2 \rangle - \langle x \rangle^2$. We must then compute

$$\langle x^2 \rangle = \sum_{n=0}^{\infty} n^2 \frac{M^n}{n!} e^{-M}. \tag{7.125}$$

We write $n^2 = n^2 + n - n = n + n(n-1)$ and substitute:

> The Poisson probability mass distribution of a process in which we observe n events, given an expectation of M, is $\frac{M^n}{n!} e^{-M}$.

> The mean of the Poisson distribution is $M = Np$. Its variance is M, too.

$$\langle x^2 \rangle = \sum_{n=0}^{\infty} (n + n(n-1)) \frac{M^n}{n!} e^{-M} = \sum_{n=0}^{\infty} n \frac{M^n}{n!} e^{-M} + \sum_{n=0}^{\infty} n(n-1) \frac{M^n}{n!} e^{-M}.$$

$$(7.126)$$

The first term is just M, while the second can be rewritten, as before, as

$$\sum_{n=0}^{\infty} n(n-1) \frac{M^n}{n(n-1)(n-2)!} e^{-M} = M^2 \sum_{n=2}^{\infty} \frac{M^{n-2}}{(n-2)!} e^{-M} = M^2.$$

$$(7.127)$$

Finally,

$$\sigma^2 = \langle x^2 \rangle - \langle x \rangle^2 = (M + M^2) - M^2 = M. \quad (7.128)$$

7

Poisson statistics apply to many counting experiments.

The Poisson distribution represents the probability of observing n low probability events in an infinitely large sample, when the expected number of observed events is M. In general, data from any experiment in which something is counted follow Poisson statistics, provided that the observed process is relatively rare.

Radioactive decay is an example of a process described by Poisson statistics.

This distribution is very useful in a number of cases. For example, radioactive decay is a process that happens on certain nuclei that spontaneously transform into a different species, emitting particles like electrons (β rays) or helium nuclei (α particles). Given a sample of ^{60}Co, for example, you can observe a certain number of decays by detecting β rays that are exiting the sample. The "activity" of the sample is defined as the number of decays per second and is measured in Becquerel (Bq, after Antoine Henri Becquerel - 1852–1908 - who discovered radioactivity). A typical value for a small radioactive source used in schools, for example, is under 200 kBq: it emits less than 200 000 particles per second, on average. The number of atoms in the sample is on the order of the Avogadro number $N_A \simeq 6 \times 10^{23}$. Radioactive decays happen randomly at a fixed rate. In other words, only a small number M of decays is observed out of an almost infinite number of possible events ($N_A \simeq \infty$), such that $p = \frac{M}{N}$ is small and constant.

With a radioactive source whose activity is 200 kBq, we expect to observe, on average, 2×10^5 counts per second, with an uncertainty of $\sqrt{N} = 450$, i.e.,

$$N = (2.0000 \pm 0.0045) \times 10^5 \text{ Bq}. \quad (7.129)$$

The real number of counts will likely be smaller, considering that the detector has an efficiency of $\epsilon < 1$, the latter being the ratio between the detected particles and those that have passed through it. Also, the source emits particles in the full solid angle $\Omega = 4\pi$, while the detector only covers part of it: ω. The ratio $A = \frac{\omega}{\Omega}$ is called the *acceptance* and the experimental number of counts is expected to be $N_{exp} = NA\epsilon$.

Consider the experiment outlined in Sect. 7.8, where we counted $N_+ = 4\,728$ and $N_- = 5\,021$ particles. We can assume that these numbers follow Poisson statistics, since the production of a particle is an event with a relatively low probability of happening during a sufficiently lengthy experiment.

The uncertainties of the numbers are thus $\sigma_+ = \sqrt{N_+} = 69$ and $\sigma_- = \sqrt{N_-} = 71$ (note that we kept two significant figures in this case, while we usually keep only one of them, in which case they are the same).

As a result, the errors on the numerator and the denominator of the asymmetry are

$$\sigma_n = \sigma_d = \sqrt{\sigma_+^2 + \sigma_-^2} \simeq 100 \tag{7.130}$$

(it is worth noting that, since uncertainties add up in quadrature, the two standard deviations are the same). The variance on the ratio is

$$\sigma_A^2 = \left(\frac{\sigma_n}{N^2}\right)^2 + \left(\frac{N_+ - N_-}{N^2}\sigma_d\right)^2 = 1 \times 10^{-12} + 1 \times 10^{-7} \simeq 1 \times 10^{-7}, \tag{7.131}$$

and the standard deviation $\sigma_A = 0.0003$, negligible with respect to the one computed in Sect. 7.8. The assumption made there, that those numbers have no errors, is justified by this observation.

Another example of experiments that follow Poisson statistics is the observation of cosmic muons: particles produced when high energy extraterrestrial protons collide with the nuclei of the atmosphere, about 10 km above the ground. Their rate is not constant, but rather varies slowly, and is well approximated by Poisson statistics. Observing cosmic muons requires expensive and complex detectors called muon telescopes, however, there are a number of universities and research centres that make data collected by these instruments publicly available, including through Apps for smartphone, providing public access to real data.

A less exotic example would be the distribution of the number of raindrops that fall on a given surface per unit time, or the number of cars that pass through a tollbooth or cross a junction, provided that, in all these cases, the events are not too overwhelming (e.g., the rain, as well as the traffic, must be light).

The rate of cosmic muons follows Poisson statistics. It can be monitored using Apps for smartphones or by searching for public data distributed through websites by laboratories around the world.

Generally speaking, most counting experiments can be described in terms of Poisson statistics.

7

Poisson statistics
applies to everyday
life, too.

Often, population statistics, political or market polls, etc., are well described by the Poisson distribution, except in a few cases in which the fraction of counted events is too high. The reliability of polls, for example, strongly depends on the size of the sample. If a party claims to have 12.2% of consensus in a poll involving 1 000 voters, it means that 122 of them are willing to vote for it. However, such a number has an uncertainty of $\sqrt{122} \simeq 11$, thus the real consensus in the population is between $89/1\,000 = 8.9\%$ and $155/1,000 = 15.5\%$, with a *confidence level* of 99%. Any change of a few percent from week to week is thus statistically insignificant, unless there is a clear trend distributed over a few weeks. In this case, the binomial statistics applies, for the purpose of computing the probability that the trend is positive or negative by chance, in a number of trials.

7.12 The Shape of the Poisson Distribution

The Poisson
distribution is
asymmetrical, but as
its mean increases, it
tends to become
wider and
symmetrical.

Plots of the Poisson distribution for various values of M are shown below.

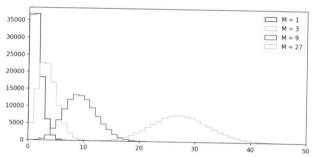

They are not much different from those of the binomial distributions. After all, they are related. As for the binomial distribution, the shape of the Poisson one becomes lower, wider and more symmetric as M increases.

The above plot has been produced using the following script:

```
import matplotlib.pyplot as plt
import numpy as np

i = 1
binwidth = 1
while i < 30:
    x = np.random.poisson(i, 100000)
```

```
    plt.hist(x, histtype = 'step',
        rwidth = 1,
        bins =range(min(x),
              max(x) + binwidth,
              binwidth),
        label ='M = {} '.format(i))
    i *= 3

plt.xlim(0,50)
plt.legend()
plt.show()
```

Using Python, one can obtain the probability and the cumulative for the given observed values N and expected value M, as in

```
import numpy as np
from scipy.stats import poisson

Ntot = 9749
N = 4728
M = Ntot * 0.5
print('N = {}'.format(Ntot))
print('Exp. value: {:.0f} +- {:.0f}'.
    format(M, np.sqrt(M)))
print('Obs. value: {:.0f} +- {:.0f}'.
    format(N, np.sqrt(N)))
d = np.fabs(M-N)
sd = np.sqrt(M+N)
print('Diff.    : {:.0f} +- {:.0f}'.
    format(d, sd))
P = poisson.pmf(N, M)
C = poisson.cdf(N, M)
print('P({:.0f}, {:.0f}) = {:.2e}'.
    format(N, M, P))
print('C({:.0f}) = '.
    format(N) + str(C))
```

`pmf()` and `cdf()` are methods common to all available distributions in the `scipy.stats` package.

The result of the script is

```
N = 9749
Exp. value: 4874+-70
Obs. value: 4728+-69
Diff. : 146+-98
P(4728, 4874)=6.28e-04
C (4728) = 0.01789369643918747
```

Distributions defined in the `scipy.stats` module offer the `pmf()` method, which returns the corresponding probability mass function, as well as `cmf()`, which returns the cumulative.

You are invited to compare the required format of the numbers and their appearance. The probability of observing 4728 events when 4874 are expected appears to be quite low, being 6.28×10^{-4}. In fact, it is not so low, considering that the Poisson distribution is quite wide. The probability of observing any number individually is low because there are many different numbers for which the chance to observe them is relatively large. In fact, the cumulative of the distribution up to 4 728 is 0.02, meaning that there is a probability of 2% of observing a number lower than that (and remember, 2% is a large probability for a physicist, corresponding to more than two standard deviations).

Summary

It is rare that the result of an experiment is just a number. In most cases, it is a set distributed over a range of values.

Data randomly fluctuate around their mean value, often taken as their *true value*. The latter is only an abstract, yet useful, concept for comparing experimental data with predictions. Both data and predictions, in fact, are always affected by uncertainties.

Fluctuations in measurements are ascribed to random effects.

The uncertainty of a linear combination of variables $c = \alpha x + \beta y$ is the square root of its variance $\sigma_c^2 = \alpha \sigma_x^2 + \beta \sigma_y^2 + 2\alpha\beta \text{Cov}(a, b)$. If the variables are uncorrelated, their covariance is null and the formula reduces to the sum in quadrature.

The mean position reached by a random walker after N steps is null, with a variance of N.

Statistics

There are various definitions of probabilities. The classical definitions are the combinatorial or analytical one and the frequentist one. In the first, the probability of an event is defined by the ratio between the favourable cases and the possible ones. In the frequentist approach, the observed frequency of events is taken as an estimation of the probability. Frequency and probabilities tend to match as the number of experiments tend towards infinity.

The Bayesian probability is subjective. It must be estimated from the current knowledge and can be updated as long as new information is available. In most cases, the subjective probability coincides, as a matter of fact, with those computed from the frequentist or combinatorial approaches.

The probability of an event occurring irrespective of other events is called the prior or marginal probability.

The probability of an event that happens with certainty is equal to 1. That of an impossible event is 0.

The joint probability $P(A \text{ and } B)$ for the occurrence of two simultaneous events is $P(A \text{ and } B) = P(A)P(B)$ if the events are independent. If not, $P(A \text{ and } B) = P(A)P(B|A)$. The latter, $P(A|B)$, is the conditional probability of A given that B has occurred.

$P(A \text{ and } B) = P(A)P(B|A)$
If A and B are independent of each other, then $P(B|A) = P(B)$.

Joint probabilities are symmetric, i.e., $P(A, B) = P(B, A)$.

If events are mutually exclusive, the probability is that $P(A \text{ or } B) = P(A) + P(B)$. If not, $P(A \text{ or } B) = P(A) + P(B) - P(A)P(B|A)$.

$P(A \text{ or } B) = P(A) + P(B) - P(A)P(B|A)$
If A and B are mutually exclusive, the probability is that $P(A \text{ and } B) = P(A)P(B|A) = 0$, thus $P(A \text{ or } B) = P(A) + P(B)$.

The posterior probability of A is updated by new evidence B according to the Bayes Theorem: $P(A|B) = \frac{P(A)P(B|A)}{P(B)}$.

Bayes Theorem:
$$P(A|B) = \frac{P(A)P(B|A)}{P(B)}$$

The expected value (or the mean) of a random variable x distributed according to the probability mass function $P(x)$ is defined as $E[x] = \sum_i x_i P(x_i)$. It can be estimated as the average. If the random variable is continuous, the sum becomes an integral and the probability mass function becomes the probability distribution function (pdf): $E[x] = \int xP(x)dx$. The pdf $P(x)$ represents the probability that the random variable lies between x and $x + dx$.

The n-th moment with respect to a value μ is defined as $M_\mu^{(n)} = \int (x - \mu)^n P(x)dx$.

The variance of the population is the expected value of the second moment of its distribution with respect to the mean and can be estimated by the variance of the sample.

The variance of a uniform distribution is the amplitude of the interval of the possible values divided by 12.

The covariance of two random variables x and y is defined as $\mathrm{Cov}(a, b) = E[(x - \bar{x})(y - \bar{y})]$, $\bar{x}$ and $\bar{y}$ being, respectively, the average values of x and y. It can be estimated as $\langle xy \rangle - \langle x \rangle \langle y \rangle$. As a consequence, the variance of x can be estimated as $\sigma_x^2 = \langle x^2 \rangle - \langle x \rangle^2$.

The covariance of two variables is always less than or equal to the product of their standard deviation: $\mathrm{Cov}(a, b) \leq \sigma_a \sigma_b$.

The binomial distribution describes events with only two outcomes: success and failure. The mean of the binomial distribution is $\langle n \rangle = Np$, where N is the number of trials and p the probability of success. Its variance is $\sigma^2 = Np(1 - p)$.

The cumulative distribution $C(a)$ is defined as $C(a) = \int_{-\infty}^{a} P(x)dx$. It represents the probability that the random variable is less than or equal to a.

The Poisson distribution represents the probability of rare events with probability p in large samples of size N. Its mean is $M = Np$, equal to its variance.

The Poisson distribution is not very different from the binomial one: it is asymmetrical and becomes wider and more symmetrical as its mean increases.

Python

Computers cannot provide random numbers, but they can provide deterministic lists of numbers having the properties of random ones: these are called pesudorandom.

The `numpy.random` package contains pseudorandom number generators.

A histogram is a plot of the number of times an event occurs versus the event itself. It can be rendered graphically using `matplotlib.pyplt.hist()`.

`scipy.stats` is a package that is useful for dealing with probability distribution.

Strings can be concatenated in Python using the + operator. Numeric data can be converted into strings by casting them using the `str()` function. Strings can be rendered according to a specified format, using the `format()` method of the `string` class.

To generate a random event with a given probability p, generate a random number $x \in [0, 1)$, then check whether $x < p$.

It is useful to decompose a problem into small chunks and write a function for each of them. Python functions are defined using the keyword `def` followed by the name given to the function with the list of needed parameters in parentheses. The body of the function is written indented on the following lines, after a colon (:).

In Python, new local variables are automatically allocated when they are found on the left side of an operator. In all other cases, the content of a variable is taken from the outer scope.

Arduino

Variables in C and C++ must be declared and exist only within the scope in which they are declared and in those nested to it. Two variables with the same name that are declared in different scopes are different and can be used as if they had different names. Variables are containers, and their names are only meaningful for the programmer.

Distributions defined in `scipy.stats` offer the `pmf()` method, which returns the corresponding probability mass function, as well as `cmf()`, which returns the cumulative.

References

1. Castro R, et al (2020) COVID-19: a meta-analysis of diagnostic test accuracy of commercial assays registered in Brazil. Braz. J. Infect. Dis. 24, pp. 180–187. ▶ https://doi.org/10.1016/j.bjid.2020.04.003
2. Laplace P (1814) Essai philosophique sur les probabilités. Cambridge library collection—mathematics. Cambridge University Press, Cambridge, pp 1–67. ▶ https://doi.org/10.1017/CBO9780511693182.001
3. The OPERA Collaboration (2012) Measurement of the neutrino velocity with the OPERA detector in the CNGS beam. J High Energ Phys 2012, 93 (2012). ▶ https://doi.org/10.1007/JHEP10(2012)093
4. "Interview of P A M Dirac" by Thomas S Kuhn and Eugene Wigner on 1962 April 1, Niels Bohr Library & Archives, American Institute of Physics, College Park, MD USA. ▶ www.aip.org/history-programs/niels-bohr-library/oral-histories/4575-1
5. Dirac PAM (1982) Pretty mathematics. Int J Theor Phys 21:603–605. ▶ https://doi.org/10.1007/BF02650229
6. Maxwell JC (1873) A treatise on electricity and magnetism. Clarendon Press, Oxford
7. "Mémoires de l'académie des sciences de Berlin" 6, 1752, pp 185–217

Counting Experiments

Contents

© The Author(s), under exclusive license to Springer Nature Switzerland AG 2021
G. Organtini, *Physics Experiments with Arduino and Smartphones*, Undergraduate Texts in Physics,
https://doi.org/10.1007/978-3-030-65140-4_8

Any measurement that consists in counting events is analysed using the statistical distributions illustrated in the previous chapter, even in fields, such as economy or sociology, that have nothing to do with physics (at least as it is traditionally intended: remember that physics is useful in any field in which we measure something). For example, understanding the way in which the number of car accidents depends on the age of the driver, his/her job, the town in which he/she resides, etc., is of capital importance for insurance companies to estimate insurance premiums. In many cases, experiments consist in counting the occurrences of an event. This chapter is devoted to this kind of experiment.

8.1 Experiments with Binomial and Poisson Statistics

We can perform some experiments with the above-mentioned distributions using data collected with PHYPHOX, as described in Sect. 6.7 using a smartphone at rest.

Data appear to be distributed around different values in the three directions. We can ask ourselves how often data lie beyond a certain threshold. In other words, we want to estimate the probability of observing $a_i > a_i^{th}$ where a_i represents the component of the acceleration measured along the i-th direction and a_i^{th} a given constant. This is an example of a binomial distribution, in which we define success as the observation of an event in which $a_i > a_i^{th}$.

In order to increase the statistics, we could consider a sample made by merging all the data from the three directions and those of the modulus of the acceleration vector. Since each subsample has its own average $\langle a_i \rangle$ and its own width σ_i, we can transform data such that they all have zero mean and a unitary standard deviation with

Normalisation consists in shifting and stretching data such that their mean is null and their standard deviation is unity. The transformation is $a_i \to \frac{a_i - \langle a_i \rangle}{\sigma_i}$.

$$a_i \to \frac{a_i - \langle a_i \rangle}{\sigma_i} . \tag{8.1}$$

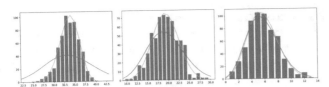

◘ Fig. 8.1 Distribution of counts of successes in experiments in which the probability of success is (from left to right) 0.84, 0.50 and 0.16

It is straightforward to show that a_i now has zero mean and a unitary standard deviation.

The collection of data now contains M measurements that can be thought of as random variables. Dividing the measurements into groups of m is equivalent to performing $N = \frac{M}{m}$ experiments in each of which we measure m times a quantity and count how many times c_n that quantity exceeds a threshold c_0, with $n = 1...N$. We then make a histogram of the counts. From the histogram, the average number of successful events is

$$\mu = \frac{1}{N} \sum_{n=1}^{N} c_n = \frac{\sum_{n=1}^{N} c_n}{\sum_{n=1}^{N} n c_n} \qquad (8.2)$$

and the probability of observing $c_n > c_0$ is $p = \frac{\mu}{m}$. If p is low, the distribution of the c_n is predicted to follow the binomial distribution with parameters N and p. As p tends to zero, the values of c_n tend to distribute as predicted by the Poisson distribution with mean μ.

Figure 8.1 shows the distributions when c_0 has been set to -1, 0 and $+1$, respectively, from left to right. Correspondingly, the expected probability p of success is expected to be 0.84, 0.50 and 0.16.

The superimposed orange curve is the shape of binomial distribution $P(n, p)$ with $N = \frac{M}{m}$, while the green one represents a Poisson distribution with mean Np. The binomial distribution always represents the data well, while the Poisson one only does so for low p.

The number of normalised data that exceeds a given threshold follows the binomial distribution. If the probability of exceeding the threshold is low, the distribution can be well approximated by a Poisson one.

8.2 Operations on Lists

Data can be **normalised**, i.e., transformed such that their mean is null and their width is unitary, as follows:

```
import numpy as np

mu = np.mean(ax)
sigma = np.std(ax)
ax = [(x - mu)/sigma for x in ax]
mu = np.mean(ay)
sigma = np.std(ay)
ay = [(x - mu)/sigma for x in ay]
mu = np.mean(az)
sigma = np.std(az)
az = [(x - mu)/sigma for x in az]
mu = np.mean(a)
sigma = np.std(a)
a = [(x - mu)/sigma for x in a]
```

where ax, ay, az and a are lists built by reading the CSV raw data file from PHYPHOX.

Lists can be joined thanks to the overloading of the + operator. Operator overloading is typical of OOP.

Merging the four lists into one unique list is as simple as

```
ax += ay + az + a
```

thanks to the **operator overloading** that consists of the fact that the result of the application of an operator to one or more operands depends on the nature of said operands. In this case, operands are lists and the + operator consists in returning a list that contains all the elements of its operands. Operator overloading is an important feature of object-oriented programming languages.

The ax list now contains normalised data from all the columns of the CSV file. We can even shuffle the elements of the list using

```
import random

random.shuffle(ax)
```

The order of the elements of the resulting list is changed randomly with respect to the original one.

8.3 Chauvenet's Criterion

The observations of the previous section is the basis for the so-called **Chauvenet criterion**, named after William Chauvenet (1820–1870), sometimes used to make a decision as to whether to discard one or more measures in a set of many.

To apply the criterion in a set of N measurements x_i, $i = 1, ..., N$, one computes

$$d_i = \frac{|x_i - \langle x \rangle|}{\sigma} \qquad (8.3)$$

and compares it with a number D taken as a threshold. If $d_i > D$, x_i is considered as an outlier and discarded.

Generally speaking, rejecting data has to be considered as a bad practice in all experiments. Outliers, in fact, can sometimes be the sign of new physics and, as such, they must be carefully scrutinised so as to be absolutely certain that they can be ascribed to known effects. Indeed, new physics is often discovered as a tiny deviation from expectations. Discarding data may lead to the loss of a big opportunity.

As usual, and thus also in the case of the Chauvenet criterion, the decision as to whether to use it or not is a matter of how confident we are in the physics with which we are dealing. A measurement of the gravitational acceleration using readily available materials can hardly lead to a discovery, and so the application of this rule is meaningful. The same rule applied to a collider experiment or to satellite observations would be totally inappropriate.

In general, it is preferable to avoid adopting arbitrary decisions. If there are a couple of outliers in a set of measurements, their weight is on the order of $\frac{1}{N}$ regarding physical results. If they are important, the experiment has serious problems that must be corrected before proceeding. Hence, in general, there is no need to use any rejection criteria. It is important to know about such criteria, because sometimes they are used, but they are strongly discouraged.

Chauvenet's criterion consists in discarding data that deviates by more than a given threshold with respect to the mean.

In general, using Chauvenet's criterion is not at all a good idea. A physicist should always identify the reasons for which data behave as outliers, unless he/she is confident enough as to the physics he/she is investigating and knows that his/her data can be affected by large systematic errors. In other cases, discarding data can deprive you of the possibility of winning a Nobel Prize.

8.4 Simulating Advanced Experiments

Many modern physics experiments, especially those involving quantum mechanics, require counting the number of events as a function of several variables (time, electric charge, direction, etc.), for which we have to use the appropriate statistical distributions. These experiments need specialised, often expensive, detectors. When those instruments are missing, one can simulate the events to practise with them.

Several isotopes undergo radioactive decays consisting in the transformation of a nucleus into another one together with the emission of one or more particles. ^{14}C nuclei, for example, transform into ^{14}N, emitting an electron and an antineutrino (^{14}C $\rightarrow$ ^{14}N $+ e^- + \bar{\nu}_e$). This process can be characterised by the number of decays per unit time, which must be measured. From these measurements, physicists developed the theory of weak interactions, originally formulated by Enrico Fermi (1901–1954) in 1933.

Our planet is constantly targeted by high-energy particles with a frequency, at sea level, of about 100 particles per square meter per second: the cosmic rays, discovered by Victor Hess (1883–1964) in 1912. This phenomenon can be characterised by the number of particles as a function of the time, the surface, their electric charge, direction, etc.

In high-energy physics laboratories, accelerated particles are focused on targets or collide head-on, resulting in a number of secondary particles produced by the conversion of the energy of the collision into matter, according to Einstein's relation $E = mc^2$. The number of the products, their angular and momentum distribution, etc., are investigated so as to understand the properties of the interactions governing the processes.

All these experiments require advanced, expensive and complex detectors and cannot be easily implemented at home or in many laboratories. In this chapter, we simulate these kinds of processes.

In this section, we show how to perform an experiment on radioactivity in environments where radioactive sources and detectors are not available, such as at home or in unequipped laboratories. This experiment, in fact, is also instructive for cases in which experiments with real radioactive sources can be conducted. In fact, while, in the case of a real experiment, the behaviour of the source can only be inferred from the results, in the case of a simulated experiment, one knows exactly how it behaves, and can then compare expectations with results, providing a way to *look inside* things that cannot be seen, such as atomic nuclei.

It is often useful to simulate an experiment, because, in this case, the behaviour of the system is perfectly known from the beginning and one can compare it directly with the results.

Radioactivity consists in the emission of particles from active atomic nuclei. There can be different types of radioactivity, historically classified as α, β and γ. α-particles are just helium nuclei. They are emitted by heavy isotopes, like bismuth (Bi), radon (Rn), radium (Ra), uranium (U), plutonium (Pu), americium (Am) and californium (Cf). α-particles are electrically charged with charge $+2$ in units of the proton charge. β-particles are nothing but electrons or positrons (their antiparticles, whose electric charge is the same as those of the proton). The most well-known β-emitter is ^{14}C, an isotope of carbon, used for dating ancient samples of living bodies. ^{60}Co, ^{137}Cs and ^{210}Bi are common radioactive sources found in laboratories. ^{22}Na is a β^+ emitter. Finally, γ-particles are photons. ^{60}Co, besides being a β-emitter, is a very common source of this type of radiation, too, often used in cancer radiotherapy.

The Geiger-Müller counter was one of the first instruments invented to detect radiation. It consists of a gas-filled tube, with a conducting wire at the center that is kept at high voltage with respect to the walls. Charged particles passing through the gas ionise it, i.e., they extract one or more electrons from the gas molecule. Electrons and ions drift towards the wire and the walls, respectively, hitting other gas molecules and producing secondary ionisation that results in a detectable electric current flowing from the wire. An electronic circuit counts the electric pulses and shows them on a display. An audible *click* is usually produced, too.

In this experiment, we simulate the radioactive source with an Arduino that produces random audible *clicks*, using a smartphone as a Geiger counter to count the latter. The details about the tools are given below. In this section, we concentrate on the measurements.

Schematic of a Geiger counter. From Wikimedia Commons, by Svjo-2.

The discovery of artificial radioactivity

In 1938, Enrico Fermi (1901–1954) was awarded the Nobel Prize for his discovery of artificial radioactivity. In his experiments, Fermi used a natural source of neutrons to irradiate samples of non-radioactive material. He then exposed recently irradiated materials to a Geiger counter to check whether they became radioactive.

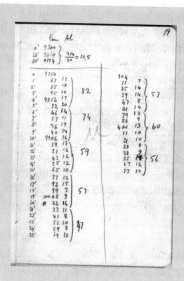

8

The figure shows the page of his logbook where he recorded the discovery, after irradiating a sample of Aℓ. The first column reports the time in minutes from the beginning of the experiment. The second column is the number shown on the counter when the measurements were performed. Fermi reported the difference in counts between adjacent lines as a third column, then summed them up every five minutes. He measured 82 counts during the first five minutes, 74 in the next interval, 59 during the third, etc. Counts drop exponentially, as expected for a radioactive source. The first three lines on the top left of the page are a measurement of the background: in 30 minutes, he measured 314 counts with no irradiated samples close to the counter, for an expectation of about 10 counts per minute, such that one would need to subtract 50 from each number in the last column to obtain the counts to be ascribed to the irradiated Aℓ sample.

It is quite surprising to see that, in fact, a Nobel Prize can be obtained with such a simple measurement. Fitting data with the techniques shown in this and the following chapters, one can easily obtain a decay time of 9 ± 3 minutes for a half-life of about 6 minutes. It is also a remarkable fact that the measurements were taken using an extremely poor set of devices: the Geiger counter was built by Fermi himself using a metallic container intended for pills. Its amplified electrical signal drove a counter used in telephony to account for the duration of the calls. The entire, original data acquisition chain is still on exhibit at the Physics Museum of Sapienza Università di Roma.

The activity of a radioactive source measures the number of decays per unit time. It is measured in becquerel (Bq).

First of all, we need to assess the decay rate, also called the **activity**. The activity of a sample is measured as the number of decays per unit time. Its unit is the becquerel (Bq), after Henri Becquerel (1852–1908), who discovered radioactivity. 1 Bq is equivalent to one count per second: $1\,\text{Bq} = 1\,\text{s}^{-1}$.

Another interesting quantity to study is how the activity changes over time.

The observed behaviour is described in terms of the **decay time** of the radioactive isotope. If N is the number of nuclei in a sample, the number of those who decay dN must be proportional to N and to the elapsed time dt, such that

Radioactive decay is yet another example of an exponential law, characterised by the decay time (also called the lifetime).

$$dN = -\alpha N dt .\tag{8.4}$$

The solution to this equation is

$$N(t) = N(0) \exp(-\alpha t) = N(0) \exp\left(-\frac{t}{\tau}\right),\tag{8.5}$$

where $\tau = \frac{1}{\alpha}$ has been introduced as the **decay time** or the **lifetime** of the radioactive sample. The meaning of τ is as follows: despite radioactivity being a random process, on average, about one third $(1/e)$ of the nuclei present at time $t = 0$ are still in the sample at $t = \tau$, as

$$N(\tau) = N(0)e^{-1} .\tag{8.6}$$

Lifetime is often expressed in terms of the **half-life** $t_{1/2}$: the time needed for a sample to reduce by one half. There is a precise connection between lifetime and half-life. In fact,

The half-life $t_{1/2}$ of an element is related to its lifetime τ by $t_{1/2} = \tau \log 2$.

$$N\left(t_{1/2}\right) = \frac{N(0)}{2}\tag{8.7}$$

and

$$N\left(t_{1/2}\right) = N(0) \exp\left(-\frac{t_{1/2}}{\tau}\right).\tag{8.8}$$

Comparing the two equations leads to

$$t_{1/2} = \tau \log 2 .\tag{8.9}$$

8.5 Using Arduino Pins

To generate an event with a fixed probability p, generate a uniformly distributed random number in the interval [0, 1), then check whether it is less than p.

In order to simulate a system in which atoms have a fixed probability p to decay within a given interval of time Δt, we can write a program that generates a random number $x \in [0, 1)$, check whether $x < p$ and, if so, let the atom decay, repeating the above process in an infinite loop.

In order to generate a *click* sound, we use an **actuator**, namely, a loudspeaker. Actuators are usually driven by Arduino digital pins. The latter can have two states: LOW, corresponding to 0 V, and HIGH, corresponding to 5 V, and can be used either as input pins or output pins. If used as input pins, their status can be read by Arduino, which can tell whether they are at low (0 V) or high (5 V) voltage. When used as output pins, Arduino can set their state. Some of them are marked with a *tilde* (~).

To prepare Arduino to drive a loudspeaker using a PWM pin, we need to configure it as an output pin and set its initial value to 0, as follows:

```
void setup() {
  pinMode(3, OUTPUT);
  digitalWrite(3, 0);
  Serial.begin(9600);
}
```

pinMode(pin, mode) assigns the mode mode to the pin pin. mode can either be INPUT or OUTPUT.

To set the state of a digital pin, use digitalWrite (pin, state), for which the state can either be LOW or HIGH.

The above piece of code shows how to set up a digital pin as an output pin. The mode of operation is set with pinMode(PIN, OUTPUT), where the first parameter is the pin number. To set a pin as an input pin, just use pinMode(PIN, INPUT). If set as output, the state of a digital pin can be set via digitalWrite(PIN, VALUE). In this case, the second parameter can either be LOW or HIGH. In the case of PWM pins, their status can be set as an integer between 0 and 255. The state of a PWM pin oscillates continuously between LOW and HIGH, with a frequency that depends on the value to which they are set.

The state of a PWM pin can be set to a value between 0 and 255. It then oscillates between the two states with a duty cycle proportional to its value.

If set to 0, they are permanently in the LOW state and a measurement of their voltage always returns 0 V. Similarly, if set to 255, they stay permanently in the HIGH state and their voltage is 5 V. Assigning a value $0 \le p \le 255$, the pin stays in the HIGH state for a fraction $f = \frac{p}{255}$ of its time. For example, if $p = 128, f = \frac{128}{255} \simeq 0.5$ and the pin remains in the HIGH state half of its time. In this case, we say that the pin's duty cycle is 50 %, and measuring the

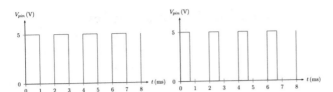

■ **Fig. 8.2** When the duty cycle of a PWM pin is set at 50 %, it oscillates between 0 V and 5 V such that it remains in the HIGH state for 50 % of its time (left). When the duty cycle is 30 %, the time spent in the HIGH state is reduced (right)

voltage of the pin as a function of time, we see something like that shown in Fig. 8.2 on the left.

A duty cycle of 30 % can be achieved by setting the pin to $p = 85$. In this case, a graph of the voltage as a function of the time is the one shown in Fig. 8.2 on the right.

The time actually spent in the HIGH state depends on the Arduino board, since it is a function of its clock, and on the pin number. For an Arduino UNO, the cycle has a duration of about 2 ms for pins 3, 9, 10 and 11. Pins 5 and 6 exhibit a frequency that is higher by a factor of two (thus, the cycle duration is half of the latter).

The time actually spent by a PWM pin in the HIGH state depends on the board and the pin number. Check the documentation, if you need to know it.

PWM pins are thus used when we need a square wave to drive an actuator, while non-PWM ones are used when we only have to switch between two values. Below, the loop() function is shown, using PWM pin 3 to produce a click sound.

```
void loop() {
    float p = (float)random
        (10000)/10000;
    if (p < 1.e-4) {
        digitalWrite(3, 127);
        delay(10);
        digitalWrite(3, 0);
    }
}
```

In order for a loudspeaker to emit a sound, it is necessary to drive it with a variable voltage. The shape of the sound waveform almost strictly follows that of the driving voltage. The latter cannot be modulated with Arduino pins that can only assume two values: 0 V and 5 V. However, exploiting PWM pins, we can put a non-constant voltage on the loudspeaker contacts, such that it reacts and emits a sound as soon as the voltage changes. If the voltage changes rapidly, our ears cannot resolve the single pulses. In order to make a click-like sound, we can drive the loudspeaker for a very short time (10 ms) with a square wave at 50 % duty cycle (the exact values are manifestly irrelevant, and you are invited to experiment with different values). After 10 ms, we silence the system by setting the voltage to zero.

In order to execute a piece of code with a probability P, it is enough to draw a random number $0 \leq p \leq 1$ and check whether $p < P$. With Arduino, random numbers can be drawn using `random()`.

The three lines of code that produce the sound are enclosed in a selection structure: they are executed only if p is less than 10^{-4}. The value of p is set randomly by means of the Arduino `random()` function. The latter returns an integer number picked randomly between 0 and the value passed as an argument (10 000, in our case), uniformly distributed. p, then, is a random number between 0 and 1. Note that, before dividing the number returned by `random()` by 10 000, we need to cast it into a `float`, otherwise, the numerator being always lower than the denominator, the ratio will always be null. In the end, the probability that $p < p_t$ is exactly p_t (note that if $p_t = 0$, the probability is null, and if $p_t = 1$, the probability is 1). Hence, in order to execute a piece of code with a probability P, it is enough to draw a random number $0 \leq p \leq 1$ and check whether $p < P$.

The above code causes a loudspeaker to emit a short sound, on average, once every 10 000 executions of the `loop()` function. The number of sounds emitted per unit time, thus, is constant, similar to what happens when a Geiger counter is placed close to a radioactive source whose activity is stable.

The activity of a radioactive source changes with time because, once decayed, nuclei can no longer contribute to it. This can happen either because the source contains a very small number of radioactive nuclei or because many of them decay per unit time such that the available number of them decreases sensibly with time. The number of nuclei contained in a source, even in a small one, is always quite large (remember that one mole of substance contains a number of elementary units equal to the Avogadro constants $N_A \simeq 6 \times 10^{23}$; even if the abundance of the radioactive isotopes is small, their number is still huge). If, however, the nucleus' lifetime is short enough, one can appreciate the reduction of the activity of the sample that follows the same exponential law given above.

We can take that into account in our Arduino sketch, producing a click sound with a given probability only for atoms that have not yet decayed. Atoms can be represented as an array of integers, which can be defined as a **global** variable. Global variables are variables that are declared outside the functions. They are shared among all the functions, unlike variables declared within them, which can only be used in the function within which they are defined.

We then add, before the `setup()` function, the following code:

Global variables are shared among functions. They are declared outside of any function block.

```
#define N 750

int n = 0; int atoms[N] = {0};
```

The `#define` directive, already encountered in Chap. 6, defines a constant N whose value is `750`. Note that, in this case, the assignment of the value to N is not done via the assignment operator = and there are no semicolons at the end of the line. A directive, in fact, is not a language statement. It is meant to simplify the program maintenance. In this case, it is used to define constants, i.e., values that are not susceptible to change during the execution of the program. Using constants has the advantage that the programmer can change their values in only one place (usually at the top of the program). For the same reason, we could add the following constants:

Constants are defined using the `#define` directive. No assignment operator, nor semicolon, should be used for this purpose. Constants make program maintenance simpler.

```
#define SPKRPIN 3
#define DUTYCYCLE 127
```

to define the pin to which the loudspeaker is attached and its duty cycle. Even the maximum integer to be drawn by `random()`, the duration of the pause and the probability with which nuclei decay can be defined through constants.

The integer variable n is used to count the elapsed time in units of "number of `loop()` executions". It will be useful to observe the exponential decay of the number of clicks. `atoms` is defined as an array of integers. Arrays are declared in the Arduino language specifying their type (`int`), their name (`atoms`) and their size (N, equivalent to 750) between square brackets. With the declaration, a number of consecutive locations is reserved in Arduino's memory to accomodate its content. In this case, since an Arduino integer takes 16 bits (2 bytes) to be represented,

Arrays are data structures composed of an ordered list of variables of the same type. Each element of an array is indexed by an integer.

$750 \times 2 = 1\,500$ bytes are used. Note that the available memory in an Arduino UNO is 2 kB. With this declaration, 75% of the available memory will be used to hold the array.

Values can be assigned to elements of an array at the declaration time as in `int array[10] = 4, 12, 8;` where the first element `array[0]` is equal to 4, the second to 12 and the third to 8. All the other elements are automatically set to zero. To access the ith component of an array, specify its index in square brackets as in `array[i]`.

Once declared, the elements of an array can be assigned at the same time by listing them within a pair of braces, separated by a comma. If the explicitly assigned integers are less than the size of the array, the rest of them are automatically assigned to be zero. Note that, if no elements are assigned, no automatic assignment is made and there is no guarantee that all the elements will be zero at the time of the declaration. This is why we need to assign at least the first element to zero to be sure that all of them are zero. In our program, each element of the array represents a radioactive nucleus. A value of zero indicates a nucleus that has not yet decayed. The `loop()` function is modified as follows:

```
void loop() {
  n++;
  int k = 0;
  for (int i = 0; i < N; i++) {
    float p = (float) random (10000)
             / 10000;
    if ((p < 1.e-4) &&
        (atoms[i] == 0)) {
      digitalWrite(3, 127);
      delay(10);
      digitalWrite(3, 0);
      atoms[i] = 1;
      k++;
    }
  }
  Serial.print(n);
  Serial.print(" ");
  Serial.println(k);
}
```

Each time the function is executed, it increases the n counter: a measure of time. Then, it loops over each atom. If atom *i* has not yet decayed, the corresponding array element is zero. It can then decay with a probability of 10^{-4}. && is a logical operator realising the AND boolean expression: the condition in the if statement is true only if both expressions (p < 1.e-4) and (atoms[i] == 0) are true (note the == operator, not to be confused with the = one). In this case, besides simulating the click of a Geiger counter, we update the corresponding array location, setting it to 1, and increase the k counter. The latter is a **local** variable (i.e., it can be used only within the function in which it is defined) used to count the number of atoms decayed within a unit time (a single loop() execution). At the end of the loop over atoms, we send both n and k over the serial line, such that we can monitor the number of decays as a function of time.

> The && operator realises the AND logical operator: an expression *A* AND *B* is true if and only if both *A* and *B* are true.

> Do not confuse the assignment operator = with the comparison one ==.

Even if the probability to decaying is constant, the fact that the number of nuclei that can decay diminishes with time produces an exponential decrease in the activity. In fact, the number of decays *dN* is proportional to time *t* and to the number of not-yet-decayed atoms *N*, such that

$$dN = -\alpha N \, dt \qquad (8.10)$$

and, as a consequence, $N(t)$ decreases as $\exp(-\alpha t)$.

8.6 The PHYPHOX **Editor**

Custom experiments can be built with PHYPHOX using a web editor available at ▶ https://phyphox.org/editor. Understanding the PHYPHOX editor is beyond the scope of this textbook, however, a brief introduction is worthwhile. Here, we briefly describe an application used to simulate the behaviour of a Geiger counter that utilizes a smartphone's microphone as the detector. The description given below cannot be exhaustive, however, looking at the way in which the experiment is realised using the editor provides enough information to start playing with it.

> Custom experiments can be added to the PHYPHOX collection using its web editor. The description of all its features is beyond the scope of this publication and we invite interested readers to refer to its documentation for further details.

The microphone is defined as the data source, defining it as an input module in the INPUT tab. The VIEWS tab contains the definition of the output modules that can be seen on the smartphone display. Each element in this tab is represented by a rectangle in the ANALYSIS tab. Data

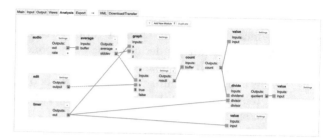

□ **Fig. 8.3** The PHYPHOX editor allows users to define their own experiments by manipulating data coming from input modules and showing the results from output modules

8

sent to their inputs is rendered on the display of the phone. The core of the custom experiment is in the ANALYSIS tab. Here, modules and operators are represented by rectangles connected by lines (called buffers) representing the flux of data.

A program for counting the number of clicks detected by the microphone is illustrated in Fig. 8.3. The example shows a possible way of realising a click counter. The "audio" input provides microphone data at its output as a list of numbers representing the amplitude of the sound waves. These data are used as input for the "average" module, which computes their standard deviation and send its value to both the graph module and an "if" module. The graph module is an output module showing a graph of its inputs on the smartphone's display. In order to plot the standard deviation of the input signal as a function of time, we send the output of a "timer" module to its x-input. The standard deviation is larger for louder sounds. The "if" module compares its a-input with its b one. The operation to be done is defined in its settings. In this case, if $a > b$, the output result is set to 1, specified as a fixed value in the "true" input of the module. Its b-input contains the value set as a threshold by the user, represented by the "edit" module, rendered as an input box on the smartphone display.

The "count" module simply counts the number of its inputs and outputs the result to the "value" module, rendered as a number on the smartphone's display. The same number n is used to compute $\frac{n}{\Delta t}$, Δt being the time elapsed since the beginning of the experiment obtained from the "timer" module. The ratio is shown on the smartphone's display, too, together with its units defined in the settings. The elapsed time is also shown on the display. The result is shown below.

The experiment described here can be downloaded from ▶ https://git.io/ clickCounter. In order to add it to your collection, load the file to the PHYPHOX web editor and generate a QR-code to be framed with your smartphone by clicking on the "plus" icon on the lower left corner of the PHYPHOX App.

With such an experiment, each time the standard deviation of the audio input exceeds the threshold, the counter is incremented by 1 and shown on the display.

The performance of the system depends on the sensitivity of the microphone and on the speed of the smartphone's processor. The higher the speed, the less time is spent in computing and showing the results, reducing the dead time. It also depends on the size of the audio input buffer: if it is too small, only a small fraction of the audio input is processed at each time; if it is too large, sound is averaged over a longer period. In the first case, the same *click* can be counted more than once, since the sound may persist for a certain time beyond the threshold. In the latter, many *clicks* may be counted just once. Such an effect can be regarded as an instrumental limitation and contributes to the finite resolution and the precision that can be attained with it.

8.7 Readily Available Particle Detectors

There exist, in fact, a number of Apps that exploit smartphone's cameras as particle detectors. Unfortunately (or, more precisely, fortunately enough), most of them suffer from the fact that the sensitive element of a smartphone camera is extremely small, hence the experiment must last for quite a considerable amount of time just to observe a few counts.

There are many sources of natural radioactivity in the environment. Even our body is radioactive, due to the presence of ^{40}K.

Most of these Apps can detect cosmic muons, but they are barely sensitive to most common products of radioactive nuclei usually found in the environment. One of the most abundant radioactive isotopes, in fact, is ^{40}K, whose activity is about 30 Bq per gram of potassium [1]. The human body contains about 180 g of potassium, hence it is a source of $180 \times 30 = 5\,400$ Bq. ^{40}K emits β-rays of 1.3 MeV of energy. Most of them are absorbed in the body of the smartphone before reaching the sensitive element and cannot be detected.

Public data about cosmic rays are released by many laboratories. Check out a list of them on the International Cosmic Day website at ▶ https://icd.desy.de/.

Public laboratories often release public data that can be used to conduct experiments similar to those described in this chapter. In particular, there are several cosmic ray observatories that publish data, sometimes together with a graphical rendering of their tracks. A comprehensive list is maintained by the DESY laboratories in Germany (▶ https://icd.desy.de/) under their PROJECTS page.

Camera sensors are good particle detectors, provided they are large enough and can be exposed for long enough.

One good alternative is a DSLR camera. They have a sensitive element about the size of their analog counterparts: $24 \times 36 = 864$ mm^2, to be compared to the typical area of a smartphone's camera on the order of 100 mm^2. Many DSLR cameras allow the user to shoot pictures with a manual exposure. Among the various exposure settings, one may find the Bulb (B) and Time (T) exposures: the first opens the shutter when the shutter button is pressed and closes it when released; the second opens the shutter at the first pressure on the shutter button and closes it once the button is pressed again. With this latter option, one can take a very long shot of a few hours, keeping the camera in the dark with the cap on the lens. Cosmic rays traversing the sensor leave a track on it. An example can be seen in Fig. 8.4, taken with an exposure of about 7 hours and saved as a JPEG image.

Each pixel in the sensor is made of three silicon photodiodes, one for each of the primary colours red, green and blue (RGB). The image is obtained as a combination of the intensities of each color. The colours of the hits in the

◻ Fig. 8.4 A small portion of a picture taken with a long exposure in the dark using a DSLR camera. With a little luck, you can even see tracks as long as the highlighted ones, corresponding to particles traversing the sensor at a small angle

picture depend on the energy deposited upon each pixel by the cosmic rays. Most of the hits appear as small clusters of pixels, because the energy released by a cosmic ray particle in the sensor is distributed over an area a bit larger than that of a single pixel and most of them cross the sensor at a relatively large angle with respect to the sensor's surface. However, if a particle enters the sensor at a small angle with respect to its surface, it can leave a relatively long track in it. These events are rare, because the sensitive thickness of the sensor is small (tens of microns) and the solid angle covered is a small fraction of 4π.

A histogram of the sum of the RGB pixels, defined as the intensity of each single sensitive element, is shown below for the full picture from which Fig. 8.4 has been extracted. The histogram is a plot of the number of times the pixel intensity is within a given interval.

Logarithmic scales are used when data span over several orders of magnitude. In logarithmic scales, one plot $\log y$, while labelling the axis with the values of y. In this scale, an exponential tail looks like a straight line.

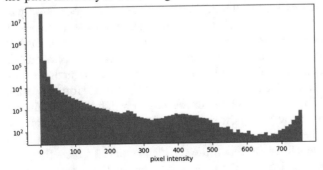

The plot is in the **logarithmic scale**. Such a scale is useful when the numbers span over a large interval. If we plotted the histogram normally, we could barely see its features: the first bins contain, in fact, about 10^7 events, while the counts in interesting ones is on the order of 10^3 or less and would be completely washed out. The scale is obtained by plotting the logarithm of the counts as a function of the intensity of the pixel, using the counts to label the axis. In other words, for each value on the x-axis, we plot $\log y(x)$, while the ticks on the y-axis are labelled as $y(x)$.

In this plot, a long exponential tail can be seen on the left side. It is worth noting that an exponential function appears as linear in a logarithmic scale and that, in fact, the tail is described by a sum of exponentials. Such a tail is due to *noise*, mostly of a thermal nature. Even if the pixel has not been hit by a particle, some current may be drawn from it. The energy gained by electrons in the pixel because of their temperature may fluctuate enough to make them free and be collected by the electronics.

> Noise consists of random fluctuations of the signal, mostly due to thermal effects.

Restricting the y-axis to values below a given value, as shown below, it is possible to better appreciate the features of the histogram, in a linear scale.

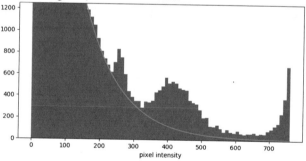

The exponential tail, shown in the picture, goes rapidly to zero. On top of that, we can measure at least three peaks. Note that the first peak is located around 255, while the last is truncated, the maximum intensity for each photodiode being set to 255 and $255 \times 3 = 765$. The intermediate peak is naively expected at $255 \times 2 = 512$, but it is found around $255 + 127 = 382$ due to the way in which the photodiodes are arranged in the sensor (as the so-called "Bayer array"). They must correspond to some physical event occurring in the silicon of which the sensor is made. In fact, they can be ascribed to the energy released in the sensor by particles. Counting only the high energy hits (those on the right whose energy in units of pixel intensity is higher than about 600) allows us to ignore the effects of the noise, which is negligible in this region.

The format of the picture is $4\,000 \times 6\,000$ pixels. Dividing the picture into chunks of 16×10^4 pixels, a distribution of the number of counts per chunk can be determined, as shown below.

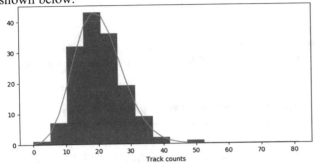

The average number of tracks found is $n = 20$, with a standard deviation of $\sigma = 7$. The process can be described by the Poisson statistics, the probability p that a cosmic ray would leave behind a track being small, while the exposure is long (a long exposure corresponds to conduct a huge number of short experiments). If the distribution of the tracks follows the Poisson statistics, the ratio between σ and n must be

$$\frac{1}{\sqrt{M}} = \frac{\sigma}{n} = \frac{7}{20} \simeq 0.35\,, \tag{8.11}$$

resulting in $M \simeq 8$. A Poisson distribution with a mean of $M = 8$, properly normalised, is superimposed onto the histogram. The distribution agrees with the data, confirming that the process is following the Poisson statistics. The fact that the number of hits whose intensity larger than the threshold is, on average, $n = 20$ means that, on average, each cosmic ray fires $\frac{20}{8} = 2.5$ pixels.

To properly normalise the distribution, we observe that the area under the histogram is given by

$$A[\text{px}] = \Delta w \sum_{i=1}^{N} n_i\,, \tag{8.12}$$

Δw being the bin width in units of pixels and n_i the number of events in bin i. In order to make A dimensionless, we need to divide it by the *calibration factor* $c_0 = 2.5$ pixels, i.e., the average number of fired pixels per track. Then, the number of events (tracks) under the histogram is

$$A = \frac{\Delta w}{c_0} \sum_{i=1}^{N} n_i . \tag{8.13}$$

Since $P(N, M)$ is properly normalised to 1, it is enough to multiply it for A to obtain the shape shown in the figure.

8.8 Image Manipulation with Python

`Pillow` is a useful Python module for working with images. It is not among the standard modules installed with the language and it has to be installed upon request. The installation of external packages depends on the operating system, but it is often straightforward. Follow the appropriate instructions for your own operating system. On UNIX-like ones, this is as simple as typing

pip3 is the package installer for Python-3 (the latest version of Python used at the time this book was written).

```
pip3 install Pillow
```

into a terminal (the command prompt on Windows). Once installed, it allows the user to play with image files. They are loaded into the memory using the `open()` method and a good deal of information about them can be easily accessed, like

```
from PIL import Image
image = Image.open('tracksmall.jpg')

print(image.format)
print(image.size)
print(image.mode)
```

A picture of $n \times m$ pixels can be thought of as an $n \times m$-matrix. Each element of the matrix represents a pixel, which, in turn, can be represented as a list of three numbers in the interval [0, 255], each one representing the intensity of the corresponding primary color, red (R), green (G) or blue (B).

Using `numpy`, it is easy to cast any picture in a list of numbers to be interpreted as above:

```
data = numpy.asarray(image)
print(len(data))
print(len(data[0]))
print(len(data[0][0]))
```

The length of `data` is *m*. It can be considered to be made of *m* arrays, each of length *n*. The first array in `data`, `data[0]`, is thus made of *n* elements, each of which is, in turn, a three-component array whose length is 3. In other words, `data[i][j]` is an array of three components containing the intensities of each primary color of the pixel at coordinates (*i*, *j*).

To loop over the pixels and determine the number of those whose intensity is large enough, we can use the following code:

```
i = 0
N2 = 16000
k = 0
c = []
p = []
threshold = 600
for row in data:
    for pixel in row:
        i += 1
        if (i % N2) == 0:
            c.append(k)
            k = 0
        h = int(pixel[0]) +
            int(pixel[1]) +
            int(pixel[2])
        p.append(h)
        if h > threshold:
            k += 1
```

Here, `row` and `pixel` are *iterators* traversing the whole data structure and returning the appropriate type. `data` is made of arrays, thus `row` is an array. As such, we can iterate over it, and since it contains arrays, too, even `pixel` is an array. The variable `i` is a local counter that holds the index of each pixel. The modulus operator % returns the remainder of the division between `i` and `N2`. Every time `i` is an integer multiple of `N2`, the remainder is zero and we reset the `k` counter after appending its last value to a list.

Lists can be traversed using iterators.

The % operator returns the remainder of the division of the operands.

`h` contains the sum of the values of each subpixel. Casting to an `int` is needed because each subpixel is, in fact, defined as a *byte*, an 8-bit integer whose maximum value is 255. Summing two or more bytes may result in a number greater than 255, and the result cannot be accommodated in a byte. All the values are appended to the list `p`, and if `h` exceeds the threshold, we increment `k` by one. The latter, then, represents the number of pixels whose intensity is above the threshold in the given chunk, whose distribution is reproduced below for convenience.

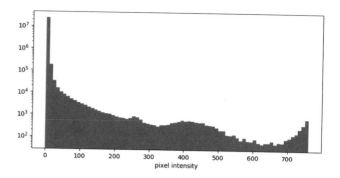

The **spectrum** of the intensity of the pixels, i.e., their distribution, can be represented in a logarithmic scale as

```
plt.hist(p, bins)
plt.yscale('log')
plt.xlabel('pixel intensity')
```

The logarithmic scale has been selected using the yscale() method of matplotlib.pyplot. The default behaviour is restored using 'linear' as the argument for yscale(). The same applies to the horizontal axis, for which the xscale() method is used.

> The axis in logarithmic scale can be produced using the xscale() and/or yscale() methods.

The relevant measures from the plot are easy to compute, remembering that c contains the number of hits found in each chunk of the picture.

```
avg = np.average (c)
sigma = np.std(c)
print('Average = {} '.format(avg))
print('RMS = {} '.format( sigma ))
m = (avg/ sigma )**2
print('Mean = {} '.format(m))
cal = avg/m
print('Calibration = {} '.format(cal))
```

Finally, the plot of the Poisson distribution on top of the experimental one can be made using the following code:

```
ch , bins , patch =
    plt.hist (c, bins = np.arange (0,
              cal*m*4, cal * 2))
binwidth = bins [1] - bins [0]
S = sum(ch)
xx = range(int(math .ceil (m*4)))
yy = poisson .pmf(xx , m)
plt. plot(xx*cal , S*yy* binwidth /cal ,
         '-')
plt.xlabel ('pixel counts')
```

The first line makes a histogram of the content of c, defining the classes as an interval of amplitude equal to twice cal, starting from 0 and up to four times m (the rightmost point on the x-axis corresponds to $4c_0 M$). The content of each bin is returned as a list ch, together with a list containing the bins' limits and a list of *patches* (a set of properties describing the graphical appearance of the histogram).

From bins, we get the bin width, and from the content of the histogram, we obtain their sum, such that we can compute the integral, to be used in the plot of the Poisson distribution.

Summary

Normalisation consists in shifting and stretching data such that their mean is null and their standard deviation is unity. The transformation is $a_i \rightarrow \frac{a_i - \langle a_i \rangle}{\sigma_i}$.

The Chauvenet criterion consists in discarding data that deviates more than a given threshold with respect to the mean. To be used with great care and *cum grano salis*.

Even if we can conduct experiments with the appropriate instruments, it is often useful to simulate them, because, in this case, the behaviour of the system is perfectly known and one can compare directly with the results of the experiment.

The activity of a radioactive source measures the number of decays per unit time. It is measured in becquerel (Bq).

Radioactive decay is yet another example of an exponential law, characterised by the decay time (also called the lifetime). The lifetime τ of a sample is related to its half-life $t_{1/2}$ by

$$t_{1/2} = \tau \log 2$$

To generate an event with a fixed probability p, generate a uniformly distributed random number in the interval $[0, 1)$, then check whether it is less than p.

There are many sources of natural radioactivity in the environment. Even our bodies are radioactive, due to the presence of ^{40}K.

Public data about cosmic rays are released by many laboratories. Check out a list of them on the International Cosmic Day website at ▶ https://icd.desy.de/.

Camera sensors are good particle detectors, provided they are large enough and can be exposed for a sufficient amount of time. They may be affected by thermal noise consisting of fluctuations of the intensity of each pixel due to thermal agitation.

Logarithmic scales are used when data span over several orders of magnitude. In logarithmic scales, one plots $\log y$, while labelling the axis with the values of y. In this scale, an exponential tail looks like a straight line.

Python

Lists can be joined thanks to the overloading of the + operator. Operator overloading is typical of OOP.

Lists can be traversed using iterators. This is more efficient than navigating the lists using their indexes.

Axes in logarithmic scales can be produced using the `xscale()` and/or `yscale()` methods.

The % operator returns the remainder of the division of the operands.

`pip3` is the package installer for Python-3 (the latest version of Python used at the time this book was written). It simplifies the management of installed packages. To install a package, use `pip3 install <package name>`.

Arduino

`pinMode(pin, mode)` assigns the mode `mode` to the pin `pin`. `mode` can either be `INPUT` or `OUTPUT`.

To set the state of a digital pin, use `digitalWrite (pin, state)`, where `state` can either be `LOW` or `HIGH`.

The state of a PWM pin can be set to a value between 0 and 255. It then oscillates between the two states with a duty cycle proportional to its value.

The time actually spent by a PWM pin in the `HIGH` state depends on the board and the pin number. Check the documentation, if you need to know it.

In order to execute a piece of code with a probability P, it is enough to draw a random number $0 \le p \le 1$ and check whether $p < P$. With Arduino, random numbers can be drawn using `random()`.

Global variables are shared among functions. They are declared outside of any function block.

Constants are defined using the `#define` directive. No assignment operator, nor semicolon, should be used to this purpose. Constants make program maintenance simpler.

Arrays are data structures composed of an ordered list of variables of the same type. Each element of an array is indexed by an integer.

Values for the elements of an array can be assigned at the declaration time as in `int array[10] = {4, 12, 8};` where the first element `array[0]` is equal to 4, the second to 12 and the third to 8. All the other elements are automatically set to zero. To access the i-th component of an array, specify its index in square brackets, as in `array[i]`.

The `&&` operator realises the AND logical operator: an expression A AND B is true if and only if both A and B are true.

Do not confuse the assignment operator `=` with the comparison one `==`.

phyphox
Custom experiments can be added to the PHYPHOX collection using its web editor. Once defined, they can be loaded onto the phone by framing the QR-code generated on the editor.

Reference

1. Samat SB, et al (1997) The ^{40}K activity of one gram of potassium. Phys Med Biol 42:407

The Normal Distribution

Contents

The original version of this chapter was revised: Typographical errors have been corrected. The correction to this chapter is available at ▶ https://doi.org/10.1007/978-3-030-65140-4_16

© The Author(s), under exclusive license to Springer Nature Switzerland AG 2021, corrected publication 2022
G. Organtini, *Physics Experiments with Arduino and Smartphones*, Undergraduate Texts in Physics,
https://doi.org/10.1007/978-3-030-65140-4_9

When the number of trials increases in the discrete distributions illustrated in the previous chapters, the probability mass distribution tends to become symmetric around a central value. For large enough numbers, the shape of these distributions is not far from what we observe in many experimental data distributions. They also become wider and the width of each single bin becomes negligible with respect to the width of the distribution, such that the latter can be considered to be a continuous distribution. In this chapter, we study the properties of the Gaussian distribution, a distribution of capital importance in statistics, mostly because it is the one to which all other distributions tend, as stated by the central limit theorem. We also learn other important results.

9.1 A Distribution Depending on the Distance

Physical data usually distribute such that the probability of a measurement x_i being different from the *true* value μ decreases with the distance $|x_i - \mu|$. Finding the distribution of random distances whose probability decreases with it is one way to understand the origin of the distribution of physical data.

Uncertainties in physics represent the average distance between the central value and the rest of the measurements. If they are ascribed to random fluctuations, they must be distributed as random distances. Let's now derive the probability density function for a random variable $r = \sqrt{x^2 + y^2}$ with null average, assuming that
- the probability depends only on r;
- the probability decreases with r.

Indicating with $P(x)$ and $P(y)$ the probability density function of x and y, respectively, the probability of finding $r = \sqrt{x^2 + y^2}$ is the product of the probabilities of finding x and y, i.e.,

$$P(r) = P(x)P(y). \tag{9.1}$$

We can always represent r in a cartesian plane as a vector starting from the origin of length r and with coordinates $x = r\cos\theta$ and $y = r\sin\theta$. Then, differentiating both sides of this equation,

$$0 = \frac{dP(x)}{dx}\frac{dx}{d\theta}P(y) + P(x)\frac{dP(y)}{dy}\frac{dy}{d\theta} = -P(y)\frac{dP(x)}{dx}r\sin\theta + P(x)\frac{dP(y)}{dy}r\cos\theta, \tag{9.2}$$

which can be rewritten as

$$xP(x)\frac{dP(y)}{dy} = yP(y)\frac{dP(x)}{dx}. \tag{9.3}$$

Separating the variables, we get

$$xP(x)\frac{dx}{dP(x)} = yP(y)\frac{dy}{dP(y)}. \tag{9.4}$$

Since the equality must be true for all x and all y, such ratios must be constant, i.e.,

$$xP(x)\frac{dx}{dP(x)} = yP(y)\frac{dy}{dP(y)} = A. \tag{9.5}$$

Defining $C = \frac{1}{A}$,

$$\frac{dP(x)}{P(x)} = Cx\,dx. \tag{9.6}$$

Integrating both members, we obtain

$$\log P(x) + c = \frac{C}{2}x^2, \tag{9.7}$$

from which we can conclude that

$$P(x) = P_0 \exp\left(\frac{C}{2}x^2\right). \tag{9.8}$$

Because the probability must decrease with r,

$$\frac{dP(x)}{dr} = \frac{dP(x)}{dx}\frac{dx}{dr} = \frac{dP(x)}{dx}\cos\theta < 0. \tag{9.9}$$

For positive x, $\cos\theta > 0$ and $\frac{dP(x)}{dx} < 0$, and, as a consequence, $C < 0$, because $P(x) \geq 0$, being a probability density function. It is convenient to write $C = -k$ with $k > 0$ and rewrite

Writing the probability of finding r in polar coordinates and differentiating, we find that $\frac{dP(x)}{P(x)} = Cx\,dx$.

$C < 0$ because the probabilities are always non-negative and $P(r)$ decreases with r.

$$P(x, k) = P_0 \exp\left(-\frac{k}{2}x^2\right),\qquad(9.10)$$

where we added k among the parameters upon which $P(x)$ depends, as its shape depends on it (it depends on P_0, too, but we show below that P_0 depends, in turn, on k). The function $P(x, k)$ attains its maximum at $x = 0$, it is symmetric with respect to $x = 0$ and its shape is shown below for $k = P_0 = 1$.

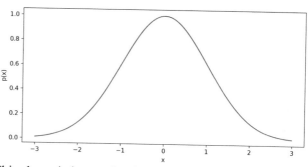

The shape of the obtained distribution, called the normal distribution, very much resembles experimental ones. It has zero mean and unitary variance.

This shape is interesting because it very much resembles the frequently observed distribution of experimental data. The mean of such a distribution is, by definition,

$$\mu = \int xP(x, k)dx = 0,\qquad(9.11)$$

because $xP(x, k)$ is an odd function. The variance is given by

$$\sigma^2 = \int (x - \mu)^2 P(x, k)dx = \int x^2 P(x, k)dx = P_0 \int x^2 \exp\left(-\frac{k}{2}x^2\right)dx.$$

$$(9.12)$$

Let's observe that

$$d\exp\left(-\frac{k}{2}x^2\right) = -kx\exp\left(-\frac{k}{2}x^2\right)dx.\qquad(9.13)$$

We can then integrate by parts, writing $z = \exp\left(-\frac{k}{2}x^2\right)$ such that

$$\sigma^2 = -\frac{P_0}{k}\int xdz = -\frac{P_0}{k}\left(xz\Big|_{-\infty}^{+\infty} - \int zdx\right).\qquad(9.14)$$

The first term in the parentheses is null, since z goes to zero faster than x, as can be clearly seen from its graphical representation and

$$\sigma^2 = \frac{P_0}{k} \int \exp\left(-\frac{k}{2}x^2\right) dx. \qquad (9.15)$$

The normalisation of $P(x)$ imposes that

$$P_0 \int \exp\left(-\frac{k}{2}x^2\right) dx = 1 \qquad (9.16)$$

and

$$\sigma^2 = \frac{1}{k}. \qquad (9.17)$$

The value of P_0 can be evaluated observing that

$$\left(\int \exp\left(-\frac{k}{2}x^2\right) dx\right)^2 = \frac{1}{P_0^2} = \int \exp\left(-\frac{k}{2}x^2\right) dx \int \exp\left(-\frac{k}{2}x^2\right) dx. \qquad (9.18)$$

Making the substitution $y = x$ in one of the integrals, we can rewrite the equation as

$$\frac{1}{P_0^2} = \int \exp\left(-\frac{k}{2}x^2\right) dx \int \exp\left(-\frac{k}{2}y^2\right) dy = \int\int \exp\left(-\frac{k}{2}\left(x^2 + y^2\right)\right) dx\, dy. \qquad (9.19)$$

We can compute the last integral in polar coordinates, observing that $dx\, dy = r\, dr\, d\theta$ such that the integral can be rewritten as

$$\frac{1}{P_0^2} = \int\int \exp\left(-\frac{k}{2}r^2\right) r\, dr d\theta, \qquad (9.20)$$

where the integrals are extended in r from 0 to ∞ and in θ from 0 to 2π. If we write $r^2 = u$, $2r dr = du$ and, remembering that

$$\int_0^{+\infty} \exp\left(-\frac{k}{2}u\right) du = \frac{2}{k}, \qquad (9.21)$$

the equation becomes

$$\frac{1}{P_0^2} = \frac{1}{2} \int \int \exp\left(-\frac{k}{2}u\right) du d\theta = \frac{1}{k} \int d\theta = \frac{2\pi}{k} \quad (9.22)$$

and

$$P_0 = \frac{1}{\sqrt{2\pi k}} = \frac{1}{\sqrt{2\pi}\sigma} . \quad (9.23)$$

The normalisation factor of the normal distribution is $\frac{1}{\sqrt{2\pi}}$.

Finally, we get

$$P(x, \sigma) = \frac{1}{\sqrt{2\pi}\sigma} \exp\left(-\frac{x^2}{2\sigma^2}\right) . \quad (9.24)$$

A normal distribution whose mean is translated at $x = \mu$ and whose width is characterised by a parameter σ^2 is called a Gaussian. Its normalisation is $\frac{1}{\sqrt{2\pi}\sigma}$.

For $\sigma = 1$, such a distribution is called the **normal** distribution. Its maximum can be easily translated from x to μ by writing

$$P(x, \mu, \sigma) = \frac{1}{\sqrt{2\pi}\sigma} \exp\left(-\frac{1}{2}\left(\frac{x - \mu}{\sigma}\right)^2\right), \quad (9.25)$$

which represents the distribution of a random variable with mean μ and width σ. In general, $P(x, \mu, \sigma)$ is more often called a **Gaussian**, after Carl Friedrich Gauss (1777–1855), though the names "normal" and "Gaussian" are sometimes used as synonyms. It is straightforward to show that, if a variable x is distributed as a Gaussian of mean μ and width σ, the variable

$$z = \frac{x - \mu}{\sigma} \quad (9.26)$$

is normally distributed. It can be proven that, for $\sigma = 1$,

$$\int_{-1}^{+1} P(x) dx \simeq 0.68 . \quad (9.27)$$

Consequently, this is the probability of finding a random variable distributed as a Gaussian of width σ in the interval $[-\sigma, +\sigma]$ around μ.

9.2 The Central Limit Theorem

Experimental data distribute as Gaussians because of the central limit theorem. Simply stated, this theorem says that the sum of independent random variables $x_1, x_2, ..., x_n$ tends to be distributed as a Gaussian whatever the distribution of x_i.

The proof of this theorem is beyond the scope of this book and will be left to more advanced courses on statistics. Here, we prove the theorem *experimentally*. Before we go any further, we observe that random fluctuations in experimentally obtained data are due to many uncontrolled causes and that, whatever their distribution, their effect sums up such that the final distribution is that expected for the sum of many random variables.

In order to experimentally prove the theorem, we use Python to generate uniformly distributed integer random numbers from 1 to 6, simulating the throw of a die. The average score is $\frac{6+1}{2} = 3.5$ and the probability of obtaining the values 3 or 4, closest to it, is constant and equal to $p = \frac{1}{6}$. Throwing two dice (i.e., summing two uniformly distributed random numbers from 1 to 6), we expect the average score to be $\frac{2+12}{2} = 7$. However, the probability is no longer uniform. In fact, while there is only one way of obtaining, e.g., $2 (1+1)$, there are three ways of obtaining 7, namely: $1+6, 2+5, 3+4$. The probability to obtain 7, then, is three times higher than that of obtaining 2. Summing the scores of three dice, the average score is $\frac{3+18}{2} = 10.5$ and the probability of obtaining 10, for example, is much higher than that of observing $3 = 1 + 1 + 1$. In fact, 10 can be obtained in a variety of ways.

The higher the number of dice to sum, the higher the probability for obtaining a number close to the mean score. At the same time, the shape of the distribution increasingly resembles a Gaussian.

Note that the same behaviour has been observed in analysing the binomial and the Poisson distribution: as the number of trials increases, these distributions tend to become Gaussians.

According to the central limit theorem, the sum of many independent random variables follows a Gaussian distribution, irrespective of the distribution of each of them.

The result of the central limit theorem can be easily predicted by observing what happens to the distribution of the sum of the scores of N dice. Starting from uniformly distributed numbers, the distribution of their sum rapidly tends to become Gaussian.

9.3 Experimental Proof of the Central Limit Theorem

In order to prove the central limit theorem experimentally, we generate uniformly distributed numbers using Python, starting with a single random variable, as in

```python
import numpy as np
import matplotlib.pyplot as plt

x = np.random.randint(1, 7, 10000)
plt.hist(x, bins = range(1,8),
         rwidth = 0.9)
plt.show()
```

randint(a, b, n) returns a list of n uniformly distributed integers between a (included) and b (excluded).

where we generate 10 000 uniformly distributed random numbers between 1 and 6 and we show the distribution as a histogram with bars whose relative width is 90 % that of each bin. The result is shown below.

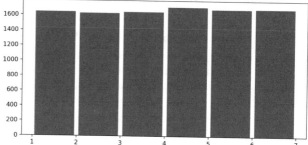

In the above code, the randint(a, b, n) method returns n uniformly distributed random numbers in the interval $[a, b)$. We also forced the histogram to have one bin per score. The elements of the bin list, in fact, are the edges of each tick in the horizontal axis: this list must contain the bin edges, including the left edge of the first bin (1) and the right edge of the last one (7, remembering that range(n, m) returns a sequence of integers from n to $m - 1$). All the bins are open on the right side, except the last. Stated differently, the first bin includes values in the $[1, 2)$ interval, the second in $[2, 3)$, etc., up to $[5, 6)$, while the latter is $[6, 7]$. Adding

```python
y = np.random.randint(1, 7, 10000)
z = [x + y for x, y in zip(x, y)]
plt.hist(z, bins = range(2, 14),
         rwidth = 0.9)
plt.show()
```

to the above script, we generate a second list of 10 000 random variables and we put the sum of them and the previously generated ones into a list called z, whose histogram is shown below.

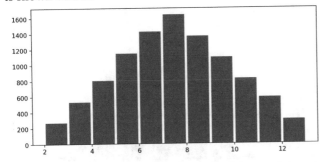

The zip() function aggregates the two lists such that the result is a list of **tuples**. A tuple is an ordered collection of data. In this case, the tuple consists of two values: each element of the resulting list consists of a pair obtained by aggregating the i-th element of list x with the i-th element of list y. The for statement applied to the list returned by zip(x, y) navigates through all the tuples and sums their values, filling a new list called z. Let's now sum the scores of ten dice.

> A tuple is an ordered collection of data. Aggregating two lists side by side using zip is a way to obtain a tuple.

```
x = []
for i in range(10):
    x.append(np.random.randint(1, 7,
        10000))
z = [sum(i) for i in zip(*x)]
plt.hist(z, rwidth = 0.9)
plt.show()
```

Here, x is now a list of 10 lists. In each component of x, we put a list of 10 000 random numbers, such that x[i][j] represents the score of die i at throw j. To sum up the scores of each die we exploit the sum() function, whose meaning is straightforward, still using zip() to build a list of tuples. Following what has been illustrated above, we would ideally write

```
z = [sum(i) for i in zip(x[0], x[1], x[2],
                         x[3], x[4], x[5],
                         x[6], x[7], x[8],
                         x[9])]
```

In order to be generic enough, we exploit the **asterisk operator** that, applied to a list, *unpacks* it in its elements. The above line of code is then equivalent to

> The asterisk operator *, applied to a list, unpacks it, i.e., splits it into a list of its elements.

```
z = [sum(i) for i in zip(*x)]
```

which is much more compact and generic. The result is shown in the histogram below, which clearly shows how the distribution tends to a Gaussian one.

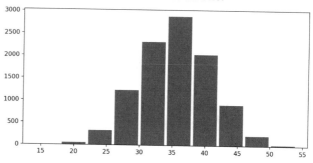

As the number of random variables increases, the distribution of their sum becomes more pronounced and peaked around their mean value. While the width of the distribution increases in absolute value, its relative value with respect to the mean becomes smaller and smaller. Correspondingly, the probability of the tails becomes negligible.

The peak has become more pronounced and data are relatively closer to the average value. While, in the case of two dice, the minimum possible value 2 still has a non-negligible probability, in this case, the minimum possible value of 10 is absent from the plot (the probability of obtaining it is lower than $\frac{1}{10\,000}$).

Repeating the experiment with 1 000 dice we obtain a distribution whose values can, in principle, span from 1 000 to 6 000, however, the histogram below shows that most of the data are distributed around the average value 3 500 within a relatively small interval.

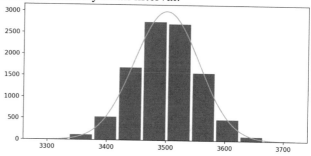

A Gaussian with the proper parameters has been superimposed onto the experimental data. Needless to say, the Gaussian almost perfectly describes the experimental distribution.

The average value of the Gaussian μ has been set to $\mu = \frac{7\,000}{2} = 3\,500$. Given that the variance σ_1^2 of the distribution of the score of one die is equal to the square of the width of the interval divided by 12, the variance of the sum of 1 000 dice is

$$\sigma^2 = \sum_{i=1}^{1\,000} \frac{6^2}{12} = \frac{36\,000}{12} \simeq 3\,000 \,. \tag{9.28}$$

The width of the Gaussian is then $\sigma = \sqrt{\sigma^2} \simeq 55$, as can be seen from the plot.

To compute the normalisation factor C in front of the Gaussian, we observe that the integral of the histogram is

$$C \int P(x, \mu, \sigma)dx \simeq \sum_{i=1}^{n} n_i \Delta w \,, \tag{9.29}$$

where n_i is the number of events in bin i, Δw is the bin width (assumed constant), and the sum is extended on all the bins, such that

$$C \simeq \Delta w \sum_{i=1}^{n} n_i = N \Delta w \,, \tag{9.30}$$

where N is the number of trials, i.e., the number of events under the histogram.

The origin of the shape of the distribution of experimental errors then, depends on the type of uncertainty. If the reading uncertainty dominates, any value between two consecutive values in the instrument scale has the same probability of being the *true* value and the corresponding uncertainty is the instrument resolution divided by the square root of 12. If random fluctuations are larger than the resolution of the instrument, many random effects sum up, giving rise to a distribution that tends to the Gaussian one and the uncertainty is given by the width of the latter.

If the reading uncertainty Δw dominates, the uncertainty about the measurement is $\frac{\Delta w}{\sqrt{12}}$. If statistical fluctuations are important, the uncertainty depends on the width of the distribution.

9.4 The Markov and Chebyschev Inequalities

Even if, in most cases, experimental data distribute as Gaussians, they sometimes exhibit different behaviours. Moreover, there can be cases in which we are not interested in attaching a value to a physical quantity like mass, time, pressure, or something similar. In certain cases, we measure distributions, in fact. In β-decays, for example, electrons are emitted from a radioactive material (e.g., ^{60}Co). These electrons carry a momentum p distributed over a continuous range from $p = 0$ to $p = p_{max}$ whose distribution is such that, indicating with $N(p)$ the number of electrons with momentum p,

$$\sqrt{\frac{N(p)}{p^2 F(Z, p)}} = a\,(Q - E)\,, \tag{9.31}$$

In a Kurie plot, the square root of the number of electrons from a β-decay with momentum p is plotted against the electron kinetic energy E. The expected distribution is a straight line with a negative slope.

where $F(Z, p)$ is called the Fermi function, accounting for the effects of the electrostatic attraction between the electron and the final state nucleus whose atomic number is Z, E is the electron kinetic energy, and Q, called the Q-value, is characteristic of the atomic species, has the dimensions of an energy and has a value on the order of 1 MeV. a is a normalisation constant, such that

$$\int_0^Q a\,(Q - E)\,dE = aQ^2 - a\frac{Q^2}{2} = a\frac{Q^2}{2} = N\,, \tag{9.32}$$

N being the number of events detected.

The distribution of the quantity on the LHS of Eq. (9.31) as a function of E is shown below and can be described by a straight line that intercepts the E-axis where $E = Q$.

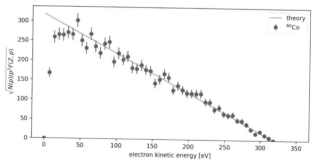

A graph like this tells us that the probability of observing an electron with a kinetic energy of, e.g., 100 eV, in cobalt β-decays is about one half of the probability of observing an electron of about 200 eV. The probability of observing electrons with energy greater than 300 eV is very low and is zero for $E \gtrsim 320$ eV.

If we knew nothing about the distribution of data, we could, in any case, perform some useful statistical tests. For example, we can estimate the probability of observing an event with $X = \sqrt{\frac{N(p)}{p^2 F(Z,p)}} \geq k$ by just counting the events n_k for which $X \geq k$ and dividing by the total number of events N:

$$P(X \geq k) \simeq \frac{n_k}{N} \,. \tag{9.33}$$

The **Markov Inequality** states that, for a random variable $X \geq 0$,

$$P(X \geq k) \leq \frac{E[X]}{k} \,. \tag{9.34}$$

The proof can be easily obtained observing that

$$E[X] = \int_0^\infty XP(X)\,dX \geq \int_k^\infty XP(X)\,dX \geq k \int_k^\infty P(X)\,dX \,. \tag{9.35}$$

The last integral is nothing but the probability that $X \geq k$. Dividing everything by k, we obtain Markov's Inequality. The **Chebyschev Inequality** follows from the latter. It states that

$$P(|X - \mu| \geq k\sigma) \leq \frac{1}{k^2} \,, \tag{9.36}$$

where μ and σ are, respectively, the mean and standard deviation of X. In fact, defining $Y = (X - \mu)^2$, $Y > 0$, and we can apply Markov's Inequality such that

$$P\left((X - \mu)^2 \geq K\right) \leq \frac{E[Y]}{K} \,, \tag{9.37}$$

but $E[Y] = \sigma^2$, thus

$$P\left((X - \mu)^2 \geq K\right) \leq \frac{\sigma^2}{K} \,. \tag{9.38}$$

The probability that $(X - \mu)^2$ is greater than K is the probability that $|X - \mu|$ is greater than $\sqrt{K}$, and defining $K = k^2\sigma^2$,

$$P(|X - \mu| \geq k\sigma) \leq \frac{1}{k^2} \,. \tag{9.39}$$

According to the Markov Inequality, $P(X \geq k) \leq \frac{E[X]}{k}$.

The Chebyschev Inequality guarantees that $P(|X - \mu| \geq k\sigma) \leq \frac{1}{k^2}$.

9.5 Testing Chebyschev's Inequality

The names of functions in Python can be assigned to a variable and still continue to represent the corresponding functions.

One of the interesting features of the Python programming language is that we can refer to functions by using their names as variables. We can, for example, build a list of statistical functions, importing the `scipy.stats` module and constructing the list as if we would with ordinary numbers or strings, as in

```python
from scipy import stats

distributions = [
    stats.binom(n = 100, p = 0.5),
    stats.poisson(mu = 5),
    stats.norm()
]
```

The list `distributions` contains three functions representing a binomial distribution with $N = 100$ and $p = 0.5$ (corresponding to the distribution of the only two values of 100 coin tosses), a Poisson distribution with $M = 5$ and a normal distribution with $\mu = 0$ and $\sigma = 1$.

We can loop over the elements of the list, generate, for each distribution, $n = 10\,000$ values and compute their average m and standard deviation s.

```python
for d in distributions:
    x = d.rvs(size = 10000)
    m = np.mean(x)
    s = np.std(x)
```

The method `rvs()` of a distribution in `scipy.stats` returns a list of random values distributed according to it.

The rvs method of the `scipy.stats` distributions returns a sample of random variables properly distributed (rvs stands for *random variables*). In order to estimate the probability that $|x_i - \mu|$ is greater than $k\sigma$, we iterate over the sample and count the number of events for which the above inequality is satisfied for a given value of k.

```python
n = 0
k = 1
for i in range(len(x)):
    diff = np.fabs(x[i]-m)
    if diff >= k*s:
        n += 1
p = n/N
```

We can then compare p, which is a function of k, with
$1/k**2$ for various values of k (only $k > 1$ makes sense,
because probabilities are always less than or equal to
one) and superimpose a graph of $1/k^2$ onto the plot of
$P(k) = P(|x_i - \mu| \geq k\sigma)$, as shown in Fig. 9.1 for the
three distributions in the list.

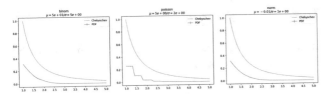

■ **Fig. 9.1**   Plots of $P(k) = P(|x_i - \mu| \geq k\sigma)$ (blu) and $1/k^2$ (orange)
as a function of k for (left to right) the binomial, Poisson and normal
distributions.

As predicted by Chebyschev's Inequality, $\frac{1}{k^2}$ is always
higher than $P(k)$.

9.6   The Law of Large Numbers

The "experiments" of Sect. 9.3 show another important
feature of random variables, known as the Law of large
numbers. Again, there are various formulations of this law
that can be formally proven. For our purposes, a good
formulation is the so-called **weak Law of large numbers**,
according to which the average $\langle x \rangle$ of a sample of N random numbers is likely to be equal to the expected value μ
of their distribution when $N \to \infty$.
It can also be expressed as

The weak Law of
large numbers states
that the average $\langle x \rangle$
of a sample of n
random numbers is
likely to be equal to
the expected value μ
of their distribution
when $n \to \infty$.

$$\lim_{N \to \infty} P(|\langle x \rangle - \mu| > \epsilon) = 0, \qquad (9.40)$$

where $P(|\langle x \rangle - \mu| > \epsilon)$ represents the probability that the
difference $|\langle x \rangle - \mu|$ is greater than ϵ. The limit operation
should not be taken as a formal limit, as in calculus. It just
means that, in order for the law to be true, N must be large
and that the larger N is, the more likely is that $\langle x \rangle \simeq \mu$. In
practice, $N \geq 30$ is considered to be large.

The uncertainty about the average $\langle x \rangle$ on N values x_i decreases as $\frac{1}{\sqrt{N}}$. If σ is the common uncertainty about x_i, the uncertainty about its average is $\frac{\sigma}{\sqrt{N}}$.

This law can easily be demonstrated using **Chebyschev's inequality**. In fact, being that x is a random variable, the average $\langle x \rangle = \frac{1}{N} \sum x_i$ is, in turn, a random variable and its variance can be computed using uncertainty propagation as

$$\sigma_{\langle x \rangle}^2 = \frac{N\sigma^2}{N^2} = \frac{\sigma^2}{N} , \tag{9.41}$$

assuming that all x_i have the same variance σ^2. According to the Chebyschev Inequality,

$$P\left(|\langle x \rangle - \mu| \geq k\frac{\sigma}{\sqrt{N}}\right) \leq \frac{1}{k^2} , \tag{9.42}$$

Taking the limit for $N \rightarrow \infty$ leads to the Law of large numbers.

In deriving the above result, we obtained another important result. Equation (9.41) states that the variance of the average value of a set of measurements decreases as $\frac{1}{N}$, thus the uncertainty about $\langle x \rangle$ decreases as $\frac{1}{\sqrt{N}}$. Averaging over many measurements, then, is good, because it decreases the uncertainty of its value.

When we take N measurements of a physical quantity x_i, $i = 1, ..., N$, we compute their average $\langle x \rangle$ and their standard deviation σ. The average then has an uncertainty that can be estimated as

$$\sigma_{\langle x \rangle} = \frac{\sigma}{\sqrt{N}} . \tag{9.43}$$

This means that, if we repeat the measurement of x, we have a 68% chance of observing a value between $\langle x \rangle - \sigma$ and $\langle x \rangle + \sigma$. However, if we take N new measurements, there will be a probability of 68 % that $\langle x \rangle$ will lie between $\langle x \rangle - \frac{\sigma}{\sqrt{N}}$ and $\langle x \rangle + \frac{\sigma}{\sqrt{N}}$.

9.7 The Uncertainty About the Average

We can test this behaviour by generating a sequence of n random variables according to any distribution, computing their average $\langle x \rangle$ and repeating the experiment N times, each time collecting the value of $\langle x \rangle$ in a list.

```python
import numpy as np

n = 10000
N = 10000

avg = []
for i in range(N):
    x = np.random.randint(1, 7,
        size = n)
    m = np.mean(x)
    avg.append(m)
```

The `randint()` method returns a list of integers that are uniformly distributed between the given limits. We are thus simulating N times n throws of a die of which we compute the average. Each set of n throws has a distribution similar to the one shown below.

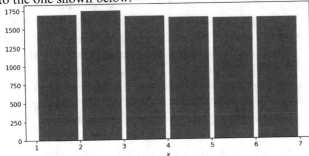

The expected value of the score of each die is 3.5, while the standard deviation of the uniform distribution is $\frac{6}{\sqrt{12}} \simeq$ 1.7. In one run, we obtained $\langle x \rangle = 3.462$ with $\sigma = 1.7$, compatible with the expectations.

In another run of n throws, we are likely going to find that $\langle x \rangle \simeq 3.5$, while the variance of the distribution is still about 3. With the above script, we make N such runs and compute N averages and variances.

We can then plot the distribution of the averages, which, as expected, is Gaussian and is shown below.

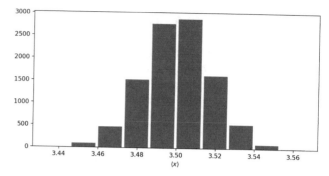

The statistics of the averages are given by

```
m = np.mean(avg)
s = np.std(avg)
print('mu = {} sigma = {}'.
    format(m,s))
```

In our run of 10 000 runs, we obtained $\mu = 3.500$ and $\sigma = 0.017$. Manifestly, the average of the *measurements* is known with better precision than the single averages. In a single run, we have an uncertainty about the value of x of 1.7, but the average over $N = 10 000$ measurements has an uncertainty of $\frac{1.7}{\sqrt{10\,000}} = 0.017$. Indeed, $3.500 - 3.462 = 0.038$, much less than 1.7 and corresponding to a bit more than two standard deviations on the average.

In other words, the standard deviation represents the spread of the measurements: in a two-σ wide interval around the average value, we will likely find 68 % of the measurements. The uncertainty about the average, however, is $\frac{1}{\sqrt{N}}$ times the latter.

As a consequence, when we take a measurement N times, the value of the measured quantity is given by the average $\frac{1}{N} \sum_i x_i$, while its uncertainty should be quoted as $\frac{\sigma}{\sqrt{N}}$, rather than the standard deviation σ of the measurements.

Summary

Physical data usually distributes such that the probability of a measurement x_i being different from the *true* value μ decreases with the distance $|x_i - \mu|$. Finding the distribution of random distances whose probability decreases with it is one way to understand the origin of the distribution of physical data. It turns out that such a distribution is a Gaussian.

When the mean value of the Gaussian is $\mu = 1$ and its variance $\sigma^2 = 1$, the distribution is called normal.
The normalisation factor is needed to make the integral of the distribution equal to 1.

$$G(x; \mu, \sigma) = \frac{1}{\sqrt{2\pi}\sigma} \exp\left(\frac{1}{2}\left(\frac{x-\mu}{\sigma}\right)^2\right)$$

In quoting the uncertainty of a measurement, if the reading uncertainty Δw dominates, the uncertainty about the measurement is $\frac{\Delta w}{\sqrt{12}}$. If statistical fluctuations are important, the uncertainty depends on the width of the distribution. The uncertainty about the average $\langle x \rangle$ on N values x_i decreases as $\frac{1}{\sqrt{N}}$. If σ is the common uncertainty about x_i, the uncertainty about its average is $\frac{\sigma}{\sqrt{N}}$.

Statistics
According to the central limit theorem, the sum of many independent random variables follows a Gaussian distribution, irrespective of the distribution of each of them.
The distribution of the sum of random variables, whatever their distribution is, becomes more pronounced and peaked around their mean value as the number of variables grows. While the width of the distribution increases in absolute value, its relative value with respect to the mean becomes smaller and smaller. Correspondingly, the probability of the tails becomes negligible.
According to Markov's Inequality, $P(X \geq k) \leq \frac{E[X]}{k}$.
Chebyschev's Inequality guarantees that $P(|X - \mu| \geq k\sigma) \leq \frac{1}{k^2}$.
The weak Law of large numbers states that the average $\langle x \rangle$ of a sample of n random numbers is likely to be equal to the expected value μ of their distribution when $n \to \infty$.

Python
`randint(a, b, n)`, in the `numpy.random` package, returns a list of n uniformly distributed integers between a (included) and b (excluded).
A tuple is an ordered collection of data. Its elements can be addressed as if they were the elements of an array. In contrast to an array, a tuple can contain objects of different types. Moreover, while lists and arrays are mutable (i.e., you can change their content), tuples are not. Aggregating two lists side by side using `zip` is a way to obtain a tuple. The asterisk operator `*`, applied to a list, unpacks it, i.e., splits it into a list of its elements.

The names of functions in Python can be assigned to a variable and still continue to represent the corresponding functions, instead of merely being represented as strings. The method `rvs()` applied to any distribution in `scipy.stats` returns a list of random values distributed according to it.

9

Kinematics

Contents

The original version of this chapter was revised: Typographical errors have been corrected. The correction to this chapter is available at ▶ https://doi.org/10.1007/978-3-030-65140-4_16

Electronic supplementary material The online version of this chapter (doi:▶ 10.1007/978-3-030-65140-4_10) contains sup-. plementary material, which is available to authorized users

© The Author(s), under exclusive license to Springer Nature Switzerland AG 2021, corrected publication 2022
G. Organtini, *Physics Experiments with Arduino and Smartphones*, Undergraduate Texts in Physics,
https://doi.org/10.1007/978-3-030-65140-4_10

This chapter is devoted to the experimental study of the motion of a projectile along an inclined plane. Despite its simplicity, characterising such a motion is very instructive and sheds some light on the very meaning of the equations of motion that can be found in physics textbooks. Those equations are always the result of (often) crude approximations and the behaviour of a real system may differ from that expected. Besides learning how to make interesting measurements in the domain of kinematics, we will then learn how to model systems in a more realistic situation.

10.1 Designing the Experiment

Experiments in kinematics require the ability to measure coordinates as a function of time.

In order to experimentally study the motion of something, we need to obtain its position $\mathbf{x}$ as a function of time t, $\mathbf{x}(t)$. Here, $\mathbf{x}(t)$ is a three-component vector whose coordinates depend on time: $\mathbf{x}(t) = (x(t), y(t), z(t))$.

The description of the motion can be greatly simplified with a clever choice of the reference frame in which it is expressed. If, for example, the trajectory of the moving body lies on a plane, choosing a reference frame such that one axis is perpendicular to that plane causes one of the coordinates to be constant. Thanks to the freedom to choose the position of the origin of the reference frame, we can always set this constant to zero and one of the coordinates becomes superfluous. Similarly, if the motion lies along a line, one can choose a 2-dimensional reference frame oriented such that one of the axes is parallel to that line. In that case, a second coordinate becomes pleonastic and the motion is fully characterised by just one number x as a function of time t, $x = x(t)$, representing the distance between the origin of the reference frame and the position of the body.

Choosing a reference frame with an axis parallel to the velocity in rectilinear motions greatly simplifies the equations.

As usual, the description of a system in terms of a mathematical model neglects many effects that exist in real systems.

It must be noted that, while the body would ideally be a pointlike particle, in practice, this cannot be true. All bodies have a non-null size along all three dimensions. $x(t)$, then, must represent the position of an arbitrary chosen point of it, provided that the body can be considered rigid enough such that all its points stay at fixed positions with respect to any of its other points. It may seem obvious, but there can be cases in which this observation is not so obvious.

Our goal is thus to measure the position $x(t)$ of a body as a function of time t as it travels along a rectilinear trajectory. A free-falling object clearly follows such a path, as well as any object falling along an incline. It is easy to realise that the speed of anything sliding along an incline moves slower than the same object freely falling vertically. An interesting experiment could thus be to characterise the motion (i.e., to measure $x(t)$) in different conditions, e.g., as a function of the inclination angle θ of the plane. We can expect that the motion when $\theta = \frac{\pi}{2}$ has something to do with free-fall. To perform such an experiment, we need something sliding along a plane whose distance from one end of the plane can be measured as long as time goes by.

If possible, always take measurements in different conditions.

10.2 Measuring Time and Distances with Arduino

Arduino provides simple, but interesting tools for performing the measurement of both time and distance.
Like any microprocessor, the Arduino ATmega 328 requires a *clock* to operate. Each elementary operation is performed at regular intervals by a device (the clock) that emits a regular sequence of electrical pulses.

Arduino can measure time with good precision.

How the CPU works

The Arduino UNO microprocessor contains a CPU (Central Processing Unit), like any computer. It provides a very clean example as to how they work, not needing an operating system. When a CPU is powered up (or reset), it always loads a fixed number of bytes (just one in the case of the Atmega 328), called a *word*, from a fixed location in its memory. Loading happens at the first clock pulse. Such a word is interpreted as an *instruction*: each CPU, in fact, has a one-to-one correspondence between a string of bits and an instruction. The instructions set is composed of operations consisting in doing basic arithmetic on internal memory locations called *registers* and moving data from one location to another in the attached memory (RAM: random access memory) and to or from memory and internal registers, as well as conducting logical operations on data in registers.
Each instruction may or may not require certain parameters to be executed. For example, an instruction to add the contents of two registers clearly needs the two registers upon which to execute the required operation. In this case, the parameters are loaded into the CPU unit at the subsequent

clock pulses, from the adjacent memory locations. The operation is actually performed at the next clock pulse (more than one if needed).
Once done, at the next clock pulse, the CPU loads the next word from the RAM and again interprets it as an instruction, such that the cycle described above restarts.
Computers appear to be able to conduct extremely complex operations, however, their CPU's simply perform only very basic operations on data that are always in the form of binary strings. The apparent complexity of their work is translated into long sequences of elementary operations by the *compilers*: software that translates programmers' code into machine code.

The Arduino UNO clock runs at 16 MHz, i.e., it provides 16 million pulses per second. Clearly, counting the number of clocks elapsed since the beginning of the run provides a way to measure time with a precision that, in principle, can be as good as

$$\sigma_t = \frac{1}{16 \times 10^6} \text{ s} = 62.5 \text{ ns} . \tag{10.1}$$

A time measurement can be used to obtain a distance using the definition of velocity.

Distances can be measured with the help of an ultrasonic device, like in a *sonar*. Sound waves travel in air at constant speed c and can be reflected by surfaces. If a sonic pulse is reflected back by an obstacle at distance d from its source, it returns to it in a time

$$t = \frac{2d}{c} . \tag{10.2}$$

A measurement of such a time, then, can be turned into a distance measurement as

$$d = \frac{ct}{2} . \tag{10.3}$$

The ability of Arduino to measure times can thus be exploited to measure both time and distances.

10.3   Ultrasonic Sensors

A rather popular ultrasonic sensor for Arduino is the HC-SR04. It comprises two piezoelectric crystals, working as a speaker and a microphone. Piezoelectric crystals change their shape and size depending on the application of a voltage drop at their ends. Applying an oscillating electric signal to a piezoelectric crystal causes it to oscillate at the same frequency. Such an oscillation is transmitted to the surrounding medium in which it propagates as a sound wave, like in a speaker.

Similarly, a piezoelectric crystal exhibits a voltage when a mechanical stress is applied, as happens when it is hit by a sound wave. Voltage variations follow the sound pressure and provide a way to record it, as in a microphone.

In an ultrasonic sensor (see Fig. 10.1), a burst of eight acoustic pulses at 40 kHz is generated by the device in response to a trigger signal with a minimum duration of $10\,\mu s$ of at least 2 V. At the same time, an electrical line is raised from 0 to 5 V and stays in this state until the microphone detects an echo of the signal.

The sensor looks like the one in the margin and has four legs: two (marked VCC and GND) are for power and must be connected, respectively, to the 5V and GND Arduino pins; the trigger can be provided to a dedicated leg (marked TRIG), while the fourth leg (marked ECHO) can be used to get the echo's digital signal, represented in red in Fig. 10.1. A variant of this sensor exists with three legs (HC-SR03) in which the same leg is used as both the trigger and the echo pin. The HC-SR05 has five pins: four are the same as the

The HC-SR04 is a popular ultrasonic sensor for measuring distances.

The HC-SR04 ultrasonic sensor. The two piezoelectric crystals are contained in the two cylindric structures on it.

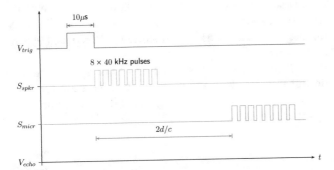

■ **Fig. 10.1**   Applying a $10\,\mu s$ voltage pulse (in blue) to the trigger pin of the HC-SR04 causes the emission of eight ultrasonic pulses from its speaker (in orange) and sets its echo pin (in red) to 5 V. The pin goes back to 0 V once the microphone (in green) detects the echo

HC-SR04, while the fifth, marked OUT, is reserved for future uses. According to the data sheet, ultrasonic sensors can provide meaningful data every 50 ms, such that we can get coordinate measurements with a rate of

$$f = \frac{1}{50 \times 10^{-3}} = 20\,\text{Hz}\,. \tag{10.4}$$

The useful range of the HC-SR04 is between 3 mm and 3 m.

The range within which these devices can measure a distance is usually within 3 mm and 3 m, with a relatively large variability among different samples.

Most colleges and universities own devices such as air tracks, low-friction cart tracks and magnetic tracks. In air tracks, air is pumped from the bottom and emerges from a series of fine holes such that a cart can glide on an air cushion with almost no friction. Low-friction carts are designed to exhibit very low friction while sliding along aluminium tracks, while magnetic tracks exploit magnetic repulsion to make magnetic carts levitate, avoiding contact between the cart and the track.

There are plenty of devices that exhibit low friction that can be used to study kinematics. There is, indeed, no need to use expensive materials to perform such an experiment. You can use the rails of an electric train as a low-friction track and a train car as a cart.

As usual in this textbook, we propose an alternative that can be realised at home using very common, readily available materials, items with which we collected the data, used below to show the performance that can be achieved with them. For this experiment, the rails of an electric train for kids can be used as the track, while one of the train cars can be used as a low-friction cart. Indeed, its friction will not be as low as in professionally designed carts, but it will be low enough to allow us to obtain satisfactory results.

The track, whose total length must be at least 1 m, must be kept flat when inclined. The rails, then, must be fixed on a rigid support like a wooden plank. They have to be fixed on it such that they are as straight as possible, with almost no bends. Mark their location with a pencil before fixing them. You can use small wide-headed nails to fix them onto the outer sides of the rails, taking care that they do not interfere with the movement of the train car. The track can be inclined by raising one of its ends using boxes or other supports.

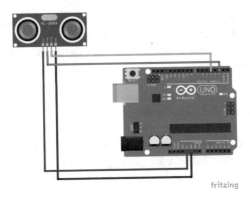

fritzing

☐ **Fig. 10.2** The TRIG and ECHO legs of the HC-SR04 sensor must be connected to digital pins, while the other two leads serve to provide power

As an alternative, you can use a duct used for electrical systems as a guide and any object capable of being slipped (without rolling) inside it. In order to reduce friction, you can even put the cart, upside down, in the freezer with a thin layer of water on its underside so that it becomes ice. When the ice starts to melt, it will make the cart quite slippery. In this case, you have to choose the cart properly such that its size efficiently reflects sound waves, while the guide does not affect their propagation.

As usual, take your time to build a stable, clean setup. Conduct tests in the early stages of the construction to be sure that the setup works well.

Another possibility is to suspend an object from an inclined smooth bar (e.g., using two metal rings around a broomstick).

The ultrasonic sensor must be fixed at one end of the rail, oriented such that the ultrasonic beam is directed towards the train car. Securing a cardboard screen on the side of the train car facing the sensor helps to reflect the waves back. The sensor must then be connected to Arduino using jumpers, as in Fig. 10.2: the VCC and GND legs must be connected, respectively, to 5V and GND pins; the TRIG and ECHO legs can be connected to any digital pin. As usual, remember that a stable, neat setup guarantees much better results: keep the sensor on a breadboard glued to the track's support and the cables running neatly behind the sensor. Fix the Arduino board firmly onto the support such that you can easily plug in the USB cables when needed.

The mechanical stability of the whole system is essential for a good experiment. Connect the sensor to Arduino properly and prevent parts from moving.

The ultrasonic beam is not very collimated (its opening angle is about 30°) and part of it will not intercept the train car. As a result, it can be reflected back by other obstacles placed along its direction. Putting a V-shaped screen

Reduce the chance of false readings.

made, e.g., out of two cardboard or polystyrene sheets at the opposite end of the track with respect to the sensor reduces this effect, improving the reliability of the measurements.

10.4 Arduino Data Acquisition

An Arduino sketch for using an ultrasonic sensor needed to trigger the device to emit the ultrasonic beam and wait for its echo.

The *sketch* for data acquisition is shown below. TRIG and ECHO are constants representing the Arduino pins to which the trigger and echo leads of the sensor, respectively, are connected.

```
#define  TRIG  2
#define  ECHO  4

#define  V  343.
#define  CONV  (1.e-6*100.)

void setup() {
  Serial.begin(9600);
  pinMode(TRIG, OUTPUT);
  pinMode(ECHO, INPUT);
  digitalWrite(TRIG, LOW);
}

void doMeasurement() {
  digitalWrite(TRIG, HIGH);
  delayMicroseconds(10);
  digitalWrite(TRIG, LOW);
  unsigned long t0 = micros();
  unsigned long ut = pulseIn(ECHO,
      HIGH);
  unsigned long t1 = micros();
  float d = V*ut*CONV/2.;
  Serial.print((t0+t1)/2.);
  Serial.print(" ");
  Serial.println(d);
}

void loop() {
  doMeasurement();
  delay(50);
}
```

The V constant represents the speed of sound in air and is set as $c = 343$ m/s. The distance is obtained as

$$d = \frac{ct}{2} \qquad\qquad (10.5)$$

and, since t will be measured in microseconds, we need to multiply it by 10^{-6} to express it in seconds. To show the distance in cm, we need to multiply the result by 100. This is done through the constant CONV. Note that its value is enclosed in parentheses, even if there is no need for them in this case. Using the parentheses is a good habit to adopt whenever a constant is defined by mathematical operations. In fact, constants are not interpreted by the compiler, i.e., the translator from the Arduino programming language into machine code for the Atmega 328. Constants are interpreted by preprocessors: software that helps the programmer manipulate their source code before passing it to the compiler. The preprocessor substitutes each occurrence of the constants found in the #define directives with the corresponding values in the source code. Suppose, then, you defined a constant as

```
#define CONV 1000+100
```

such that CONV is supposed to represent the number 1 100. Writing

```
x = CONV*3;
```

will not result in $x = 3\,300$, but rather $x = 1\,300$, since such a line is interpreted by the compiler as

```
x = 1000+100*3;
```

due to the fact that the string CONV is simply substituted to its value by the preprocessor. Enclosing the value of any constant in parentheses prevents such a harmful effect. On the other hand, an excess of parentheses does not harm your *sketch*.

Constants are useful for several reasons.

Preprocessors literally substitute each occurrence of a constant found in the source code with its value, before passing it to the compiler.

Always enclose the value of a constant written as an expression in parenthesis, even when unnecessary.

A variable, in computer programming, is a data container. Constant variables are containers whose content never changes.

Many tutorials on Arduino programming use global variables or *constant variables*, instead of constants. The term *constant variable* is only apparently an oxymoron. In computer language, the term *variable* simply indicates a container for data. If such data are constant or do not depend on how the variable is used, the variables that represent them are said to be constant. Ordinary variables, contrastingly, can contain data that are subject to change. A variable can be forced to be constant using the `const` qualifier, as in

A variable is declared as constant by adding the `const` qualifier.

```
const float v = 343.;
```

which makes the content of a variable constant. Any attempt to modify the content of a constant variable results in an error at compile time.

Constants help in reducing the memory occupation of a compiled *sketch*.

That said, it is clear that any variable in the *sketch* may require that more memory be allocated so as to hold the data it contains. As a result, a compiled *sketch* in which variables are used instead of constants requires a bit more memory, as you can check by compiling it and looking at the report at the bottom of the IDE window (Fig. 10.3). As the memory onboard is very limited, it is not a good idea to waste it.

Always try to write your programs in such a way that they can be easily maintained, even by other programmers.

Programming does not just mean writing code that is technically correct. It is also a literary exercise, in which you must exploit your communication skills, writing code that is easy to maintain and whose interpretation is crystal clear, unless you are participating in *The International Obfuscated C Code Contest*—ioccc.org—in which you are supposed to write the most incomprehensible, but still functional, computer program.

Get into the habit of writing constants' names in all capitals.

Constant names, unlike those of the variables, are, by convention, completely written in capitals. Just looking at how a string is capitalised tells the reader whether or not he/she is looking at a constant.

Constants help maintaining code.

Moreover, constants help in maintaining code in a simple, yet effective way. Suppose, for example, that you are going to use pin 9 for the trigger, instead of pin 2. Without constants, you have to look for every occurrence of the character 2 in the code and decide whether you need to change it into 9 or not. For example, you may want to change the line containing `pinMode(2, OUTPUT)`, but you don't want to change the 2 in the formula `float d = V*ut*CONV/2`. Using the `TRIG` constant allows you to change just one line to have a perfectly working *sketch*.

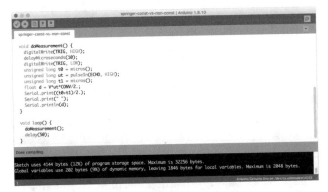

Fig. 10.3 The size of the compiled *sketch* is shown in the black part at the bottom of the window of the IDE

In the `setup()` function, besides initialising the serial connection, we define the mode of operation of the digital pins. Remembering that digital pins can act both as input and output pins, the trigger pin is set as an OUTPUT pin using

```
pinMode(TRIG, OUTPUT);
```

while the echo pin is set as an input one. We also set the state of the trigger pin to LOW with

```
digitalWrite(TRIG, LOW);
```

such that there are 0 V on it.

The `loop()` consists in repeating two instructions: doMeasurement() sends the result of the execution of a distance measurement over the serial line, then Arduino waits 50 ms before attempting to run another `loop()`. Here, doMeasurement() is a function, as shown by the parentheses at the end of its name. Such a function is not part of the Arduino programming language, however, we can define our own functions if needed. In order to define a function, we need to assign a name (doMeasurement()) and a type to it. doMeasurement() is not supposed to

The programmer can define his/her own functions.

In function declaration, the returned type must be specified together with the name.

return something and is defined as `void`. This is the reason for defining the function as

```
void doMeasurement()  {
    . . .
}
```

Note that its definition must precede its usage in the program. Calling `doMeasurement()` results in the execution of all the code enclosed in braces. The first three lines of code are needed to generate the trigger pulse. As illustrated in the previous section, an ultrasonic pulse is emitted when a voltage pulse of at least $10\,\mu s$ duration is applied to the TRIG leg of the sensor. Writing `HIGH` on that pin causes it to exhibit a voltage of 5 V. The `delayMicrosecond()` function pauses the execution of the *sketch* for the number of microseconds passed as its argument (10), after which putting the pin back to `LOW` produces the desired result. At this time, the ultrasonic trail of pulses is emitted and the ECHO pin is set to `HIGH`. We then get the current time with `micros()`, record it in `t0` and then call the `pulseIn()` function. The latter continuously monitors the pin passed as the first argument and, when the pin switches from the state passed as the second argument (`HIGH`) to the complementary one, it returns the time elapsed since its beginning in microseconds. Such a number is put in the `ut` variable defined as an `unsigned long`, i.e., a 32-bit-long non-negative integer. After getting the current time and recording it in `t1`, the value of `ut` is used to compute the distance in cm as

Digital pins can be used to generate rectangular signals.

`delayMicro seconds()` suspends the execution of the *sketch* for a given amount of μs.

`pulseIn(pin, state)` monitors the indicated pin. It returns the time elapsed since its call as soon as it is found in a different state with respect to the given one.

$$d = \frac{v\ [\text{m/s}]\,t\ [\mu s]}{2} \times 10^{-6}\frac{s}{\mu s} \times 100\frac{\text{cm}}{\text{m}}\,. \tag{10.6}$$

Once time and distance are measured, we send them over the serial line to be stored on a computer.

With the last three lines, we send over the serial line: the average time between `t0` and `t1`, corresponding to the time at which the ultrasonic beam hit the obstacle before being reflected, a blank and the distance d. In this way, Arduino provides a measurement of the position of an obstacle, with respect to a reference frame whose origin is on the ultrasonic sensor, as a function of time every 50 ms.

10.5   **Collecting Data**

We can now collect the data by letting the train car slide along the inclined track. First of all, we need to measure the angle of the track, for which we can choose among a protractor, a ruler or a smartphone. The first gives us the angle directly; with a ruler, one can measure the height h of the end of the track and, knowing its length ℓ, obtain the angle as

$$\theta = \sin^{-1}\frac{h}{\ell}. \tag{10.7}$$

If the plane supporting the track is not horizontal, both methods can be affected by a systematic error. This error can be mitigated using a smartphone, whose accelerometer can be exploited to determine the inclination of a plane by measuring the components of the gravity acceleration along the device axis. Apps like PHYPHOX provide an easy way to measure θ.

Angles can be easily measured using a smartphone with PHYPHOX or similar Apps.

In order to avoid any push or pull on the train car when launched, we can tie it to the upper part of the board using sewing thread. Burning the thread with a lighter will make the cart free to move with null initial speed.

For a simpler analysis, we need not give any initial momentum to the cart.

Using the `arduinoReader.py` script developed in Chapter 4, we get the data from the serial line and record them on a file for each inclination angle θ. In order to keep only the interesting data, we can set up the Arduino reader to keep only data such that the distance is within a reasonable interval.

Collect data for each angle in separate files, named after the angle for easier bookkeeping.

Data must be collected at angles that are not too large, nor too small. If angles are too large, the cart can go fast enough to be damaged as a result of the collision at the end of the track; if it is too small, static friction prevents the cart from moving.

Avoid risks to the equipment. Keep conditions within a range such that they meet the assumptions of your model.

10.6   **Data Analysis**

Data are represented such that we can easily appreciate their features. Figure 10.4 shows the coordinate x of the cart along the track as a function of time t: $x(t)$. The overall path followed by the cart was a bit more than 60 cm and

The position data versus time distribute as a parabola.

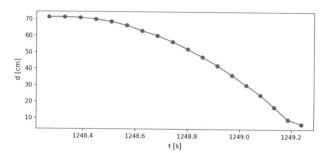

◘ Fig. 10.4 The position of the cart along the track, in cm, as a function of the time, in μs

the duration of the descent was about 1 s. The plot appears to be parabolic, so we can easily argue that

$$x(t) = At^2 + Bt + C\,, \tag{10.8}$$

where $C \simeq 60$ cm, while $B \simeq 0$ because, with an appropriate translation of the origin of the axes, the vertex of the parabola is at coordinates ($t = 0$, $x = 60$). The last point may not be usable: it is probably too close to the sensor to give a reliable result.

The derivative of the position with respect to time $v = \frac{dx}{dt}$ is, by definition, its velocity. It can be obtained through data, remembering that

> The derivative of a physical quantity can be obtained from data, taking the difference quotient.

$$v = \lim_{\Delta t \to 0} \frac{\Delta x}{\Delta t} \simeq \frac{x_{i+1} - x_i}{t_{i+1} - t_i}\,, \tag{10.9}$$

where x_i and t_i are, respectively, the ith coordinate of the cart and its corresponding time. Similarly, one can get the acceleration as

$$a = \lim_{\Delta t \to 0} \frac{\Delta v}{\Delta t} \simeq \frac{v_{i+1} - v_i}{t_{i+1} - t_i}\,, \tag{10.10}$$

with the meaning of the symbols being obvious. Results are shown in Fig. 10.5 (after removal of the last point). From the figure on the left, one can see that the absolute value of the velocity of the cart increases almost linearly with time, as expected from the fact that $x(t)$ is a parabola, i.e.,

> Velocity appears to increase linearly with time.

$$v(t) = 2At + B\,, \tag{10.11}$$

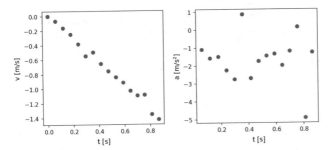

□ Fig. 10.5 Velocity and acceleration of the cart along the track, in SI units, as a function of the time

with $B \simeq 0$ consistent with what can be inferred from the graph of $x(t)$. A first guess for A can be easily obtained as

$$2A \simeq \frac{\Delta v}{\Delta t} = \frac{-1.4}{0.8} \simeq -1.8 \, \frac{m}{s^2} . \tag{10.12}$$

A better estimation can be obtained by averaging over various pairs of points, as seen in Sect. 4 of Chap. 5. Here, we learn a new technique for estimating the parameters of a function that best fits the data. We assume a linear model, i.e., $v = \alpha t + \beta$, and use the "least squares" method, by means of the Python `curve_fit` from `scipy.optimize`, illustrated in the next section. In short, the least squares method consists in finding the parameters $\mathbf{p}$ of a function $y = f(\mathbf{p}, t)$ that minimises

> The least squares method provides a convenient way of estimating function parameters from data, given a model for them.

$$\chi^2 = \sum_{i=1}^{N} \left(\frac{y_i - f(\mathbf{p}, t_i)}{\sigma_i} \right)^2 \tag{10.13}$$

(pronounced ki-square), where y_i is the experimental data collected when $t = t_i$ (in the example above, $y_i = v_i$, $\mathbf{p} = (\alpha, \beta)$ and $f(\mathbf{p}, t) = \alpha t + \beta$), N is the number of points and σ_i the uncertainty about y_i. The uncertainty about t_i is supposed to be negligible. It is easy to recognise in each term of the sum the squared distance between each data point y_i from the fit function, measured in units of their uncertainty. Finding the best fit function means, in fact, minimising the sum of such distances.

It would be quite natural to minimise the sum of the distances between experimental points and the theoretical curve, but it is not convenient.

The reason why we minimise the sum of the squares of the distances and not that of the distances themselves is that, in the latter case, we would be forced to use the modulus operator, defining the total distance as

$$\chi = \sum_{i=1}^{N} \left| \frac{y_i - f(\mathbf{p}, t_i)}{\sigma_i} \right|. \tag{10.14}$$

In order to perform operations on this expression, we should always consider the sign of each term in the sum, and this is very uncomfortable. The squares, instead, are defined as positive and, if the distance is minimal, so is its square.

Finding the minimum of χ^2 is relatively easy, since, in this case, χ^2, seen as a function of α and β, is a parabola. One can then find the vertex of such parabolas to compute the values of $\mathbf{p}$. Details about this technique are given below (see Sect. 10.10). Here, we discuss the results of the minimisation.

Taking $\sigma_i = 1, \forall i$ does not affect the result of the minimisation, provided that all the uncertainties are equal to each other (or, at least, of the same order of magnitude).

We have not evaluated the uncertainties about experimental data, however, we can assume that they are all of the same order of magnitude and put $\sigma_i = 1 \forall i$. On the other hand, if all the uncertainties are equal to each other, the result of the minimisation will be independent of the value of σ_i and we can get rid of them for the moment. The result of the fit is

$$\alpha = -1.61 \frac{m}{s^2}$$
$$\beta = 0.017 \frac{m}{s}, \tag{10.15}$$

where α is not far from our first raw estimation and β is relatively close to zero (however, think about how reasonable this statement is). As a result of a measurement, both α and β must be affected by some uncertainty that, in turn, depends on σ_i.

Few points, in fact, appear to fluctuating around the best fit, such that the best fit itself can be seen as susceptible to fluctuate around each point. If we trust the fit, data points give us the opportunity to estimate the uncertainty of the measurements. The difference between the velocities at $t \simeq 0.36$ s and at $t \simeq 0.42$ s (those that exhibit the largest deviation from a straight line) is on the order of 0.05 m/s. Given that $v \simeq 0.5$ m/s, the relative uncertainty is about 10 %.

From our model, we expect $a \simeq const$, while, from the plot on the right in Fig. 10.5, it seems that a is far from being constant. However, we should not forget that data are affected by a relatively large uncertainty: if the uncertainty of v, σ_v, is on the order of 10 %, σ_a, the uncertainty about a can be as large as 20 %. We found that

$$\langle a \rangle = -1.65 \,\mathrm{ms}^{-2}$$
$$\sigma_a = 1.3 \,\mathrm{ms}^{-2} \qquad (10.16)$$
$$\frac{\sigma_a}{\sqrt{N}} = 0.32 \,\mathrm{ms}^{-2} \,.$$

The latter is the uncertainty about the average value, so, assuming $a = const$,

$$a = -1.65 \pm 0.32 \,\frac{\mathrm{m}}{\mathrm{s}^2} \,, \qquad (10.17)$$

which is still compatible with the fit result, for a relative uncertainty of 20 %,

Using the least squares method, one can take a similar approach to evaluating the uncertainties about parameters that take into account the shape of data distribution. Assuming that the model is correct, one can suppose that the uncertainty is such that data fluctuate around the best fit line according to a Gaussian distribution, inferring the uncertainty from the width of the distribution of "residuals", defined as

$$r_i = y_i - f(\mathbf{p}, t_i) = y_i - \alpha x_i - \beta \,. \qquad (10.18)$$

The result, obtained with this technique, detailed in Sect. 10.7, is

The effective uncertainty about data can be inferred by looking at the plots.

Assuming $a = const$, its value can be taken by averaging the data. The uncertainty about the average is the standard deviation divided by the square root of the number of points.

Using the least squares method, it is possible to exploit data distribution to infer parameters.

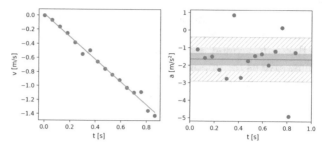

■ **Fig. 10.6** Velocity and acceleration of the cart along the track, in SI units, as a function of the time, with the results of the fits superimposed. The green and yellow bands represent, respectively, the one- and two-sigma uncertainty intervals as in the linear fit. The hatched band shows the one-standard deviation interval amplitude

$$\alpha = -1.61 \pm 0.05 \, \frac{m}{s^2}$$
$$\beta = 0.017 \pm 0.023 \, \frac{m}{s} \,, \tag{10.19}$$

which is consistent with the above results. If we fit the acceleration data with a constant, i.e., with a polynomial of grade 0, using the least squares method, we obtain the same exact value $a = -1.65 \, \text{ms}^{-2}$ found averaging the slopes. It can be shown, in fact (see Sect. 10.10), that finding the minimum of the χ^2 in this case is equivalent to taking the average of the data.

Figure 10.6 shows the same data as in Fig. 10.5 with the results of the fits superimposed. The one- and two-sigma intervals are shown in the case of the acceleration, in the form of coloured bands.

10.7 Evaluating the Goodness of a Fit

Once the parameters **p** have been determined, the χ^2 is a number that can be computed by substituting the parameters where appropriate. With the proper normalisation of the terms in the sum, the value of χ^2 gives an indication of how good the fit is. Minimising the χ^2 assuming $\frac{\sigma_i}{v_i} \simeq 10\%$ leads to the following result:

An indication of the quality of the fit is given by the value of χ^2, if it is properly normalised.

$$\alpha = -1.61 \pm 0.06 \, \frac{m}{s^2}$$
$$\beta = 0.017 \pm 0.033 \, \frac{m}{s} \,. \tag{10.20}$$

Substituting $\alpha = -1.61$ and $\beta = 0.017$ in the expression of χ^2, its resulting value is $\chi^2 = 6.8$. As expected, the result of the weighted fit (the one in which we normalised the χ^2 using the σ_i) is compatible with that of unweighted one.

Data points are random variables distributed as Gaussians, thus each term in the χ^2 is a random variable distributed as a Gaussian with zero mean and unitary sigma. As a consequence, the χ^2 itself is a random variable distributed as a sum of Gaussians. Needless to say, it follows the χ^2-distribution: the distribution of the sum of normally distributed random numbers. As with any other distribution, the χ^2-distribution has an expected value that can be proven to be $E[\chi^2] = v$, and a variance $\sigma^2(\chi^2) = 2v$, where v, called the number of "degrees of freedom", is given by $v = N - m$, m being the number of parameters estimated from the data. The distribution, in fact, depends upon v. In the above example, $v = 16 - 2 = 14$.

> The χ^2 is a random variable with $E[\chi^2] = v$ and $\sigma^2 = 2v$, where $v = N - m$, m being the number of parameters estimated from the fit. v is called the number of degrees of freedom.

Given the PDF of the χ^2, $P\left(\chi^2, v\right)$, one can compute

$$P\left(\chi^2 > \chi_0^2, v\right) = 1 - \int_0^{\chi_0^2} P\left(\chi^2, v\right) d\chi^2, \qquad (10.21)$$

i.e., the probability that a randomly picked χ^2 for v degrees of freedom is greater than a given value χ_0^2. This probability is also called a p-value. The integral on the RHS of the equation is the cumulative distribution function and gives the probability that the variable is between the integral limits (note that, by definition, $\chi^2 \geq 0$, thus $P\left(\chi^2 < 0, v\right) = 0 \, \forall v$).

> The goodness of the fit, also called a p-value, is given by the area of the tail of the χ^2-distribution, computed after its cumulative distribution function.

The value of the cumulative can be computed numerically and, as usual, there are functions in Python that return it. In our case, the p-value of the fit is 0.94: the probability that our experiment leads to a $\chi_0^2 = 6.8$ or higher is thus quite high. This usually gives physicists confidence that the model well describes the data. As a rule of thumb, a *good* fit is one returning a "reduced" χ^2, i.e., the ratio $\frac{\chi^2}{v} \simeq 1$. In fact, $E\left[\chi^2\right] = v$, because each term in the sum is a normally distributed random variable and, on average, contributes about 1 to the sum. Since there are v independent terms in the sum (there are N terms, of which m are constrained by the values of **p**), the latter is expected to be v. A much larger value of χ^2, as well as an extremely low value, should

> As a rule of thumb, a good fit returns a reduced χ^2, $\frac{\chi^2}{v}$, close to 1.

draw your attention. In summary, a *good* fit should return a reduced χ^2

$$\frac{\chi^2}{\nu} = \frac{\nu \pm \sqrt{2\nu}}{\nu} = 1 \pm \sqrt{\frac{2}{\nu}}. \tag{10.22}$$

Statistics can only provide the probability that a given result is consistent with random fluctuations. It cannot tell us how good a model is, nor how data are consistent with the model. It can only tell us how consistent the deviations from the model are with random fluctuations.

It is important to observe that statistics cannot give absolute answers about the goodness of a fit. It can only evaluate the probability that the observed χ^2 can reasonably be ascribed to statistical fluctuations. It is always possible that the χ^2 of a fit is large because of unfortunate circumstances or, conversely, that the χ^2 of a fit is extremely low because the model *overfit* data (any $(n-1)$-degree polynomial fits a set of n data points perfectly).

When we affirm that a fit is *good* because of its χ^2, we implicitly assume that the model is correct. Quoting the *p*-value of a fit as in the above paragraphs, in fact, we implicitly provide the conditional probability that, given a certain hypothesis H_0 (the acceleration is constant), the data distribute according to our expectations: $P(\text{data}|H_0)$. As a matter of fact, the *p*-values are often misinterpreted as the probability of a hypothesis H_0 given the data: $P(H_0|\text{data})$. Remember that conditional probabilities are not symmetric!

Bearing that in mind, there is nothing really wrong in quoting *p*-values as above, even if care must be taken in using this metric as a measure of the quality of the fit itself. On the other hand, observing very large or very low χ^2, especially when the model is manifestly correct, should draw your attention: often, it is a signal that either you made a mistake or that you are over/underestimating uncertainties or neglecting some systematic effect.

Never take rules as laws

Statistics often provides rules for making decisions that, however, cannot be taken as universal laws. Usually, they can be applied more or less blindly in most situations, but in subtle cases, they can lead to completely wrong conclusions.

Consider, for example, the case in which you found a value $x = 5.0$ during an experiment and you want to compare this value with two alternative models. In one model, x is distributed such that it peaks around $x \simeq 3$ (in orange in the plot below); in the alternative model, the x distribution is *bimodal* (in blue), i.e., it has two modes.

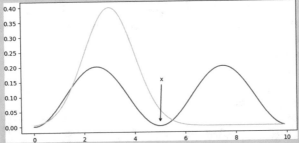

Manifestly, the p-value of the alternative is much larger than that of the other. Naively adopting the p-value rule, we would conclude that $x = 5$ is compatible with the hypothesis that it is distributed according to the blue curve. Unfortunately, the probability of $x = 5$ in this case is predicted to be zero!

It is instructive to work out a goodness-of-fit test by comparing two models in a poorly modelled experiment, considering the acceleration as a function of time. Our hypothesis H_0, stemming from mechanics, is that $a = const$, which is further corroborated by the fact that the observed velocity is linearly dependent on time.

The hypothesis is also compatible with the result of the measurement of $\langle a \rangle$ leading (Eq. (10.16)) to a relative uncertainty $\frac{\sigma_a}{a} \simeq 20\,\%$ consistent with the error on v.

Manifestly, however, we could have obtained the same results even if $a \neq const$. An appropriate change of a with time would lead to similar results, in fact. So, fitting the a versus t distribution is worthwhile and permits us to test the hypothesis directly, exploiting the way in which data distribute over time. Performing a weighted least squares minimisation with $f(\mathbf{p}, t) = a = const$, the result is

$$a = -1.65 \pm 0.08\,\text{ms}^{-2}, \qquad (10.23)$$

A large χ^2, corresponding to a low p-value, is a signal that either the model does not describe the data, or that uncertainties about the data are underestimated. Conversely, an excessively high p-value probably means that uncertainties are overestimated or that the model overfit the data.

with a $\chi_0^2 = 219$ for $\nu = 14$ degrees of freedom. The result is perfectly compatible with the one obtained by simply averaging the data. In fact, minimising the χ^2 is equivalent, in this case, to taking the average, as can be seen by just imposing that

$$\frac{\partial \chi^2}{\partial a} = 0 = 2 \sum_i \frac{y_i - a}{\sigma_i^2} \tag{10.24}$$

satisfied for

$$Na = \sum_i y_i , \tag{10.25}$$

i.e., for $a = \langle y_i \rangle$. The uncertainty is a bit larger, due to the fact that a few data points deviate significantly from the predicted behaviour. The fact that the χ^2 of the fit is so large ($\chi^2 = 219$) tells us that the hypothesis H_0 that the acceleration is a constant is unlikely to be supported by the data. The p-value of the fit is practically zero, in fact.

Such a result would naively lead to the conclusion that the hypothesis is wrong, so we could test a different hypothesis. Since any function $f(\mathbf{p}, v)$ can always be rewritten as a Taylor series, we can try with a truncated series. A fit with a straight line gives a similar result, with $\chi^2 = 216$. Even if the χ^2 is smaller than the previous one, the difference is not significant (remember that the variance on its values is about twice the number of degrees of freedom) and, in fact, even in this case, the p-value is, in practice, null. Increasing the order to which the Taylor series is truncated is not worthwhile because it is unlikely that a reasonable function describing the data cannot be approximated by a low degree polynomial and, on the other hand, a 14th-degree polynomial will lead to a *perfect* fit with $\chi^2 = 0$.

The correct way to interpret the result is as follows: a *bad* χ^2 may have a different interpretation.

- The hypothesis H_0 is wrong. In the case we are considering, this is very unlikely, since H_0 stems directly from Newtonian mechanics that has been extensively tested over centuries and, unless we are conducting an extremely precise experiment in very unusual conditions, no one is going to *bet* on that. Following Bayes' definition of probability, the probability that H_0 is wrong is almost null.

- Data are affected by large statistical fluctuations. Such a hypothesis cannot be ruled out completely, because the probability that the χ^2 is large is never 0, although small. After all, the p-value is greater than zero for any $\chi^2 \in (0, \infty)$. However, this is unlikely to happen. If this is, in fact, a fluctuation, performing the experiment a second time should lead to a much better result.
- Data do not follow the predicted distribution. Such a hypothesis can be true, in fact. However, the experiment we conducted is simple enough that, even considering the presence of friction, we would expect the data to be compatible with H_0, still according to our previous (*a priori*) knowledge of the laws of mechanics. On the other hand, the velocity measured during the same experiment is compatible with H_0.
- Data are affected by large systematic errors that are not properly included in the χ^2.

The last explanation is, indeed, the most credible one. Data about acceleration clearly exhibit large fluctuations at selected points along the track. Those points coincide with those in which the values of v also show larger fluctuations at times corresponding to points at which two rails join. It is then possible that, when the train car crosses the joints, its motion is affected in a way that is difficult to predict. The result is that its position changes abruptly, and that reflects on both v and a.

In summary, while a low χ^2 is often translated into "data follows the predicted model" or "the model is correct", one should never forget that statistics can only predict probabilities for purely random variables. In particular, it cannot predict how good a model is. Hence, we cannot ask ourselves how large the probability is that a fit is good, but rather how large the probability is that the fit is not too bad.

10.8 Data Processing

A least squares fit can be performed using Python thanks to the `curve_fit()` function defined by `scipy.optimize`, as follows:.

```
from scipy.optimize import
    curve_fit
...
```

`curve_fit()` imported from `scipy.optimize` is used to perform a best fit to data with a user function.

```
res,  cov  =  curve_fit(f,  t,  v)
```

where res, cov, t and v are arrays and f is the fitting function. When the latter is a straight line, it is defined as

```
def f(x, alpha, beta):
    return alpha*x+beta
```

curve_fit()
returns the list of
parameters estimated
by the fit and their
covariance matrix.

The returned objects res and cov contain, respectively, the parameters **p** of the fit, in the order in which they are listed in the function definition, and the covariance matrix. The slope of the line is then given by res[0], while the intercept is given by res[1]. The variances of parameters are given by cov[0][0] and cov[1][1]. Off diagonal elements cov[0][1] and cov[1][0] are the covariances $Cov(\alpha, \beta)$ and $Cov(\beta, \alpha)$ (and they are equal to each other, the matrix being symmetric).

Data collected by Arduino consists of the list of positions x collected at times tx. From them, the list of velocities v can be obtained as

```
for i in range(len(x) - 1):
    v.append((x[i+1]-x[i])/(tx[i+1]-tx[i])
        /100.)
```

0

(don't forget that x contains positions in cm). When producing the plot of *v* versus *t* and the corresponding fit, we have to take into account that the length of v is that of x minus one. The corresponding list of times must thus be shorter. In particular, the last time must not be included in the list. A copy of the list of times can be obtained as

To create a copy of
an object, use the
copy() method.
The = operator does
not create a copy of
the operand on its
right, but rather a
reference to it.

```
t = tx.copy()
```

From the latter, the last element can be removed with

```
t.pop()
```

It is interesting to observe that a statement like t = tx does not actually create a list t whose elements are equal to those of tx, but just a reference to the original list. In this case, a change in the elements of tx is reflected in the elements of t, and vice versa, so t.pop() would remove the last element of tx, too. The copy() method, contrastingly, creates a *shallow copy* of the list, constructing a new list and inserting the elements found in the original one into it (there is also a deepcopy() method that is useful when the list contains complex objects).

curve_fit(f, t, v) performs an unweighted fit for which $\sigma_i = 1$ is assumed for all data. The elements of the covariance matrix are evaluated from the width of the distributions such that, when the parameters change by one standard deviation, the χ^2 increases by the corresponding number of degrees of freedom.

In order for the fit to take into account the proper normalisation, we need to pass an array w containing the values of σ_i to curve_fit(), as in

By default, curve_fit() performs an unweighted fit. To properly weigh the terms in the sum of the χ^2, a list of the uncertainties must be passed as an optional argument.

```
res, cov = curve_fit(f, t, v, sigma = w)
```

In this case, the minimisation is performed weighing each term in the sum by $\frac{1}{\sigma_i^2}$, such that the points with smaller uncertainties dominate while we look for the minimum. However, the uncertainties are still evaluated as above, assuming that the model is correct and looking for the values of the parameters that cause the χ^2 to increase by v. In order to compute the uncertainties from σ_i, we need to state that explicitly, as in

```
res, cov = curve_fit(f, t, v, sigma = w,
                     absolute_sigma = True)
```

After a weighted fit, the χ^2 of the observed data can be used to evaluate its goodness. Unfortunately, there is no way to ask Python to return that value, and we need to compute it explicitly by ourselves, as in

```
def chisquare(x, y, fun, p, sigma=None):
    chi2 = 0.
    for i in range(len(x)):
        s = 1
        if sigma:
            s = sigma[i]
        fi = fun(x[i], p[0], p[1])
        chi2 += ((y[i]-fi)/s)**2
    return chi2
```

Parameters in a function can be optional if they are assigned a default value when the function is defined. Arguments to function are either passed in the order in which it expects them or as a comma-separated list of keyword/value pairs.

When defining a user function, besides positional parameters, one can define optional parameters that can be omitted when calling the function; in this case, those parameters get their default value specified after the = sign (it can be None). In the case of chisquare, the parameter sigma can be omitted. If it is, its value is None and the weight of each term in the sum is taken to be 1. Otherwise, it is assumed to be a list of as many elements as x and y. Note that positional parameters, contrastingly, are mandatory and their values are assigned taking into account the order in which arguments are listed in the function call, such that, in order to call the above function to compute the weighted χ^2 of the fit to v versus t with a function linear using res as **p**, we write

```
chi2  =  chisquare(t,  v,
    linear,  res,  wv)
```

However, values to parameters can be assigned in any order by explicitly indicating the corresponding parameter name (also called the keyword) together with the argument, such as in

```
chi2  =  chisquare(y  =  v,  x  =  t,
                   sigma  =  wv,
                   fun  =  linear,
                   par  =  res)
```

A mix of the techniques is allowed when there is no ambiguity. This technique is used when the first arguments are passed in the default order, as in

```
chi2  =  chisquare(t,  v,  par  =  fitres,
                   fun  =  linear,
                   sigma  =  wv)
```

For a clean code, we recommend that one always call the functions using positional arguments when they are mandatory and keyword arguments when they are optional. The above function returns

$$\chi^2 = \sum_i \left(\frac{x_i - f(x_i, \alpha, \beta)}{\sigma_i} \right)^2 , \qquad (10.26)$$

where fun is supposed to be the name of a function that accepts two arguments, besides x, contained in the par list. The returned χ^2 can be used to evaluate the goodness of the fit.

The χ^2-distribution is defined in the `scipy.stats` package and is called `chi2`. The corresponding cumulative distribution function is given by `scipy.stats.chi2.cdf(X2, NDF)`, where `X2` represents the value of χ^2 and `NDF` ν.

10.9 The χ^2-Distribution

Suppose we have ν normally distributed random variables x_i, $i = 1...\nu$. Their sum $\chi = \sum_i x_i$ is distributed as a Gaussian with zero mean and standard deviation $\sigma = \sqrt{\nu}$. Figure 10.7 on the left side shows the distribution of 10 000 random variables χ for $\nu = 1$, $\nu = 5$ and $\nu = 25$, with their PDF superimposed as a continuous curve.

The distribution of $\chi^2 = \sum_i x_i^2$ is shown on the right side of the same figure, for various ν. Manifestly, $\chi^2 > 0$ and its distribution is asymmetric. It exhibits a peak around ν and a tail on the right side. The tail becomes less pronounced as ν increases, as predicted by the central limit theorem. In fact, for $\nu = 100$ (Fig. 10.8), the difference between the χ^2 distribution and a Gaussian is hard to spot. The number ν is called the "number of degrees of freedom" (NDF). From the figures, we can easily spot a few features of the χ^2 distribution that can be rigorously proven. The mean of an χ^2 variable with ν degrees of freedom is ν, while its variance is 2ν. From these properties, it is easy to show that the mean of the reduced χ^2

> The sum of ν normally distributed variables is distributed as a Gaussian with $\sigma = \sqrt{\nu}$. The sum of the squares of ν normally distributed random variables is distributed as a χ^2 with ν degrees of freedom.
>
> The mean of the χ^2 variable with ν degrees of freedom is ν, while its variance is 2ν.

$$\chi_\nu^2 = \frac{\chi^2}{\nu} \tag{10.27}$$

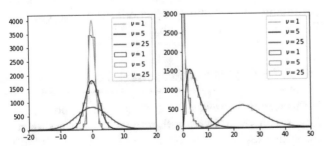

◘ Fig. 10.7 The distribution of 10 000 random variables obtained by summing up ν normally distributed variables (left) and that of as many variables obtained by summing up their square (right)

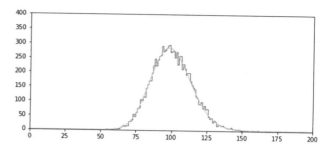

Fig. 10.8 The distribution of an χ^2 random variable obtained by summing up $v = 100$ normally distributed variables. One can hardly distinguish it from a Gaussian distribution with mean 100

is 1, while its variance is $\frac{2}{v}$. The PDF of χ^2 can be obtained with Python from `scipy.stats.chi2.pdf(x, NDF)`. Its analytical expression is

$$P\left(\chi^2, v\right) = \frac{\left(\chi^2\right)^{\frac{v}{2}-1} e^{-\frac{\chi^2}{2}}}{2^{\frac{v}{2}} \Gamma\left(\frac{v}{2}\right)}, \tag{10.28}$$

$\Gamma\left(\frac{v}{2}\right)$ being the Gamma function, defined as

$$\Gamma(x) = \int_0^\infty y^{x-1} e^{-y} dy, \tag{10.29}$$

for which

$$\Gamma(n) = (n-1)!. \tag{10.30}$$

In most cases, there is no need to memorise it. Values of $P\left(\chi^2, v\right)$ are either tabulated in printed books or can be obtained using library functions such as those from Python's `scipy.stats.chi2.pdf()`.

10.10 The Least Squares Method

In this section, we review the least squares method in a more formal and general way with respect to what is shown in the examples.

Having collected data of $x(t)$, $v(t)$ and $a(t)$, as shown in Sects. 10.5 and 10.6, we need to find the function $f(\mathbf{p}, t)$ that best describes our data. One possible approach is to find the function for which the distance between the function itself and the experimental points is the minimum. Such a distance can be defined as the sum of the distances of each experimental point to the corresponding value of the function of the variable upon which it is supposed to depend, e.g., for $x(t)$,

> The parameters of the function that best describes data can be found by imposing that its distance from experimental points is the minimum possible one.

$$d_i = |x_i - f(\mathbf{p}, t)| \tag{10.31}$$

such that we want to minimise

$$D = \sum_{i=1}^{N} d_i = \sum_{i=1}^{N} |x_i - f(\mathbf{p}, t_i)|. \tag{10.32}$$

The need for the absolute value $|...|$ comes from the fact that each experimental point can lie below or above the function. However, such an operation is complicated to handle, so we should try to avoid it. One possibility is to minimise the sum of the squares of the distances d_i. In fact, if $\sum d_i$ is the minimum, so is $\sum d_i^2$. The advantage, in the latter case, is that we don't care about the sign of each term in the sum, all of them being non-negative by definition. We then find the best fit by minimising

> The absolute value is difficult to manage, but if the distance is at its minimum value, then so is the distance squared.

$$D^2 = \sum_{i=1}^{N} d_i^2 = \sum_{i=1}^{N} (x_i - f(\mathbf{p}, t_i))^2. \tag{10.33}$$

However, we should take into account that each x_i is affected by an uncertanty σ_i and each term in the sum should thus be weighted by their uncertainty. We take that into account in minimising the sum of the squared distances in units of their uncertainties

$$\chi^2 = \sum_{i=1}^{N} \left(\frac{d_i}{\sigma_i}\right)^2 = \sum_{i=1}^{N} \left(\frac{x_i - f(\mathbf{p}, t_i)}{\sigma_i}\right)^2, \tag{10.34}$$

such that the minimum location is determined mostly by more precise points (those whose σ_i is lower) and less by the less precise ones.

Finding the best fit means finding the values of parameters $\mathbf{p}$ such that χ^2 is the minimum. Since each term in the sum is a normal distributed random variable, the sum is distributed as a χ^2 variable with ν degrees of freedom, where $\nu = N - m$ and m is the number of parameters $\mathbf{p}$ estimated from the fit. In fact, for each parameter estimated from the data we introduce a constraint among the variables such that the number of random variables decreases by one.

Fitting with a constant is equivalent to taking the average of the data.

Consider, for example, the fit to $a(t)$, where we supposed that $f(\mathbf{p}, t) = C = const$. In this case, $\mathbf{p} = (C)$ is a vector with just one component (i.e., a scalar) and the χ^2 can be computed as

$$\chi^2 = \sum_{i=1}^{N} \left(\frac{a_i - C}{\sigma_i} \right)^2 . \tag{10.35}$$

Clearly, χ^2, as a function of C, attains its minimum when

$$\sum_{i=1}^{N} a_i - C = 0 , \tag{10.36}$$

i.e., for

$$C = \frac{1}{N} \sum_{i=1}^{N} a_i . \tag{10.37}$$

Indeed, we have already observed that finding the best fit in this case is equivalent to taking the average of the measurements. We can also realise that, once C is determined by the above formula, the N normally distributed random variables represented by the data are no longer independent of each other. They are constrained such that C is given by the equation above. That is why the number of degrees of freedom of such a χ^2 is not N, but $N - 1$.

The best fitting for $v(t)$ is more interesting. In this case, $f(\mathbf{p}, t) = \alpha t + \beta$ and $\mathbf{p} = (\alpha, \beta)$, thus

$$\chi^2 = \sum_{i=1}^{N} \left(\frac{v_i - (\alpha t_i + \beta)}{\sigma_i} \right)^2. \tag{10.38}$$

In this case, the minimum of χ^2 as a function of α and β is attained when its derivatives with respect to α and β vanish. In fact, expanding the square in the sum, we have

In general, χ^2 is regarded as a function of **p** and reaches its minimum when the derivatives with respect to the components of **p** vanish.

$$\chi^2 = \sum_{i=1}^{N} \frac{v_i^2 + (\alpha t_i + \beta)^2 - 2v_i (\alpha t_i + \beta)}{\sigma_i^2}$$

$$= \sum_{i=1}^{N} \frac{v_i^2 + \alpha^2 t_i^2 + \beta^2 + 2\alpha \beta t_i - 2\alpha v_i t_i - 2\beta v_i}{\sigma_i^2}. \tag{10.39}$$

In this simple case, we can avoid taking derivatives by observing that the numerator is a parabola as a function of α and β. Finding the minimum means finding the abscissa of the vertex of each parabola. Writing the expression above by ordering in powers of α gives

Fitting with a straight line is equivalent to finding the abscissa of the vertex of two parabolas.

$$\chi^2 = \alpha^2 \sum_{i=1}^{N} \frac{t_i^2}{\sigma_i^2} + 2\alpha \sum_{i=1}^{N} \frac{\beta t_i - v_i t_i}{\sigma_i^2} + \sum_{i=1}^{N} \frac{v_i^2}{\sigma_i^2} + \beta^2 \sum_{i=1}^{N} \frac{1}{\sigma_i^2} - 2\beta \sum_{i=1}^{N} \frac{v_i}{\sigma_i^2}. \tag{10.40}$$

We can improve readability by introducing the following definitions:

Building the sums of the data (t and v), their product vt and their squares (t^2 and v^2) is useful for simplifying notation.

$$S_{tt} = \sum_{i=1}^{N} \frac{t_i^2}{\sigma_i^2} \quad S_{vv} = \sum_{i=1}^{N} \frac{v_i^2}{\sigma_i^2} \quad S_{11} = \sum_{i=1}^{N} \frac{1}{\sigma_i^2}$$

$$S_{1t} = \sum_{i=1}^{N} \frac{t_i}{\sigma_i^2} \quad S_{vt} = \sum_{i=1}^{N} \frac{v_i t_i}{\sigma_i^2} \quad S_{1v} = \sum_{i=1}^{N} \frac{v_i}{\sigma_i^2} \tag{10.41}$$

such that

$$\chi^2 = S_{tt}\alpha^2 + 2\alpha (\beta S_{1t} - S_{vt}) + \left(S_{vv} + \beta^2 S_{11} - 2\beta S_{1v} \right). \tag{10.42}$$

The abscissa of the minimum (the vertex) of this parabola is at

$$\alpha = \frac{S_{vt} - \beta S_{1t}}{S_{tt}}.$$

(10.43)

Seen as a function of β, the χ^2 has a minimum at

$$\beta = \frac{S_{1v} - \alpha S_{1t}}{S_{11}},$$

(10.44)

then, substituting the latter in the expression of α:

$$\alpha = \frac{S_{vt}S_{11} - S_{1v}S_{1t}}{S_{tt}S_{11} - S_{1t}S_{1t}},$$

(10.45)

Dimensional analysis provides an effective way to remember formulas.

which is not so difficult to remember, taking into account that α should have the dimensions of a velocity divided by a time. The numerator is made up of a combination with the dimensions of vt (ignoring the σ_i^2 part that is common to the numerator and the denominator), and to obtain the right dimensions, we need to divide by something that has the dimensions of t^2.

Knowing α, it is straightforward to obtain β from Eq. (10.44). Since we constrained the data such that they satisfy both equations for α and β, the number of degrees of freedom of the χ^2 is $N - 2$.

We can also fit $x(t)$ to the data. In this case, $f(\mathbf{p}, t) = At^2 + Bt + C$ and $\mathbf{p} = (A, B, C)$. Finding the values of A, B and C is a bit trickier, but, in principle, we just need to compute the derivatives of the χ^2 with respect to A, B and C and set them to zero. The same happens for any function $f(\mathbf{p}, t)$.

In these cases, we profit from the power of a computer to numerically solve equations. There are plenty of modules for finding the coordinates of the minimum of a function of q variables. Some of them adopt a classic, deterministic method consisting in exploring the q-dimensional space *moving* in the direction along which the gradient of the function diminishes (the "gradient descent" method). Such a method is simple to implement, but it relies on the fact that the function exhibits just one minimum within the search interval. More powerful techniques use stochastic methods to find the minimum, often stolen from physics. Optimisation methods are discussed in the next chapter.

Fitting data can be difficult when the fit function is not as simple as a straight line. Computers can help, thanks to their ability to perform a great number of computations in a short amount of time.

Coming back to the problem of evaluating the parameters of a function, we still need to find the uncertainties with which we know them. Given that data are affected by uncertainties, so are the parameters derived from them. The uncertainty of each term in the numerator in the expression of χ^2 is σ_i. As seen above, since each term has a σ_i in the denominator, each term contributes 1 to the sum. Since there are N terms in the sum, a one-sigma fluctuation of the data causes χ^2 to fluctuate by $\sqrt{2N}$. However, two parameters are constrained when we choose the slope and the intercept of the line, thus, in fact, only $\nu = N - 2$ of the terms in the sum may actually fluctuate, i.e.,

When data fluctuate by one sigma, the corresponding χ^2 increases.

$$\chi^2 \to \chi^2 \pm \sqrt{2\nu}, \tag{10.46}$$

upon a one-sigma fluctuation of the data. It can be shown that the error with which we know the parameters of the fit function is such that it increases χ^2 by 1. The way in which this is done is clearly illustrated when **p** actually only has one component, as in the case of $a = C = const$. In this case, the χ^2 as a function of the C has a minimum for $C = \langle a \rangle$. In a relatively narrow interval around the minimum, the χ^2 has the shape of a parabola. Finding the intersection between $\chi^2(C)$ and a horizontal straight line $y = \chi^2_{min} + 1$ identifies two points whose abscissa determine the uncertainty about C.

See Sect. 12.3 for a proof of this result.

In the case of two free parameters, the χ^2, as a function of the parameters, is a function of two variables: $\chi^2 = \chi^2(A, B)$, and is graphically represented as a paraboloid. To find the uncertainties about A and B, we find the plane at $z = \chi^2_{min} + 2.4$ whose intersection with the χ^2 gives an ellipse. The intersection of such an ellipse with the A and B-axis determines the uncertainties about the parameters.

When the χ^2 cannot be well approximated by a Taylor series truncated at the second order, it may happen that errors are asymmetric. If, for example, we find that the minimum is obtained for $C = 1.65$ in some units, while the interceptions between the χ^2 and the line $\chi^2 + 1$ are found at $C = 1.60$ and at 1.82, we express our result as

$$y = 1.65^{+0.17}_{-0.05}. \tag{10.47}$$

We exploit such a property to evaluate the uncertainties of the fit function parameters.

Then, when the uncertainties of the data are not known, assuming that the model is a good description of the data we can estimate the uncertainties about the parameters **p** by finding the values of **p** that cause χ^2 to increase by a number that depends on the length of **p** (1, 2.4, 3.5, etc). This is how the uncertainties of the parameters are found in the residuals method.

10.11 Discarding Bad Data

Never delete data without having identified the causes of any discrepancies.

When the goodness of fit test returns a low p-value, students are often tempted to just discard data that behaves as outliers. The first reaction is to assume that they made some mistake in collecting them, thus it is better to delete them such that the professor cannot ask them why those points behave that way.

This habit, in fact, can turn out to be a big mistake. First of all, instructors are often smart enough to spot this simply by performing statistical tests on the distribution, often by eye. Secondly, and most importantly, maintaining this habit may lead to miss big opportunities.

Important discoveries are often made while trying to identify the origin of the discrepancy between data and models.

If, in 1967, Jocelyn Bell Burnell, observing the signals of the radio telescope she was operating, had simply ignored the small deviations from the average signal (as shown in Fig. 10.9), she would not be remembered today as the discoverer of pulsars.

Unfortunately, at that time, concerns about gender equality were not yet widespread and, indeed, the contributions of female scientists were often ascribed to their (mostly male) supervisors. For that discovery, the Nobel Prize was awarded to Antony Hewish. Bell Burnell herself was circumspect about the matter, stating that, as Hewish was the supervisor of the project, he would also be held responsible for any failures, so he should be similarly celebrated for its success. The major role that she played in the discovery was ultimately only recognised recently, through the attri-

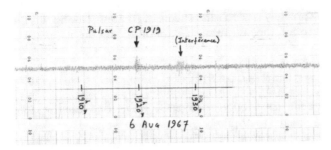

◘ Fig. 10.9 The original chart of the recordings of Jocelyn Bell's radiotelescope in Cambridge, with her annotations

bution of the Breakthrough Prize in Fundamental Physics in 2018 and the Copley Medal in 2021.

We include this story here in the hope that none of you, in the near future, will miss an opportunity due to carelessness in data analysis or because of his/her gender, race or political views. The message is: never discard data you do not understand. Try hard to explain their origin. Even if you do not succeed, report them to your community. It is very possible that you will be blamed for that, but if those data turn out to be of interest and/or value, your career will get a boost. On the other hand, do not place too much trust in the results of your data analysis, based only on p-values. Statistics can only provide the probability that some effect can be produced by a random fluctuation. It can never make predictions about the reliability of your models. A very low probability that an observation is due to statistical fluctuations is not equivalent to a high probability that the same observation is due to a real effect!

> At the same time, never put too much trust in the frequentistic approach to statistics.

10.12 Measuring Gravity Acceleration

Newtonian mechanics predicts that an object sliding along an incline falls with an acceleration

$$a = g \sin \theta . \qquad (10.48)$$

Knowing θ, one can easily measure g from the data collected in the experiment described above. The result could be disappointing, if g, expected to be close to 9.8, turns out to be quite far from it. The reason is that the above formula completely neglects the effects of friction. Taking this into account, the acceleration is found to be

$$a = g \left(\sin\theta - \mu\cos\theta\right) , \tag{10.49}$$

Taking friction into account is important for obtaining reasonable results from an experiment conducted using readily available materials.

μ being the friction coefficient. We need at least two measurements to obtain both g and μ, so we need to repeat the experiment twice at different angles. Repeating the experiment several times at several angles allows us to fit the distribution of a versus θ to obtain both g and μ with better precision. It is interesting to observe that, if θ is small enough, $\sin\theta \simeq \theta$ and $\cos\theta \simeq 1$, such that

$$a \simeq g \left(\theta - \mu\right) . \tag{10.50}$$

For small angles θ, the acceleration is a linear function of g and μ.

This way, a is a linear function of θ, whose slope is g, while μ can be computed after the intercept, dividing it by g. In conducting such an experiment, θ cannot be too small, otherwise static friction prevents the object from sliding, nor too high, otherwise the approximation is no longer valid. For θ below 35°, however, $\sin\theta \simeq \theta$ is still valid within a few percentage points, being that $\theta \simeq 0.61$ and $\sin\theta \simeq 0.57$.

Summary

Experiments in kinematics require measuring the position of an object as a function of time. A clever choice of the reference frame may lead to great simplifications.

Even if a physical quantity can be obtained by a single measurement, exploiting the relationship that exists between different quantities helps in reducing the uncertainties and identifying possible sources of systematic errors. For example, in the case of the motion along an incline, the gravity acceleration g can be inferred, in principle, by a single measurement. However, measuring the accelerations of the object by varying the angle of the incline leads to a much better result.

The position of an object falling along an incline as a function of time can be well described as a parabola.

Knowing the position at different times, velocities can be computed by taking the numerical derivative $\frac{\Delta x}{\Delta t}$. Similarly, the acceleration can be computed as $\frac{\Delta v}{\Delta t}$.

The least squares method provides a convenient way of estimating function parameters from data, given a model. It consists in minimising

$$\chi^2 = \sum_{i=1}^{N} \left(\frac{y_i - f(x, \mathbf{p})}{\sigma_i} \right)^2$$

Once the minimum is found, the χ^2 is a number that can be computed by substituting the values of $\mathbf{p}$ in its expression. The value of the χ^2 is a measurement of the goodness of the fit.

The χ^2 is a random variable with $E[\chi^2] = \nu$ and $\sigma^2 = 2\nu$, where $\nu = N - m$, m being the number of parameters estimated from the fit. ν is called the number of degrees of freedom.

The goodness of the fit, also called the p-value, is given by the area of the tail of the χ^2-distribution, computed after its cumulative distribution function.

$$p \text{ - value} = 1 - \int_0^{\chi_0^2} P\left(\chi^2, \nu\right) d\chi^2$$

As a rule of thumb, a good fit returns a reduced χ^2, defined as $\frac{\chi^2}{\nu}$, close to 1 with a standard deviation of $\sqrt{\frac{2}{\nu}}$.

Statistics can only provide the probability that a given result is consistent with random fluctuations. It cannot tell us how good a model is, nor how data are consistent with the model. It can only tell us how consistent the deviations from the model are with random fluctuations.

A large χ^2, corresponding to a low p-value, is a signal that either the model does not describe the data or that uncertainties about data are underestimated. Conversely, an excessively high p-value probably means that uncertainties are overestimated or that the model overfit the data.

Arduino

With Arduino, times can be measured by exploiting the fact that its operation is regulated by a clock running at 16 MHz.

Distances can be measured with Arduino using ultrasonic sensors, like the HC-SR04. It is composed of two piezoelectric crystals: one emits ultrasonic pulses, the other detects their echo by working as a microphone. Measuring the time t needed for the ultrasonic beam to travel back and forth, we can infer the distance d, knowing the speed of sound c.

$$d = \frac{ct}{2}$$

An Ultrasonic sensor can measure distances over an interval of a few meters. It produces a train of ultrasonic pulses when triggered by a positive voltage signal on its TRIG lead whose duration is at least $10\,\mu$s. It then sets its ECHO lead to HIGH and puts it back to LOW when an echo is detected. The function `pulseIn(pin, state)` monitors a given pin so as to change its state and returns the time elapsed since it was called, in microseconds.

Constants in Arduino sketches are useful for maintenance. They are introduced by the `#define` directive, interpreted by the preprocessor that substitutes its value to any occurrence of them in the program before sending it to the compiler. Remember that the substitution happens literally and before compilation.

Variables, in programming, are data containers. Each variable corresponds to one or more memory locations. Variables can be made constant (i.e., immutable) by adding the `const` qualifier to their declaration. This way, the compiler can detect any place in the code at which the content of a variable can potentially be modified and exit with an error.

It is important to write programs in a clean, fashion. When writing a program, always try to transmit implicit messages to other programmers. Adopt conventions like writing variables in lowercase and constants in uppercase.

Phyphox

PHYPHOX has a function for measuring angles with respect to its axis. The function exploits the accelerometer and the angles are measured comparing the components of the gravity acceleration.

Python

`curve_fit()`, imported from `scipy.optimize`, is used to perform a best fit to data with a user function. It returns the list of parameters estimated by the fit and their covariance matrix.

To create a copy of an object, use the `copy()` method. The = operator does not create a copy of the operand on its right, but rather a reference to it.

By default, `curve_fit()` performs an unweighted fit. To properly weigh the terms in the sum of the χ^2, a list of the uncertainties must be passed as an optional argument.

Parameters in a function can be optional if they are assigned a default value when the function is defined. Arguments to a function are either passed in the order in which they are expected, or as a comma-separated list of keyword/value pairs.

Statistics

The sum of the squares of ν normally distributed random variables is distributed as a χ^2 with ν degrees of freedom. The mean of an χ^2 variable with ν degrees of freedom is ν, while its variance is 2ν.

In general, χ^2 is regarded as a function of $\mathbf{p}$ and reaches its minimum when the derivatives with respect to the components of $\mathbf{p}$ vanish. The parameters that best describe data are taken to be those that lead to the minimum possible χ^2.

Fitting data to a constant is equivalent to averaging data. Fitting to a straight line consists in finding the vertex of the parabola represented by the χ^2 as a function of the fit parameters.

There exist explicit formulas for computing the fit parameters in the case of a linear fit (or regression). In general, a computer finds the minimum of the χ^2 numerically.

Uncertainties can be computed by propagating them to the fit parameters or using the method of residuals, i.e., finding the values for which the χ^2 increases by ν.

Oscillations

Contents

© The Author(s), under exclusive license to Springer Nature
Switzerland AG 2021
G. Organtini, *Physics Experiments with Arduino and Smartphones*, Undergraduate Texts in Physics,
https://doi.org/10.1007/978-3-030-65140-4_11

This chapter is devoted to the experimental study of the motion of a spring, resulting in the Hooke Law. In fact, there is nothing particularly interesting about the motion of the end of a spring, except that it has a certain regularity. On the other hand, flipping through a physics textbook, it seems that physicists are almost obsessed by them. The reason why physicists are so interested in springs is that the elastic force that governs their motion is, in the first approximation, similar to the forces that govern other much more interesting phenomena. Understanding how a spring works allows us to investigate a variety of physical phenomena, from subnuclear to cosmological scales.

11.1 An Experiment to Study Elasticity

Elasticity can be studied using very simple tools: a smartphone with PHYPHOX, a flask and one or two rubber bands. You may have noticed that there are no springs in the list. In fact, the role of the spring is taken, in this case, by the rubber bands. On the other hand, they behave almost like springs, at least qualitatively. From now on, then, "spring" and "rubber band" will be synonymous in the text.

A very simple experimental setup is shown in the margin, where we show a flask suspended from a clothes line by means of a rubber band.

In order to start a quantitative study of the spring motion, we need to identify the relevant physical quantities. The state of the system under investigation is given by the spring length $y = y(t)$, which is a function of time t. The length of the spring varies according to the external force applied to one of its extrema, keeping the other one fixed. Its shape, the material of which it is made, its weight, etc., have some influence on its motion. However, given a spring, they all remain constant under all circumstances, thus they can be modelled by simply identifying one or more constants with the proper dimensions.

From Newton's second Law, we can write that

$$\frac{d^2y}{dt^2} = \frac{F}{m},$$ (11.1)

where m is the mass attached to one of the spring's ends. The spring itself is considered as massless in this model. In order to reproduce this condition in the real setup, we need to make sure that the mass of the flask is much higher than the mass of the spring. Integrating this equation twice, we should obtain $y(t)$, as experimentally observed when the spring is subject to force F, i.e.,

To simplify the analysis of the setup, we introduce approximations that allow us to neglect the effects of parameters that are difficult to be taken into account.

$$y(t) = \int dt \int \frac{F}{m} dt' . \tag{11.2}$$

Here, F can be a function of time, too, i.e., $F = F(t)$ and, since $y = y(t)$, $F = F(y)$. In principle, we know almost nothing about F.

Even if we do not know much more than the fact that F must be continuous and derivable, we can go further in our investigation by assuming that, if F is smooth enough (and we have no reason to believe it is not), we can approximate it using Taylor expansion and write

Irrespective of how complicated the expression of the elastic force is, for small elongation, it can always be written as being linearly dependent upon it. The elastic force is the Taylor expansion of any force $F(y)$ truncated at the first order.

$$F(y) = F(y_0) + F'(y_0)(y - y_0) + \frac{F''(y_0)}{2}(y - y_0)^2 + \cdots , \tag{11.3}$$

$F'(y_0)$ and $F''(y_0)$ being, respectively, the first and second derivative of $F(y)$, computed for $y = y_0$. If we stretch the spring only a little, $y - y_0$ is small, thus $(y - y_0)^2$ can be considered as negligible and

$$F(y) - F(y_0) \simeq F'(y_0)(y - y_0) . \tag{11.4}$$

This formula can be rewritten by defining y_0 as the length of the spring for $F = 0$, such that $F(y_0) = 0$. Being that $F'(y_0) = k$ is a constant and given that the force exhibited by the spring is always opposite to the external force,

$$F(y) = -k(y - y_0) . \tag{11.5}$$

We can further simplify its expression by defining the elongation $\Delta y = y - y_0$, such that

The Hooke Law can be written as $F = -k\Delta y$.

$$F(y) = -k\Delta y , \tag{11.6}$$

also known as the Hooke Law, after Robert Hooke (1635–1703), who first formulated such a Law as *ut tensio, sic vis*: the force (*vis*) depends (*sic*) on the elongation (*tensio*, from which we derive the term *extension*). The constant k is called the "elastic constant" and represents the spring's stiffness: the higher k, the stronger the force needed to change its length by Δy. The elastic constant's dimensions can be easily computed as

$$[k] = \left[\frac{F}{\Delta y}\right] = \left[\frac{MLT^{-2}}{L}\right] = \left[MT^{-2}\right] . \tag{11.7}$$

It can then be measured in units of kg/s^2, but it is usually given in N/m (the unit of a force divided by the unit of a length).

As can be seen, the Hooke Law is nothing but a law that states that, in first approximation, every force that depends on the position y is proportional to the displacement from the equilibrium position. A toy model of a solid, for example, is made as a regular mesh of points representing atoms attached to each other by springs. It is manifest that physicists do not, in fact, believe that atoms are attached to each other by microscopic springs. In fact, they believe that there must be a force that keeps the atoms close to each other in a solid. There is no way to measure such a force directly using a dynamometer, however, we can argue that the force must depend on the distance Δx between atoms (at least, because we can break apart a solid if we separate its two ends by sufficient distance with ample intensity). Repeating the above reasoning, we can conclude that, for small displacements Δx, such as those presumably taking place in normal conditions, the interatomic force F must be proportional to them, and $F = -k\Delta x$.

In solids, atoms are held together by a force that manifestly depends on the interatomic distance Δx. Irrespective of how complicated it is as a function of Δx, the latter being small, the force can be written as a truncated Taylor series and behaves like an elastic force.

In order to find the equation of motion, we need to integrate Newton's second Law and solve the following differential equation:

$$\frac{d^2 y}{dt^2} = -\frac{k}{m}(y - y_0) . \tag{11.8}$$

In other words, we need to find a function $y(t)$, whose second derivative is proportional to the function itself. Trigonometric functions have this property, i.e.,

$$\frac{d^2}{dt^2} \sin \omega t = -\omega^2 \sin \omega t$$
$$\frac{d^2}{dt^2} \cos \omega t = -\omega^2 \cos \omega t \,, \tag{11.9}$$

A combination of trigonometric function is always a solution of differential equations in which the second derivative of a function is equal to the function itself.

and since they differ only by a constant angle (the "phase"), we can try with

$$y(t) = y_0 + A \sin (\omega t + \phi) \,. \tag{11.10}$$

Substituting in the differential equation, we obtain

$$-\omega^2 \sin (\omega t + \phi) = -\frac{k}{m} \sin (\omega t + \phi) \,, \tag{11.11}$$

which is satisfied for

$$\omega = \sqrt{\frac{k}{m}} \,. \tag{11.12}$$

The phase ϕ depends on our choice concerning the origin of times. For $t = 0$,

$$y(t) = y_0 + A \sin \phi \,. \tag{11.13}$$

If we choose $t = 0$ when the spring is at rest, $y(0) = y_0$ and $\phi = 0$. If, instead, $t = 0$ when the spring reaches its maximum elongation A, $y = y_0 + A$ and $\phi = \frac{\pi}{2}$.
We are now ready to conduct an experiment to test our theory. We can, for example, measure y as a function of t and compare with predictions. Since the second derivative of y is proportional to y itself, we can also measure the acceleration of the weight as a function of time, to see if it behaves as expected.

11.2 A Study of Spring Dynamics with Smartphones

A smartphone's accelerometer can be exploited to measure the acceleration of the weight attached to the spring. Among the PHYPHOX experiments, in fact, the "Spring" one uses it to compute the period of oscillations.

To collect data from the accelerometer, use a "Timed run", setting the delay and the experiment duration to a few seconds (e.g., 3 s and 6 s, respectively), such that we can avoid measuring the acceleration during the preparation of the experiment and we can catch a few oscillations of the spring. Data can be collected using either the "Acceleration" experiments (both with and without g) or using the dedicated "Spring" experiment. In this case, the experiment collects data and tries to determine the period of oscillation, shown at the end of the experiment in the main tab. It also shows its inverse (the frequency).

A PHYPHOX timed run automatically starts after the given delay and stops after the specified time has elapsed.

Start the run, stretch the spring, then let it go just before the data acquisition starts. Wait for the chosen duration, then export the data as a CSV file. Try to avoid giving boosts to the weight when the experiment starts and make sure, as much as possible, that the system oscillates in the vertical direction only, avoiding oscillations on the other axis. Repeat the experiment several times, such that you can later choose the one that is less affected by systematic effects, deleting the previous data from the phone's memory each time after exporting them.

A smartphone accelerometer can be used to collect data about the spring's acceleration.

Ideally, we should observe the horizontal components of the acceleration (a_x and a_z) to be zero, while the vertical one (a_y) is expected to oscillate as

$$a_y = \frac{d^2 y}{dt^2} = -\frac{k}{m} A \sin(\omega t + \phi).$$ (11.14)

Let's have a look to the data collected using the "Spring" experiment with PHYPHOX. The raw data CSV file contains four columns: t, a_x, a_y and a_z. A plot of a_y versus t is shown in Fig. 11.1 for a 5 s run.

We can clearly spot a few interesting features. The acceleration, in fact, oscillates, but the amplitude of the oscillations is not constant: it decreases with time. Oscillations do not seem to be quite as we expected: they are not harmonic.

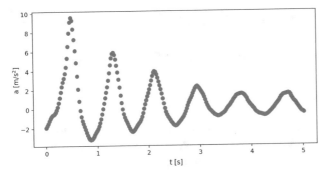

Fig. 11.1   Raw data collected during a PHYPHOX run using the "Spring" experiment with a rubber band and a flask with $m = 0.930 \pm 0.001$ kg. The acceleration of the mass is shown as a function of time

Our simple model of a harmonic oscillator only included the elastic force $F = -k\Delta y$. In a real experiment, non-conservative forces such as dissipative ones can be relevant and must be taken into account. Usually, this is a difficult task, however, in this case, it is relatively easy to include them in our model. In fact, ideally, the mass attached to the spring moves as

$$y(t) = y_0 + A \sin(\omega t + \phi).$$ (11.15)

The speed of the mass is then

$$v_y(t) = \frac{dy}{dt} = \omega A \cos(\omega t + \phi).$$ (11.16)

Neglecting its gravitational potential energy, the mechanical energy E of the mass can be written by taking its elastic potential energy $U = 0$ when $y = y_0$, such that

$$E = U + K = \frac{1}{2}ky^2(t) + \frac{1}{2}mv_y^2(t) =$$
$$\frac{k}{2}A^2 \sin^2(\omega t + \phi) + \frac{m\omega^2 A^2}{2}\cos^2(\omega t + \phi).$$ (11.17)

Substituting $\omega^2 = \frac{k}{m}$,

$$E = \frac{k}{2}A^2.$$ (11.18)

Even if the acceleration is not well described by a harmonic function, it is periodic. The Fourier Theorem guarantees that its shape can be represented by a series of harmonic functions with appropriate frequencies. Despite its anharmonicity, a rubber band behaves like a set of springs with different elastic constants acting together.

The decrease in amplitude is due to friction. The latter subtracts mechanical energy from the system.

If the energy of the system is conserved, $A = \sqrt{\frac{2E}{k}}$ is constant, as expected. If not, the mechanical energy is reduced by the work ΔL done by dissipative forces and

$$\Delta E = \Delta L . \tag{11.19}$$

These forces can only subtract a fraction of the system's energy at any time, i.e., $\Delta E = \alpha E$. Moreover, the longer the time it takes them to act, the larger ΔE, thus

$$\Delta E = -\alpha E \Delta t , \tag{11.20}$$

where the minus sign tells us that E decreases with time. As a result, E decreases with time as an exponential with a characteristic time $\tau = \frac{1}{\alpha}$, thus making the amplitude, i.e.,

$$A = A(t) = \sqrt{\frac{2E(t)}{k}} = \sqrt{\frac{2E(0)}{k}} \exp\left(-\frac{t}{2\tau}\right). \tag{11.21}$$

Taking $t = 0$ when the amplitude is at its maximum $y(0) = A_0$, $E(0) = \frac{1}{2}kA_0^2$, $\phi = \frac{\pi}{2}$ and, including non-conservative forces, the theoretical prediction about the amplitude of the oscillations as a function of time is modified into

$$y(t) = y_0 + A_0 \exp\left(-\frac{t}{2\tau}\right) \cos(\omega t) . \tag{11.22}$$

The net result is that the acceleration oscillates sinusoidally with an exponentially decreasing amplitude. Velocity and position follow this behaviour.

The predicted velocity is then

$$v_y(t) = \frac{dy}{dt} = -\frac{A}{2\tau}e^{-\frac{t}{2\tau}} \cos \omega t - A\omega e^{-\frac{t}{2\tau}} \sin \omega t , \tag{11.23}$$

while the acceleration is

$$a_y(t) = \frac{d^2y}{dt^2} = Ae^{-\frac{t}{2\tau}}\left(\frac{1}{4\tau^2} \cos \omega t - \omega^2 \cos \omega t + \frac{\omega}{\tau} \sin \omega t\right) , \tag{11.24}$$

which can be written as

$$a_y(t) = Ae^{-\frac{t}{2\tau}} (C \cos \omega t + S \sin \omega t) \tag{11.25}$$

with

$$C = \frac{1}{4\tau^2} - \omega^2 \qquad (11.26)$$

and

$$S = \frac{\omega}{\tau}. \qquad (11.27)$$

The latter is an oscillating function whose period is $T = \frac{2\pi}{\omega}$ and whose amplitude decreases exponentially, as observed in the experimental data.

The reason why the oscillations do not appear to be harmonic is that the conditions under which we derived the solutions are not perfectly matched in the real setup. We assumed the validity of the Hooke Law; however, as seen, it is a first order approximation of the expression of the elastic force, which can be much more complex. Hooke Law is valid only for small elongation and, at the beginning of the experiment, that condition may not be valid. In fact, as the amplitude diminishes, the shape of the curve is increasingly approximated by a harmonic motion.

Using a spring, rather than a rubber band, causes the experimental data to align with our expectations to a far greater degree, because, with a spring, the effective elongation is much less than in a rubber band. Imagine the turn of a spring as being composed of flexible curved sections, like those shown on the left in Fig. 11.2

Springs behave much as predicted by Hooke Law because their length practically changes infinitesimally: they become longer or shorter because the turns rotate around the wire axis.

▢ Fig. 11.2 Stretching a spring does not change its length. In fact, it causes the turns of the spring to rotate slightly

Stretching the spring does not, in fact, cause the length of the spring to change. It causes its turns to rotate slightly such that a tiny increase in the angle between turns causes a relatively large change in its effective length. As a result, the conditions for the validity of the Hooke Law are well satisfied with springs.

Observing the data closely, we can see that minima and maxima show a sort of cusp. This happens because, when the length of a rubber band becomes too long (or too short), it becomes more rigid and does not behave like a harmonic oscillator at all.

Despite the fact that the conditions in which the experiment has been conducted do not match the mathematical model well, it is instructive to analyse systems like these. They provide many useful insights about the correct interpretation of analytical models, as we have already seen above. Performing the experiment with a spring that has been carefully selected to behave as expected, in fact, is not that interesting.

11.3 Obtaining Parameters from Data

In order to test our model against the data we can do a *best fit* of the model against the data using the technique introduced in Chap. 10.

Even if linearisation is not possible (or difficult), with the help of a computer, it is possible to find the minimum of a χ^2 with almost no effort.

In this case, it is not possible to linearise the data and we must rely on the ability of computers to perform a lot of elementary operations in a small amount of time, so as to find the parameters that minimise the χ^2 defined as

$$\chi^2 = \sum_{i=1}^{N} \left(\frac{a_i - a_y(t_i)}{\sigma_i} \right)^2, \tag{11.28}$$

where the sum is extended to the N points collected during the experiment and a_i is the experimental values of the acceleration of the mass; $a_y(t_i)$ is the values computed from Eq. (11.25), giving the acceleration as a function of the time, for $t = t_i$, the times at which data were collected. σ_i is the uncertainties about a_i. In order to find the parameters (A, ω and τ) that best describe the data, we need to minimise the value of χ^2.

The fit procedure returns the fit parameters, together with the covariance matrix: a 3×3 symmetric matrix, whose diagonal elements represent the variances on the free parameters of the fit.

Indeed, the latter can be taken as such only if the parameters are uncorrelated and the χ^2 is properly weighted. This is probably not the case. In fact, obtaining a reasonable estimation of σ_i is difficult in this case. Moreover, even if it were possible to evaluate the statistical error on each individual point, the systematic one is, in any case, going to dominate the uncertainty. As already observed, initially, the motion is far from harmonic, and even if the fit converged, the reliability of the parameters has to be carefully evaluated.

The uncertainties evaluated by taking the square root of the diagonal elements of the covariance matrix are thus not very meaningful from a statistical point of view. In fact, they are evaluated under the assumption that the uncertainties about the data have been correctly estimated. Nevertheless, we can extract meaningful information from the procedure.

Comparing uncertainties estimated with different methods provides a mean for evaluating systematic uncertainties due to the adopted procedure. Moreover, comparing the results of different fits provides a mean for evaluating their relative goodness, even if an absolute figure of merit regarding their quality cannot be established. We can then perform an iterative procedure in which we start by fitting the whole range of data, then we progressively reduce the number of periods included in the fit. At each iteration, we can fit only the data collected after each maximum into the plot. At each step, the conditions of the experiment increasingly match those in the model and the fit is expected to become better and better.

We don't need the granularity of the original data to perform a good fit. In Sect. 7.7, we show that averaging is good for reducing fluctuations in data. If σ is the uncertainty about a single point, the uncertainty about the average of N points is $\frac{\sigma}{\sqrt{N}}$. From Fig. 11.1, we can estimate a period on the order of 1 s. Reducing the density of points such that we keep about ten points per period is then possible, keeping the main features of the model clearly visible. To reduce data by a factor of n, each data point can then be defined as $\left(t'_k, a'_k\right)$, where t'_k is the average of the time of a few adjacent points as

Especially in the initial experiments on a given topic, systematic uncertainties often dominate and the corresponding uncertainty about the results are affected. Physicists refine their experiments over time, trying to eliminate any source of systematic error, either by refining the mathematical model of the system or improving the correspondence between the model and the experimental apparatus.

Data reduction is a way to improve statistics. It does not really reduce the information. Instead, it aggregates data such that their statistical fluctuations are reduced by averaging them.

$$t'_k = \frac{1}{n} \sum_{i=(k-1)n+1}^{kn} t_i .$$ (11.29)

Similarly

$$a'_k = \frac{1}{n} \sum_{i=(k-1)n+1}^{kn} a_i .$$ (11.30)

For example, for $n = 5$, t'_1 is the average of t_1, t_2, t_3, t_4 and t_5. Similarly, a'_2 is the average of a_6, a_7, a_8, a_9 and a_{10}. This way, we can attribute an uncertainty to each point of the fit as the standard deviation of the group of n measurements.

Having observed that the system behaves like an ideal spring only for relatively large times, we perform several fits considering only the data collected at $t > t_{min}$.

Fitting our data, we find, for ω, the following values, depending on the starting point of the fit:

run no.	t_{min} [s]	ω [s^{-1}]	χ^2/N_{dof}
0	0.44	7.47 ± 0.04	12.3
1	1.35	7.42 ± 0.07	13.2
2	2.16	7.39 ± 0.10	13.2
3	2.96	7.29 ± 0.12	7.5
4	3.77	7.11 ± 0.10	1.9

The corresponding graphs are shown in Fig. 11.3 for the first iterations. According to our rule of thumb, only the last row in the table provides a *good* fit, and we should take this value as $\omega = 7.11 \pm 0.10$ Hz. It is worth noting that the various values of τ are not compatible with each other and become systematically lower. The reduced χ^2, too, becomes lower and lower, as a consequence of the fact that the model tends to represent the data better and better, as long as the conditions under which it was assumed are matched in the real system.

With time, the system matches the model better and better, because the elongation of the rubber band becomes shorter and shorter. Discarding the first period improves the precision of the data.

Figure 11.3 shows clearly how bad the fit is when starting from the very beginning of data acquisition. Ignoring the initial oscillations the fit becomes better and better. The result of the last iteration is shown below.

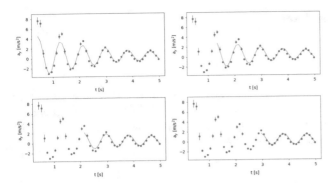

Fig. 11.3 Iterative fits made on data. The result of the fit is shown as an orange curve

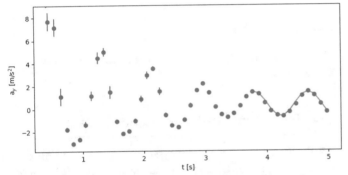

We can associate a systematic uncertainty with ω, observing that, even when the model is not so good, the value of ω is not too far from the one obtained in ideal conditions. The evaluation of a systematic uncertainty is somewhat more arbitrary than the statistical one and depends on how confident we are in the apparatus, too. A conservative estimation could be given by the difference between the biggest and smallest values of ω obtained under different conditions, i.e.,

$$\sigma_{sys} \simeq \frac{\omega_{max} - \omega_{min}}{2} = \frac{7.47 - 7.11}{2} = 0.18\,\text{Hz}, \qquad (11.31)$$

When a systematic error is quoted, the first number represents the statistical error, the second the systematic one. The latter has no statistical content, even if it is sometimes added in quadrature with the statistical uncertainty for convenience.

and we can write

$$\omega = 7.11 \pm 0.10 \pm 0.18 \, \text{Hz}. \tag{11.32}$$

In this case, we keep two significant figures, so as to mitigate the effects of rounding when determining further data. Indeed, knowing the relationship between ω, the period T and the elastic constant k, we can derive

$$T = \frac{2\pi}{\omega} \simeq 0.884 \, \text{s} \tag{11.33}$$

and, for $m = 930$ g,

$$k = m\omega^2 \simeq 47 \, \text{Nm}^{-1}. \tag{11.34}$$

The evaluation of the corresponding uncertainties is left as an exercise.

11.4 Extracting and Manipulating Data

As discussed in the previous section, when looking for the parameters of a function that best fit the data there is no need for the latter to be much denser than necessary. Data should be enough to represent the main features of the model and, for sure, more than m, the number of free parameters in the fit.

Data look like the following.

```
"Time (s)","Acceleration x (m/s^2)","Acceleration y (m/s^2)","Acceleration z (m/s^2)"
0.000000000E0,3.376618028E-1,-1.893230915E0,9.337148070E-1
2.011108400E-2,2.025120258E-1,-1.701057911E0,7.484606504E-1
4.022216800E-2,8.397972584E-2,-1.461338520E0,5.465144515E-1
6.033325200E-2,-2.084195614E-2,-1.195132732E0,3.279784322E-1
8.047485300E-2,-8.745026588E-2,-9.496812820E-1,1.668487489E-1
1.005859370E-1,-6.927591562E-2,-7.615351677E-1,9.249490499E-2
```

From the first column, we get that data are collected every 20 ms. Averaging every 5 points will make the shape smoother without affecting the capability of catching the features of the model: the acceleration oscillates and its amplitude decreases.

Using `pandas`, we extract the data from the CSV file as follows:

```
f = pd.read_csv('rawdata-930g.csv',
                usecols=['Time (s)',
                'Acceleration y (m/s^2)'])
print(f.columns)
t = f['Time (s)']
a = f['Acceleration y (m/s^2)']
```

The usecols optional parameter in read_csv() allows us to extract only those columns from the file in which we are interested. Note that the name we gave to the file contains some relevant information, such as the mass of the weight. This way, we do not need to look at our logbook each time we analyse these data. We can easily spot the most important information from the file itself. The names of the columns to be extracted must match those given in the file and included in a comma-separated list defined by the pair of square brackets [...]. The following line simply prints the list of column names on the screen, as a cross-check.

There is often no need to load the whole set of data into the memory. We can select interesting columns using the usecols parameter in read_csv().

Time and corresponding acceleration can then be defined as Series simply by indicating the corresponding column name in the DataFrame. A Series is almost like a list, in pandas. Series can be thought of as lists of key-value pairs. By default, the keys of a Series returned by read_csv() are an integer from 0 to $N-1$, N being the number of rows in the DataFrame. As such, in most cases, Series and lists can be confused. A plot like the one shown in Fig. 11.1 is done with

```
plt.plot(t, a, 'o')
plt.xlabel('t [s]')
plt.ylabel('a [m/s$^2$]')
plt.show()
```

In order to average data in groups of 5, we can exploit the groupby(), mean() and std() methods as follows:

The groupby() method applied to a list returns a list of lists grouped such that each sublist shares the same value passed as an argument.

```
N = 5
t = t.groupby(t.index // N).mean()
e = a.groupby(a.index // N).std() / np.sqrt(N)
a = a.groupby(a.index // N).mean()
```

The // operator is the floor division operator. It returns the integer part of the division between the operands. t.index is a list containing the row numbers of a Series, and hence can be thought of as a list of integers from 0 to $N-1$, N being the number of rows in the Series. A list divided by a scalar is a list, too, and t.index // N is a list containing the floor division of each row number by N. The first five elements of the resulting list will thus

The integer part of the division between two integer numbers is given by the floor division operator //.

be equal to zero, the following group of five elements are equal to one, and so on.

The `groupby()` method returns a list of lists. Each element of the resulting list is, in turn, a list of five numbers extracted from the original list based on the fact that they share the same value of `t.index // N`. If the original list `t` contained 250 rows, the resulting one contains 50 elements, each of which contains five consecutive data points. Applying the `mean()` method to them results in a list containing 50 numbers, each of which is the average of each group of five. We can bring the result back to the original list, such that `t` and `a` are a factor five shorter and contain the data collected during the experiment averaged every 0.1 s. The `e` list contains the standard deviations of each group of five accelerations.

`find_peaks()`, provided by `scipy.signal`, is a method for identifying peaks in a distribution and returning their indexes.

As described in the previous section, the fit is performed iteratively on a subset of the available data. Let's then start by dividing the time interval into chunks, starting from the moment at which the weight reaches its maximum acceleration per oscillation. In other words, we need to identify the positions of the maxima of Fig. 11.1. This is as simple as

```
p, pdict  = find_peaks(a)
print('==== peaks found at ====')
print(t[p])
```

`find_peaks()` is a function defined in the `scipy.signal` module, included by using

```
from scipy.signal import find_peaks
```

which looks for local maxima in the list passed as an argument (the accelerations) and returns an array containing their index and a dictionary: a data structure composed of key-value pairs. Given that `p` contains the indexes of maxima, the times at which they occur can be selected as `t[p]`, which returns the list of corresponding times. The result is something like

When the index of a list `lst` is a list itself, `indx`, the result `lst[indx]` is still a list containing the elements of `lst` whose indices are listed in `indx`.

```
==== peaks found at ====
4       0.443048
13      1.349774
21      2.155756
29      2.961743
37      3.767725
46      4.674457
```

Compare the list with the position of the maxima in Fig. 11.1. For each interval starting at the time given in the list, we fit the data using

```
for k in range(len(p) - 1):
    res, cov = curve_fit(spring, t[p[k]:],
                         a[p[k]:], maxfev=100000)
```

`curve_fit()` can be used if imported from the `scipy.optimize` module. The first argument (`spring`) is the name of a user-defined function representing the model against which we are fitting the data. The data follow as a pair of arrays of the same size, the first containing the values of the independent variable (t_i) and the second those of the dependent one (a_i). The optional argument `maxfev=100000` sets the maximum number of calls for the function, used either to limit or to extend the number of iterations performed by the minimisation algorithm to find the minimum.

In the above example, the first fit is performed within the interval starting at $t = 0.443048$ s up to the last data point. Then, we repeat the fit, considering only the data with $t \geqslant 1.349774$ s, and so on. In total, we are going to make five fits, indexed from 0 to 4 (the last maximum is not used: we cannot fit our model using only the data after it). The model function is defined as follows:

```
def spring(x, b, A, tau, omega, t0):
    C = 1/(4*tau**2)-omega**2
    S = omega/tau
    t = x-t0
    y = b + A*np.exp(-t/(2*tau))*
        (C*np.cos(omega*t) + S*np.sin(omega*t))
    return y
```

The function takes the independent variable as the first argument and the free parameters as separate remaining arguments. In this function, we define `C` and `S` as the corresponding symbols in Eq. (11.25). Since, in this equation, $t = 0$ at the beginning of the motion, we need to translate the time axis by an offset `t0`. We consider the possibility of a constant bias `b` to the measured acceleration, due to the limited accuracy of the smartphone's accelerometer. Trigonometric functions `sin` and `cos` are defined in the numpy module aliased to np.

The χ^2 of the fit is computed *manually* as the result of the function

```
def computeChi2(y, e, f, dof):
    chi2 = 0
    i = 0
    y = y.to_list()
    f = f.to_list()
```

The function to be used in a fit must have the independent variable as the first argument. Other arguments represent the free parameters and their values are returned by `curve_fit()` in the same order.

```
e = e.to_list()
for i in range(len(y)):
    chi2 += ((y[i]-f[i])/e[i])**2
return chi2
```

to which we pass, at each iteration, a[p[k]:] and
e[p[k]:]. f is computed as

```
f = spring(t[p[k]:], res[0], res[1], res[2],
           res[3], res[4])
```

As is often the case throughout this textbook, there are probably better and more concise ways, in Python, to perform the same operations described here. We use these examples mainly to show features that otherwise may give rise to difficulties.

where res is the list of the free parameters of the fit returned by curve_fit(). It is interesting to note that, in order to loop on data so as to compute the χ^2, we need to transform Series into lists by means of the to_list() method. In fact, this is a case in which a Series cannot be considered as equivalent to a list. When extracting the subset of the data from the original Series, as in a[p[k]:] (which selects only the components whose key follows p[k]), the resulting one keeps the same list of keys. Moreover, t[p[k]:] being a series, f is a series too. Consider, for example, the Series resulting from a[p[0]:]. The first element of p is 4, thus a[p[0]:] is equivalent to a[4:] and only the data whose key is greater than or equal to 4 is kept in the resulting Series. The first key of the resulting Series is then 4 and, when looping on i from 0 to the length of y, the first four elements (whose keys are expected to be from 0 to 3) are not found in the Series, resulting in a runtime error. Converting the Series into a list, the indexing is based on the position of the value, rather than its key.

The errorbar() method of matplotlib. pyplot shows the error bars together with the data points.

To plot the error bars on the data, together with the result of the fit, as in Fig. 11.3, we use the following code:

```
plt.errorbar(t[p[0]:], a[p[0]:], e[p[0]:],
             fmt = 'o')
plt.plot(t[p[k]:], spring(t[p[k]:], res[0],
         res[1], res[2], res[3], res[4]), '-')
plt.xlabel('t [s]')
plt.ylabel('a$_y$ [m/s$^2$]')
plt.show()
```

The errorbar() function plots the values in the list passed as its second argument as a function of those in the first, taking the values of the third argument as the corresponding uncertainties. The format of the plot (circles) is chosen using the optional parameter fmt = 'o'.

11.5 Optimisation Methods

Computer algorithms minimise the χ^2, considering it a surface in a $(m + 1)$-dimensional space, m being the number of the free parameters in the fit, namely, A, ω and τ in our example. This way, χ^2 is a function of $\mathbf{x} = (x_1, x_2, \ldots, x_m)$: $\chi^2(\mathbf{x})$. A common minimisation scheme is the so-called "gradient descent method". It consists in evaluating $\chi^2(\mathbf{x_0})$ at a random point $\mathbf{x_0} = (A_0, \omega_0, \tau_0)$ in that space, comparing it with the value of the χ^2 evaluated in $\mathbf{x_1} = \mathbf{x_0} - \delta\nabla\chi^2(\mathbf{x_0})$ and iterating the procedure such that

Optimisation algorithms consist of techniques for minimising (or maximising) a *cost* function. They are divided into two categories: deterministic, exploiting the mathematical properties of the cost function, and stochastic, looking for minima by randomly sampling the space of the parameters, with some deterministic guidance.

$$\mathbf{x_{n+1}} = \mathbf{x_n} - \delta\nabla\chi^2(\mathbf{x_n}) \tag{11.35}$$

until $f(\mathbf{x_{n+1}}) < f(\mathbf{x_n})$. δ is the amplitude of each step that can be adjusted during the minimisation process. We can easily visualise the method in three dimensions, representing the surface $z = f(x, y)$ as a set of contour levels as in the figure below.

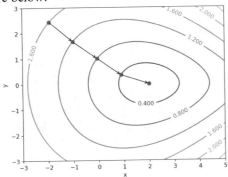

Suppose that we start from the top left point at coordinates (x_0, y_0). We compute the gradient of the function at the red point

The gradient of a function is null at its minima.

$$\nabla f(x, y) = \left(\frac{\partial f}{\partial x}, \frac{\partial f}{\partial y}\right)\Bigg|_{x=x_0, y=y_0} \tag{11.36}$$

and find the new point according to the rule above. Moving in the direction of the gradient consists in moving along the arrow shown in the figure to the next red point. Here, we compute the new values for the gradient, lower it along the next arrow and reach the third red point, and so on, until we reach the minimum value for which the gradient is null.

The gradient descend method is also called the hill-climbing method, especially when the problem is reversed, i.e., we look for the maximum rather than the minimum.

Many stochastic methods are inspired by physical processes, often in the realm of thermodynamics. They usually look for minima using techniques similar to deterministic methods, but there is a non-null probability of jumping on points randomly selected in the given space. This guarantees that a wider region of parameters is explored by the algorithm, increasing the chance of finding the absolute minimum in less time.

Sometimes, stochastic methods are more efficient, especially if the function to be minimised is not so smooth. Stochastic methods explore the space of the parameters randomly, guided by certain criteria. In "simulated annealing" methods [1], for example, the point at iteration n, $\mathbf{x}_n$ is tested against the one obtained at the previous iteration $n-1$, $\mathbf{x}_{n-1}$. Before accepting it as the new starting point, the algorithm selects another random point $\mathbf{x}_{rnd}$. If $f(\mathbf{x}_{rnd}) < f(\mathbf{x}_n)$, the randomly chosen point is selected as the new starting point instead of $\mathbf{x}_n$. Otherwise, the random point is chosen instead of $\mathbf{x}_n$ with a probability $P \propto \exp\left(-\frac{\Delta f}{kT}\right)$, where k is a constant, $\Delta f = f(\mathbf{x}_{rnd}) - f(\mathbf{x}_n)$ and T is a parameter that decreases with a given scheme as the algorithm proceeds.

The method is inspired by physics, where $\exp\left(-\frac{\Delta E}{k_B T}\right)$ is the Boltzmann distribution that gives the PDF of finding a system in an energy state ΔE at temperature T, k_B being the Boltzmann constant. To understand how the method works, consider a very simple case of a function depending on only one parameter $f(x)$, as shown in Fig 11.4. The walker is represented by the red point and starts on the rightmost part of the figure. Initially, it descends the gradient of the function towards the local minimum. However, there is a certain non-null probability of jumping on the left valley, from which point it can descend to the real minimum. The probability of returning to the local minimum is

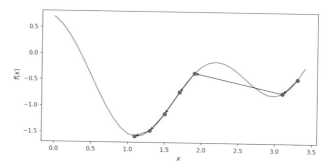

■ **Fig. 11.4** In simulated annealing, there is a non-null probability of jumping from a position x_n to another position x_{n+1} where $f(x_{n+1}) > f(x_n)$. This allows the walker (the red point) to jump from a local to an absolute valley

lower, because, as the walker proceeds, the *temperature T* reduces and the difference Δf increases.

Various methods are implemented in `curve_fit()` and are selectable by the user, the default one being a variant of the gradient descend method.

11.6   A Harmonic Oscillator with Arduino

Rather than being a problem, the opportunity to study a physical system that behaves only partially as a harmonic oscillator is welcome, as it allows us to understand that modelling a real system can be a tough task, yet it is an invaluable tool for understanding why physicists need to learn computing and programming.

Only a few extremely simple systems can be solved analytically in physics. Most of the time, numerical methods are employed. Being able to write a computer program is thus as important as having a solid mathematical background.

Most interesting problems in real life applications cannot be solved analytically: only numerical solutions are possible. In fact, solving physics problems numerically is not necessarily worse than solving them analytically, nor are they less instructive. In many cases, numerical approaches can be even more illustrative than analytical ones, and we encourage you to compare numerical solutions with known analytical solutions.

Nevertheless, the realisation of experiments that are as close as possible to the simplest models is important for the correct identification of the relevant parameters affecting the dynamics. For example, it is difficult to realise the validity of the Newton's second Law without trying to eliminate friction from the experiments.

On the other hand, it is important to realise systems that behave as expected. This ability, too, is one of those that a physicist must develop. In this section, we will thus try to build a system that behaves much like a harmonic oscillator. We already discussed the fact that springs, in fact, do not really stretch when a (moderately strong) force is applied to their ends: their coils just twist and the conditions under which the Hooke Law is valid are fulfilled.

We can thus realise a system by suspending a slinky spring with a screen on its free end aimed at reflecting the ultrasound signal emitted by an ultrasonic sensor connected to Arduino, as shown in the figure above.

The ultrasonic sensor must be oriented towards the screen. We can then measure the distance of the screen as a function of the time after pulling the spring and letting it move under the elastic force. A typical plot of the distance versus time obtained with the Arduino serial plotter is shown below.

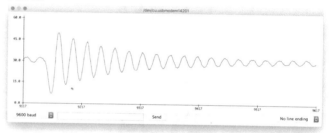

The extinction of the amplitude of the oscillations due to dissipative forces is clearly visible in this case, too. With respect to the rubber band case, the shape of the graph is smoother and more similar to the theoretical predictions. The analysis of the motion of the spring end is almost identical to that made in the case of the rubber band, in which we measured the acceleration with the smartphone. In this case, we measure the position y of the oscillating body and we expect that

$$y = y(t) = y_0 + A_0 \exp\left(-\frac{t}{2\tau}\right)\cos(\omega t).\qquad(11.37)$$

Finding the values of y_0, A_0, τ and ω consists in minimising the χ^2 defined as

$$\chi^2 = \sum_{i=1}^{N} \left(\frac{y_i - y(t)}{\sigma_i} \right)^2, \qquad (11.38)$$

where y_i is the measured distances and σ_i the corresponding uncertainties. According to the manufacturer of the sensor, its resolution is about 3 mm, thus we can assume $\sigma_i \simeq 0.3$ cm $\forall i$.

It is instructive to perform the fit with a different starting point t_{min}, as in the previous section. The table below shows the main results of various fits with various t_{min}.

i	t_{min} [s]	A_0 [cm]	τ [s]	ω [Hz]	χ^2/ν	$P(\chi_\nu^2 > \chi_0^2)$
0	1.2	21.7 ± 0.2	3.54 ± 0.05	4.904 ± 0.002	3.73	0.00
2	2.6	20.0 ± 0.2	3.81 ± 0.05	4.914 ± 0.002	1.65	0.00
3	3.8	19.3 ± 0.3	3.94 ± 0.05	4.920 ± 0.002	1.27	0.01
4	5.1	18.4 ± 0.3	4.10 ± 0.07	4.925 ± 0.002	1.09	0.19
5	6.4	16.7 ± 0.4	4.4 ± 0.1	4.926 ± 0.003	0.88	0.86
6	7.7	14.8 ± 0.5	4.9 ± 0.1	4.931 ± 0.003	0.71	1.00
7	9.0	12.7 ± 0.6	5.5 ± 0.2	4.935 ± 0.004	0.61	1.00
8	10.2	10.2 ± 0.7	6.8 ± 0.5	4.947 ± 0.006	0.50	1.00
10	11.5	9 ± 1	8 ± 1	4.967 ± 0.009	0.46	1.00
12	12.8	7 ± 2	12 ± 5	4.97 ± 0.02	0.52	1.00
13	14.0	8 ± 5	9 ± 7	5.1 ± 0.1	0.54	0.94

A typical fit is shown in the figure below.

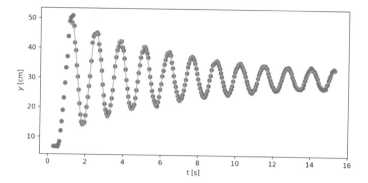

The values of ω are consistent among them, contrary to what happens with rubber bands. The extinction time τ becomes longer with time. This behaviour can be interpreted in two ways: either the fit is not sensitive enough when the starting point moves to the right or, in fact, the dissipative forces are not constant. In this case, we can lean towards the second hypothesis, because we observe a rather clear pattern towards larger values of τ. The drift in the value of A_0 is a consequence of the latter effect.

Another feature that can be observed in the table is that the uncertainty about the free parameters increases with t_{min}. This is reasonable, because, with increasing t_{min}, there are fewer points to exploit to fit the data.

The quality of the fit can be inferred from the reduced χ^2, which is almost always as good as expected. A more quantitative evaluation of the quality of the fit can be done by considering the cumulative distribution function of a χ^2-distributed variable. The fit appears to be quite good only for $i \geqslant 4$. The number in the last column of the table represents the probability that a variable distributed as a χ^2 with ν degrees of freedom will attain the value found in the column on its left simply because of statistical fluctuation: in other words, the p-value of the fit. If this probability is high, it means that we cannot distinguish between a statistical fluctuation and a genuine, systematic deviation from the model, and we can ascribe the value of χ^2_ν to both them. Usually, we prefer to quote the best result obtained using the highest possible number of data that provide a good fit. In this case, for $i = 4$,

> If the p-value is high, we cannot distinguish between a statistical fluctuation and systematic deviations from the model, hence we ascribe the differences to statistics.

$$\omega = 4.925 \pm 0.002 \, \text{Hz} . \tag{11.39}$$

For the mass, we can take that of the system composed of the spring and the cardboard screen, $m = 40 \pm 1$ g, such that

$$k = m\omega^2 = 40 \times 10^{-3} \times 4.925^2 \simeq 0.970 \, \text{Nm}^{-1}. \quad (11.40)$$

The computation of the uncertainty is left as an exercise, too. With Arduino, we can make further observations. We measure $y(t)$, then we can compute $v(t)$ as

$$v(t_i) \simeq \frac{y(t_i) - y(t_{i-1})}{t_i - t_{i-1}}. \quad (11.41)$$

Velocity and acceleration can be computed numerically from position data.

The figure below shows how $v(t)$ and $y(t)$ depend on time. Both are normalised such that they are divided for their maximum $v_M = \max_i v(t_i)$, $y_M = \max_i y(t_i)$.

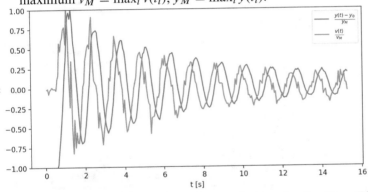

As expected, the graph of $v(t)$ is out of phase by $\frac{\pi}{2}$ with respect to $y(t)$. Similarly, we can obtain a measurement of $a(t_i)$ as

$$a(t_i) \simeq \frac{v(t_i) - v(t_{i-1})}{t_i - t_{i-1}}. \quad (11.42)$$

The plot of the velocity versus time of a harmonic oscillator is out of phase by $\frac{\pi}{2}$ with respect to that of the position. Acceleration is out of phase by π with respect to the latter.

Due to the relatively large statistical fluctuations, the graph of $a(t)$ is much noisier. As usual, averaging over a few points mitigates the effects of fluctuations. The result is shown below.

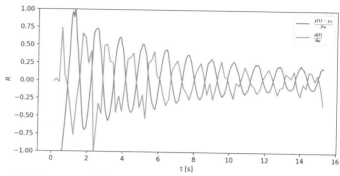

In this case, $a(t)$ is out of phase by π, as expected from the fact that $\frac{d^2 y}{dt} \propto -y(t)$. A plot of the energy versus time can also be made, computing

Energy is easy to measure starting from the velocity and position of the oscillator.

$$E(t) = \frac{1}{2}mv^2 + \frac{1}{2}k\,(y - y_0)^2 \;, \tag{11.43}$$

and is shown below.

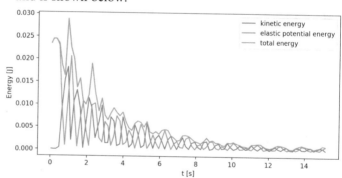

The work done by friction is proportional to time and to the energy of the system, thus the latter decreases exponentially.

Neglecting fluctuations, the total energy appears to decrease exponentially, as expected from the fact that dissipative forces subtract part of it from the system. The total energy is the sum of the kinetic and potential energies that oscillate with opposite phase. If there were no dissipative forces, the total energy was conserved and was equal to $\frac{1}{2}kA(0)^2$. Fitting the total energy with an exponential

$$E(t) = E_0 \exp\left(-\frac{t}{\tau}\right) \tag{11.44}$$

returns $E_0 \simeq 0.026$ J and $\tau \simeq 3.1$ s. The fitting curve is shown below.

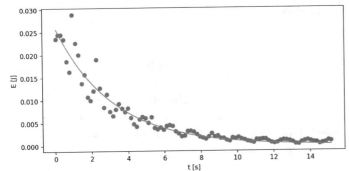

The extinction coefficient τ is consistent with the value that can be predicted from the fit to $y(t)$. In fact, assuming that $\tau = \tau(t)$, we can extrapolate to $t = 0$, assuming that, at first approximation, $\tau \simeq \tau_0 + \eta t$, and find that $\tau_0 \simeq 3.3$ s. From $E(0)$, we can infer $A(0)$ to be

$$A(0) = \sqrt{\frac{2E(0)}{k}} \simeq 23 \, \text{cm} . \tag{11.45}$$

Data for $y(t)$ can be extrapolated to $t = 0$, too. Identifying the position t_i of the peaks in $y(t)$, one can fit $y(t_i)$ versus t_i with an exponential

$$A(t_i) \simeq A(0) \exp\left(-\frac{t}{2\tau}\right), \tag{11.46}$$

obtaining $A(0) \simeq 22$ cm and $\tau \simeq 4$ s, not far from the values guessed from the analysis of energy. In the real system, gravity is not negligible, however, including its effects, the solution to Newton's second Law is

$$y(t) = A(t) \cos(\omega t) + \frac{g}{\omega^2} + y_0 \tag{11.47}$$

and, for $t = 0$,

$$y(0) = A_0 + \frac{g}{\omega^2} + y_0 . \tag{11.48}$$

Note that

$$\frac{g}{\omega^2} = \frac{mg}{k} \simeq \frac{40 \times 10^{-3} \times 9.8}{0.970} \simeq 0.40\,\text{m} \qquad (11.49)$$

is the length of the spring when subject to its own weight, $m = 40 \times 10^{-3}$ kg, which was, in fact, about 40 cm. Correspondingly, neglecting friction, the energy can be written as

$$E(0) = \frac{1}{2}m\omega^2 A_0^2 - mg\left(\frac{g}{2\omega^2} + y_0\right). \qquad (11.50)$$

It only differs from the previous one for a constant term. Being that the energy is defined up to an arbitrary constant, the previous analysis remains valid: it is enough to add the constant

$$C = -mg\left(\frac{g}{2\omega^2} + y_0\right) \qquad (11.51)$$

to the expression of the energy to make it consistent with the one computed above.

In summary, using Arduino, it is possible to perform detailed experiments on the topic of oscillations.

11.7 Newton's Laws

Data collected with Arduino give us the opportunity to test Newton's Laws experimentally.

Using data collected with Arduino, we can easily make an experimental verification of Newton's first and second Laws.

According to the first Law, a particle at rest stays at rest unless some force acts upon it. As a consequence of the arbitrariness in the choice of the reference frame, even a particle moving at constant speed continues moving at the same speed unless a force acts upon it.

The validity of such a statement is clearly shown in the plot with both $y(t)$ and $v(t)$.

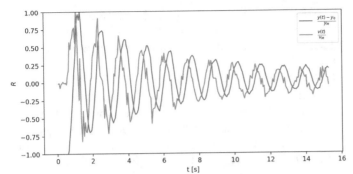

A spring can be viewed as a dynamometer: knowing that $\Delta y = y - y_0$, y_0 is the value of y when the dynamometer is in the equilibrium position, we can obtain the intensity of F as $F = -k\Delta y$.

In this plot, $y(t) = 0$ corresponds to the case in which the spring is in equilibrium. Here, $F = 0$. As soon as $y(t) \neq 0$, the force $F = -ky \neq 0$ and the free end of the spring starts moving. The highest force intensity is reached where $y(t)$ attains its maximum or its minimum. In these positions, the speed is null. In fact, the elastic force progressively reduces the speed of the spring tip until it becomes zero. Due to the elastic force, the speed starts increasing soon after, in the opposite direction. At a certain point, $y(t) = 0$ and $F = 0$. Despite $F = 0$, $v(t)$ reaches its maximum in absolute value and, according to Newton's first Law, the spring tip continues moving such that it compresses or stretches the spring, which reacts by applying a force to it.

Having measured ω from oscillations, we can obtain the values of

A plot of $y(t)$, for a harmonic oscillator, is equivalent to a plot of $F(t)$. When $F(t) = 0$, $v(t)$ attains its maximum and the oscillator keeps moving, as predicted by Newton's first Law.

$$\frac{F}{m} = -\frac{k}{m}\Delta y = -\omega^2 (y - y_0) \tag{11.52}$$

and make a plot of this quantity as a function of a, as below.

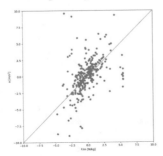

The orange line is not a fit, but rather the prediction made using Newton's second Law, according to which $a = \frac{F}{m}$. The system, then, allows for a direct experimental verification of Newton's Laws.

Indeed, fitting the data with a straight line of the form $a = \alpha \frac{k\Delta y}{m} + \beta$, we obtain $\alpha \simeq 0.67$ and β compatible with zero. The slope of the effective line is lower, indicating that only part of the energy transferred to the system by the elastic force is used to accelerate the mass (a bit less than 70 %). The rest is lost via dissipative forces.

11.8 A Widely Applicable Model

The Hooke Law simply states that a force that depends on a parameter can be written as a truncated Taylor series: when the truncation is done at first order, the solution to Newton's Law is harmonic.

The very simple model studied in this chapter, despite its simplicity, can be applied to a variety of systems. As shown in the chapter opening, the Hooke Law is nothing but Newton's second Law applied to forces that depend on the distance when the latter is small, i.e., when the function describing the dependency of the intensity of the force on distance can be expanded in a Taylor series and the expansion is truncated at the first order.

In general, the same behaviour is exhibited by any force F described by a function of some parameter x, $F(x)$, such that

$$F(x) \simeq F(x_0) + F'(x_0)(x - x_0). \tag{11.53}$$

For example, Newton's Law for a pendulum, neglecting friction, is

$$\mathbf{a} = \frac{m\mathbf{g} + \mathbf{T}}{m}, \tag{11.54}$$

T being the tension of the suspension wire. It must be noted that the mathematical model of a pendulum consists of a pointlike particle of mass m attached to a massless wire of length ℓ. Gravity, then, applies only to the suspended mass and ℓ simply represents the distance between the latter and the points to which is suspended. Equation (11.54) can be cast in three scalar equations, writing the components of the vectors in a reference frame where the x-axis is horizontal and lying in the oscillation plane, the y-axis is vertical and upward, while z is perpendicular to both x and y. In this frame, $\mathbf{g} = (0, -g, 0)$ and $\mathbf{T} = (-T \sin \theta, \ T \cos \theta, \ 0)$, θ being the angle made by the suspending rope and the vertical direction. Equation (11.54) reads as

Pendulums are often imagined as the familiar model: a ball attached to a string. In fact, a pendulum is really any system subject to both gravity and a source of tension, in which the suspension device has a negligible mass with respect to the rest of it.

$$
\begin{cases}
a_x = -\dfrac{T}{m} \sin \theta \\[2mm]
a_y = -g + \dfrac{T}{m} \cos \theta \\[2mm]
a_z = 0 .
\end{cases}
\tag{11.55}
$$

If θ is small, $\sin \theta \simeq \theta$, $x \simeq \ell \theta$ and $\cos \theta \simeq 1$, such that

$$
\begin{cases}
a_x = -\dfrac{T}{m} \dfrac{x}{\ell} \\[2mm]
a_y = -g + \dfrac{T}{m} .
\end{cases}
\tag{11.56}
$$

For $\theta \simeq 0$, we can write $\mathbf{a} = a(\cos \theta, \sin \theta, 0) \simeq (a, 0, 0)$, from which we obtain

$$
\frac{T}{m} \simeq g
\tag{11.57}
$$

and

$$
a_x \simeq -g \frac{x}{\ell} .
\tag{11.58}
$$

It is straightforward to show that this equation must have exactly the same solution for that obtained in the case of the Hooke Law, thus the pendulum oscillates with a frequency

$$
\omega = \sqrt{\frac{g}{\ell}}
\tag{11.59}
$$

corresponding to a period

$$T = \frac{2\pi}{\omega} = 2\pi \sqrt{\frac{\ell}{g}} \, . \tag{11.60}$$

An experiment similar to those performed above can be conducted by suspending a mass M from a "wire" of mass $m \ll M$. To avoid rotations, the weight can be suspended using at least two wires. In this case, ℓ is the distance between the suspension axis and the center of mass of the weight. Note that one can always write $\ell = \ell_0 + d$, where ℓ_0 is the distance between the axis from which M is suspended and its upper edge, while d is the distance between the latter and its barycentre. Then,

$$\frac{T^2}{4\pi^2} = \frac{1}{g} (\ell_0 + d) \, . \tag{11.61}$$

Taking the measurement of T for different ℓ_0 allows us to compute both g and d, respectively, from the slope and the intercept of T^2 versus ℓ_0.

The precision that can be achieved with Arduino is sufficient to appreciate the deviations from the predicted approximated solution. The solution found for Newton's equation is such that the period of the pendulum is independent of the angle. This is known as the pendulum's isochronism and is the basis for the construction of mechanical clocks. However, this is only a first order approximation solution, valid until $\sin \theta \simeq \theta$. It can be shown that, for larger angles,

Pendulums are isochronous only in the first approximation. Their period depends, even if only slightly, on the amplitude of oscillations.

$$T \simeq 2\pi \sqrt{\frac{\ell}{g}} \left(1 + \frac{\theta_0^2}{16} \right) \, . \tag{11.62}$$

This relation, too, is not exact, though it is valid for a wider interval of angles (it is obtained by expanding $\sin \theta$ to the next order). In this formula, θ_0 is the starting angle that is supposed to remain constant because of energy conservation. In practice, it will not be such and must be determined by extrapolating the amplitude to $t = 0$.

Another model for which the Hooke Law turns out to be useful is a model for a solid or a liquid, described as an array of pointlike particles (the atoms) bound together by springs, as shown below for a 2D arrangement.

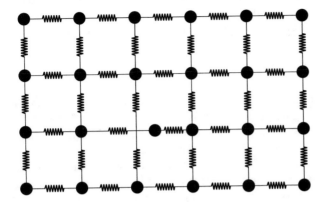

The idea behind this model is that the forces that keep the atoms bound together in a solid clearly depend on the distance between the atoms. In fact, keeping two solids far apart, no interactions are observed. On the other hand, if we press them one against the other with sufficient force, we can cause them to merge. As long as there is no way to measure the force between the atoms, the only conclusion at which we can arrive is that $F = F(\Delta x)$, Δx being the interatomic distance. If Δx is small,

A solid is composed of atoms held together by forces that depend on the interatomic distances. At the first order, such forces behave like elastic ones, and a toy model of a solid consists of point-like sites connected by springs. Despite the approximation being rather crude, it describes many observed phenomena, such as the increase in size when objects are heated.

$$F \simeq F(\Delta x_0) + F'(\Delta x_0)(x - x_0) + \frac{F''(\Delta x_0)}{2}(x - x_0)^2 + \cdots$$
$$(11.63)$$

Truncating the series at the first order, whatever the force between atoms looks like, it exhibits a behaviour close to that of a spring. This explains why ordinary materials increase in size when heated. Heating a body is equivalent to transferring energy to it. The increase in energy to the springs modelling interatomic forces causes them to stretch and, as a result, we observe a macroscopic increase in the size.

There are a few materials (e.g., water) that do not *obey this law*, meaning that the above approximation is not appropriate for them.

Summary

Objects exhibiting the property of elasticity can be treated, at least at first approximation, like a spring.
The elastic force is described by the Hooke Law.

$$F = -k\Delta x$$

In fact, the Hooke Law describes any force $F(x)$ that depends on a parameter x, when x is small enough to allow $F(x)$ to be written as a truncated Taylor series. It is strictly valid only for small Δx.
For example, in solids, atoms are held together by a force that manifestly depends on the interatomic distance Δx. Irrespective of how complicated it is as a function of Δx, the latter being small, the force can be written as a truncated Taylor series and behaves like an elastic force.
Rubber bands, for example, do not behave exactly like springs, however, it is interesting to perform experiments aimed at determining their dynamics. Though the motion of an object attached to a rubber band is not harmonic, it is periodic and can be described by a series of harmonic terms.
Contrastingly, springs behave much as predicted by Hooke Law, because their length practically changes infinitesimally: they do not really stretch or compress; they become longer or shorter mostly because the turns rotate around the wire axis.
The solution to Newton's Law for elastic forces is always a combination of trigonometric functions, being, in this case, the second derivative of the position proportional to the position itself.
The decrease in amplitude observed in real oscillating systems is due to friction. The latter subtracts mechanical energy from the system, proportional to time and energy. The amplitude, then, decreases exponentially.
Even if the system under investigation is not well described by the simplest models, the results obtained by conducting experiments on it help physicists in identifying the parameters to be optimised in order to make the system increasingly similar to that represented by the mathematical model.
One way to improve statistics is to reduce data. This does not consist in reducing the amount of information. Instead, it aggregates data such that their statistical fluctuations are reduced by averaging them.

When a systematic error is quoted, the first number represents the statistical error, the second the systematic one. The latter has no statistical content, even if it is sometimes added in quadrature with the statistical uncertainty for convenience.

If the p-value of a statistical test is high, we cannot distinguish between statistical fluctuation and systematic deviations from the model, hence we ascribe the differences to statistics.

From position measurements, velocity and acceleration are computed numerically. Also, energy can be easily computed and studied as a function of time and other variables.

Measuring the position of a mass attached to a spring is equivalent to measuring the force applied by the spring to the mass. This gives us the opportunity to directly measure a force through use of a dynamometer. This way, a study of a harmonic oscillator can be turned into an experiment about Newton's Laws.

Never take a model literally: pendulums are often imagined as the familiar model, i.e., as a ball attached to a string. If you try to realise a pendulum like this, however, you often run into various kinds of trouble. In fact, a pendulum is really any system subject to both gravity and a source of tension, in which the suspension device (the system applying the tension) has a negligible mass with respect to the rest of it.

Pendulums are isochronous only in the first approximation. Their period depends, even if only slightly, on the amplitude of oscillations.

Phyphox

A PHYPHOX timed run automatically starts after the given delay and stops after the specified time has elapsed.

A smartphone's accelerometer can be used to study the dynamics of springs.

Python

Computer programming is important, because, among other reasons, only a few extremely simple systems can be solved analytically in physics. Most of the time, numerical methods are employed.

When linearisation is not possible (or difficult), with the help of a computer, it is possible to find the minimum of an χ^2 with almost no effort. Computers look for minima adopting numerical techniques that can be either deterministic or stochastic. Python's `curve_fit()` adopts, by default, a modified gradient descent method.

Reading a CSV file, there is often no need to load the whole set of data into the memory. We can select interesting columns using the `usecols` parameter in `read_csv()`. The `groupby()` method applied to a list returns a list of lists grouped such that each sublist shares the same value passed as an argument.

The integer part of the division between two integer numbers is given by the floor division operator `//`.

`find_peaks()`, provided by `scipy.signal`, is a method for identifying peaks in a distribution and returning their indexes.

When the index of a list `lst` is a list itself, `indx`, the result `lst[indx]` is still a list containing the elements of `lst` whose indices are listed in `indx`.

The `errorbar()` method of `matplotlib.pyplot` shows the error bars together with data points.

Reference

1. Kirkpatrick S, Gelatt CD Jr, Vecchi MP (1983) Optimization by simulated annealing. Science 220(4598):671–680

1

Maximum Likelihood

Contents

Electronic supplementary material The online version of this chapter (doi:▶ 10.1007/978-3-030-65140-4_12) contains sup-. plementary material, which is available to authorized users

G. Organtini, *Physics Experiments with Arduino and Smartphones*, Undergraduate Texts in Physics, https://doi.org/10.1007/978-3-030-65140-4_12

Experimental data often distribute as Gaussians. The purpose of measuring quantities is to increase our confidence about the knowledge of a particular phenomena. It is important to recognise that we do not always need a measurement to know the value of a quantity. For example, the period T of a pendulum can be easily estimated by eye, and we would never consider the possibility of using an hourglass to obtain it. If we want to increase our knowledge about its value, we must choose an instrument, like a stopwatch, able to provide us with a better resolution, i.e., with less uncertainty, increasing the probability that the measured value T_0 is closer to its *true* value. Thanks to the Bayes Theorem, the probability attached to each estimation, though subjective, can be formally computed, and even if it remains subjective, its value tends to be agreed upon by everyone. Most importantly, asymptotic results will be independent of our prior subjective knowledge. This chapter deals with the review of the meaning of the measurement process, leading to a formal justification of formulas used in the previous chapters.

12.1 Application of the Bayes Theorem to Measurements

The reason why we believe that we are able to measure the *true* value of a quantity resides in Bayes Theorem, according to which the probability of obtaining a quantity T from data is proportional to the probability of obtaining that data when T is its *true* value. In fact, Bayes Theorem defines what we intend when we speak of a *true* value.

According to the Bayes Theorem, we may know something (the period T of a pendulum) with an a priori knowledge to which we assign a subjective probability $P(T)$ that can be updated through a measurement, making it less subjective. The Bayes Theorem states that

$$P(T|\text{data}) \propto P(T)\, P(\text{data}|T).\qquad(12.1)$$

Here, $P(\text{data}|T)$ is the probability of observing the collected data, given T as the *true* value. $P(T|\text{data})$ is the probability that the period is, in fact, T, given the observed data. In this textbook, when the words "true value" are used, the word "*true*" is always in italics, because the existence of a *true* value is a human prejudice: we are not even able to define it for most measurements. Bayes Theorem provides a clear definition of that for which we should be striving: a *true* value is a parameter of the system under investigation that determines the data we can collect during a measurement. The probability of obtaining that value for

the parameter from our data is proportional to the probability that it will exactly determine our set of data.

Measuring the period of a pendulum with a stopwatch
whose resolution is σ results in a distribution of times that
likely resembles a Gaussian,

$$P(x|T) = Ce^{-\frac{1}{2}\left(\frac{x-T}{\sigma}\right)^2}, \tag{12.2}$$

representing the probability of obtaining x in a trial, while
the period is T. The probability that the period is T, given
a single measurement T_1, is then

$$P(T|T_1) \propto P(T)\, Ce^{-\frac{1}{2}\left(\frac{T_1-T}{\sigma}\right)^2}. \tag{12.3}$$

We do not know $P(T)$, but, in fact, we do not need to
know it. Manifestly, the period of a 1 m-long pendulum
cannot be in ms or hours, neither can it be in minutes. It is
enough to look at it to realise that T is on the order of two
seconds. This constitutes our prior knowledge, from which
we can conclude that $P(T)$ is roughly constant within a
range of a few seconds centred on 2 s ($P(T) = \alpha = const$ is a
uniform distribution). On the other hand, there is no reason
to prefer one value to another within the given range, thus

$$P(T|T_1) \propto Ae^{-\frac{1}{2}\left(\frac{T_1-T}{\sigma}\right)^2}, \tag{12.4}$$

with $A = \alpha C$, whose value is fixed by the normalisation
condition. The measurement is then expressed as

$$T = T_1 \pm \sigma. \tag{12.5}$$

It is worth noting that the roles of T_1 and T are now
inverted: T is the unknown period of the pendulum, while T_1
is the known value of its measurement. In other words,
the value assigned to T is the one (T_1) that maximises the
probability that the period is the chosen one given that we
measured it as T_1.

Clearly, with such an estimation, we cannot include any
statistical effect leading to measurement fluctuations and σ

When we take a
measurement of a
quantity, we update
our knowledge about
that system thanks to
Bayes Theorem. The
(known) value of the
measurement of
something is that
which we assign to
the (unknown) *true*
value with a given
probability. We
choose to assign the
measured value to a
parameter such that
we maximise the
probability of
measuring it.

Measurements are repeated to increase our confidence in the results. The probability content of a measurement is always subjective, but adding information makes its estimation more reliable.

could be underestimated. In this sense, the evaluation of T is subjective, though it is the result of a well-defined procedure. To increase our confidence in its value, we need to perform more measurements.

Again using the Bayes Theorem, after taking a second measurement T_2, we can write the probability that the period is T given T_2 as

$$P(T|T_2) \propto P(T) P(T_2|T) . \tag{12.6}$$

Now, our prior $P(T)$ is more confidently known to be

$$P(T) = Ae^{-\frac{1}{2}\left(\frac{T-T_1}{\sigma}\right)^2} \tag{12.7}$$

and

$$P(T|T_2) \propto Ae^{-\frac{1}{2}\left(\frac{T-T_1}{\sigma}\right)^2} A'e^{-\frac{1}{2}\left(\frac{T_2-T}{\sigma}\right)^2} . \tag{12.8}$$

A formal proof is a bit tedious and laborious, but it is not too difficult to convince yourself that the maximum probability is attained when

$$T = \frac{T_1 + T_2}{2} \tag{12.9}$$

(see next section, too) and that the resulting probability density function (PDF) is a Gaussian whose variance is twice the variance of each, supposed to be the same. In practice, the best estimate of T is

$$T = \frac{T_1 + T_2}{2} \pm \frac{\sigma}{\sqrt{2}} . \tag{12.10}$$

Adding a third, a fourth, a fifth, an Nth measurement leads to a sequence that, not surprisingly, provides the average of the measurements as the best estimate of the *true* value of the period, with an uncertainty given by $\frac{\sigma}{\sqrt{N}}$. Interestingly enough, the asymptotic result is independent of the choice of the initial prior. If we started with a different prior with respect to a uniform PDF, we would arrive at the same conclusion.

In summary, the reason why we average data is that the average is the best estimator of the *true* value of a quantity, where the latter is defined as the one having the highest probability of representing it. This principle is called the "maximum likelihood principle". According to it, the sequence of measurements is the most probable one given the *true* value of the quantity and the boundary conditions (instruments used, environmental conditions, technology adopted, etc.).

According to the maximum likelihood principle, the result of a series of measurements is the one with the highest possible probability of happening.

12.2 An Experimental Proof

Instead of manipulating Eq. (12.8) algebraically to prove that the resulting distribution is a Gaussian centred on $T = \frac{T_1+T_2}{2}$, we can set up a numerical *experiment* to observe how the product behaves upon the parameters being changed.

After defining a function gauss(x, mu, sigma) that returns

$$P(x; \mu, \sigma) = \frac{1}{\sqrt{2\pi}\sigma} e^{-\frac{1}{2}\left(\frac{x-\mu}{\sigma}\right)^2}, \qquad (12.11)$$

a plot of the product of two Gaussians with mean $\mu_1 = 1$ and $\mu_2 = 2$, respectively, and with unitary σ, can be produced by defining the function

```
def plot(mu2):
    mu1 = 1
    mu2 += 1
    mu = (mu1+mu2)/2
    x = np.arange(0, 10, .1)
    y = gauss(x, mu1, 1)*gauss(x, mu2, 1)
    label = f'$\\mu_1={mu1};\\,\\mu_2={mu2}$'
    fig.clf()
    plt.plot(x, y, label = label)
    plt.legend()
    arr = dict(facecolor='orange')
    M = np.max(y)*0.2
    plt.annotate('f$\\mu = {mu}$',
                 xy = (mu, M/0.2*0.95),
                 xytext = (mu, M/0.2*0.70),
                 arrowprops = arr, ha = 'center')
    plt.show()
```

and using it as

```
plot(1)
```

f-strings are evaluated at runtime. They are preceded by an f-character. Curly brackets contain the variable part of the string expressed as the name of the variable whose content has to be displayed.

Here, we introduce a new technique for formatting strings, especially useful when the format of the variables to be displayed does not require specifying the length of the field and/or the precision, as in this case. *f*-strings are preceded by the f character and are evaluated at runtime. Curly brackets contain the variable part of the string as the name of the variable whose content must be displayed when printed. In the above example, the string is rendered as

$$\mu_1 = 1 \, , \, \mu_2 = 2$$

Note that double backslashes ($\backslash\backslash$) are interpreted as a single backslash ($\backslash$), the latter being a meta-character used as an escape character. It allows the subsequent character to be interpreted non-literally.

In turn, this string is interpreted by Python's LaTeX engine and rendered as

$$\mu_1 = 1 \, \mu_2 = 2. \tag{12.12}$$

A dictionary is a collection of key value pairs.

The $\backslash$, characters are used in LaTeX to typeset a short space in math mode. The arr dictionary is used to define the properties of the arrows used to indicate the relevant points on the plot (their color), while ha = 'center' tells Python to draw the annotation as centred with respect to the given coordinates.

The plot can be made for various values of μ_2, and to see how $P(x; \mu_1, \sigma)P(x; \mu_2, \sigma)$ evolves with μ_2, one can animate the plot by showing its content for a short time, cleaning the figure and drawing a new plot with different μ_2.

The top level container of all the plots is called a Figure and can be obtained via the gcf() (get current figure) method.

In Python's matplotlib, a Figure is the top level container for all the plot elements. It can be accessed via the gcf() (get current figure) method of a plot object as

```
fig = plt.gcf()
```

We can then benefit from the animation subpackage of matplotlib. It includes a methodFuncAnimation()

that requires at least two parameters: the first is the figure object to be updated at each step of animation; the second is the name of a function that is called before each step. The latter must have at least one integer parameter representing the animation step. The maximum number of steps (frames) can be specified as an optional parameter, as well as the interval between two frames, expressed in ms. For example,

```
import matplotlib.animation as animation
anim = animation.FuncAnimation(fig, plot,
                    frames = 10,
                    interval = 1000)
```

repeatedly calls the function `plot()` defined above by passing, as its arguments, the integer numbers from 0 (included) to 10 (excluded), such that $\mu_2 \in [1, 10]$ (this explains why we used mu2+=1 in the function code). Each step in the animation lasts for 1 s. Including the `fig.clf()` (clear figure) statement in the `plot()` function, the figure content must be cleared before drawing a new plot, such that the user has the impression that its content actually moves within the figure.

A single step of the animation is shown below for $\mu_2 = 6$.

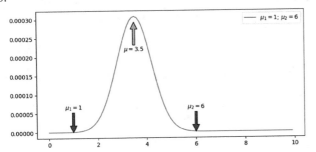

We leave as an exercise the evaluation of the resulting width of the Gaussian in the cases above and a study of what happens when more than two Gaussians are multiplied by each other.

FuncAnimation() repeatedly calls a function with a variable argument. That argument can be used to draw something in a figure. Cleaning the figure with clf() before redrawing it gives the impression that its graphical elements move on the screen.

The product of two Gaussian PDF with respective means μ_1 and μ_2 and with the same variance is a Gaussian with mean $\frac{\mu_1 + \mu_2}{2}$.

12.3 Parameter Estimation

The maximum likelihood principle is equivalent to the least squares method in searching for the parameters that best describe a set of data. The probability $\mathcal{L}$ of observing a set of data is, in fact, the product of the probabilities of obtaining each single value in the given measurement. Maximising it is equivalent to minimising $-2\log\mathcal{L}$, i.e., to minimising the χ^2 if the PDF of each measurement is a Gaussian.

The maximum likelihood principle can be used to estimate unknown parameters from a set of observations x_1, $x_2, \ldots, x_N$. Consider the law relating the length ℓ of a pendulum to its period T,

$$\frac{T^2}{4\pi^2} = \frac{\ell}{g}, \tag{12.13}$$

and suppose we measured T as a function of ℓ for $\ell \in \{\ell_1, \ell_2, \ldots \ell_N\}$. The value of g can be obtained as the inverse of the slope of the line that describes the data. Using the principle of maximum likelihood, we can say that the set of measurements is such that their probability is the highest possible under the given conditions. The PDF of each measurement is a Gaussian centred at $\frac{\ell}{g}$, and is thus

$$P(T_i|\ell_i, g) = A_i e^{-\frac{1}{2\sigma^2}\left(\frac{T_i^2}{4\pi^2} - \frac{\ell_i}{g}\right)^2} \tag{12.14}$$

such that the probability of observing the whole set of data is the product of each single probability:

$$\mathcal{L} = P\left(\{T_i\}|\{\ell_i\}, g\right) = \prod_{i=1}^{N} P(T_i|\ell_i, g). \tag{12.15}$$

More generally, if $y = f(\mathbf{p}, x)$, the likelihood $\mathcal{L}$ is

$$\mathcal{L} = \prod_{i=1}^{N} A_i e^{-\frac{1}{2}\left(\frac{y_i - f(\mathbf{p}, x_i)}{\sigma_i}\right)^2}, \tag{12.16}$$

$\mathbf{p}$ representing the unknown parameters (in this case, $\mathbf{p} = (g)$ is unidimensional). Taking the logarithm of both members, the product of exponentials turns into a sum of their exponents and

$$\log\mathcal{L} = \sum_{i=1}^{N}\log A_i - \frac{1}{2}\sum_{i=1}^{N}\left(\frac{y_i - f(\mathbf{p}, x_i)}{\sigma_i}\right)^2. \tag{12.17}$$

The first term is a constant and is irrelevant in the search for the maximum of $\mathcal{L}$, attained where $-2\log\mathcal{L}$ is the minimum, thus it can be dropped and the maximum of $\mathcal{L}$ is found, where

$$\chi^2 = \sum_{i=1}^{N}\left(\frac{y_i - f(\mathbf{p}, x_i)}{\sigma_i}\right)^2 \tag{12.18}$$

is the minimum. The latter can be found by exploring the space of the parameters upon which $f(\mathbf{p}, x)$ depends. For example, if $y = Ax$, as in the case in which $y = \frac{T^2}{4\pi^2}$ and $x = \ell$, a plot of

$$\chi^2 = \chi^2(A) = \sum_{i=1}^{N}\left(\frac{y_i - Ax_i}{\sigma_i}\right)^2 \tag{12.19}$$

reveals that χ^2 is a parabola as a function of A. Locating its vertex is then equivalent to finding the value of A that maximises the probability of having observed the given sequence, under the assumption that $y = Ax$. Substituting A with $A \pm \sigma_A$,

The χ^2, equal to $-2\log\mathcal{L}$, as a function of the free parameters, is a paraboloid in an $(m+1)$-dimensional space. For $m = 1$, looking for the values of the parameters for which $\chi^2 \rightarrow \chi^2 + 1$ allows us to find the uncertainties about the latter.

$$-2\log\mathcal{L} = \sum_{i=1}^{N}\left(\frac{y_i - (A \pm \sigma_a)x_i}{\sigma_i}\right)^2 = L_0 + \sum_{i=1}^{N}\left(\frac{\sigma_a x_i}{\sigma_i}\right)^2 \pm 2\sum_{i=1}^{N}\frac{y_i - Ax_i}{\sigma_i^2}\sigma_a x_i, \tag{12.20}$$

and, observing that the last term in the sum averages to zero and that $\sigma_A = \frac{\sigma_i}{x_i\sqrt{N}}$, each term contributes $\frac{1}{N}$ to the sum such that

$$-2\log\mathcal{L} = L_0 + 1, \tag{12.21}$$

L_0 being the minimum of $-2\log\mathcal{L}$. Locating the abscissa of the points in which $-2\log\mathcal{L} = L_0 + 1$, we can identify the values of A that differ by σ_A from the minimum, as shown below.

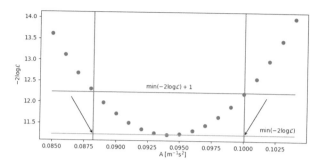

In this figure, the value of $-2 \log \mathcal{L}$ is plotted against A. It follows a parabola whose minimum is reached for $A_{min} \simeq$ 0.094, corresponding to $g \simeq 10.6 \ \mathrm{ms}^{-2}$. The arrows indicate the abscissa of the points defining σ_A. Since the shape of $-2 \log \mathcal{L}$ is a parabola, they are symmetric with respect to A_{min} and are found at a distance of about 0.006 in SI units. Finally, then,

$$A = 0.094 \pm 0.006 \ \mathrm{m}^{-1}\mathrm{s}^2 , \tag{12.22}$$

corresponding to

$$g = 10.6 \pm 0.7 \ \mathrm{ms}^{-2} . \tag{12.23}$$

Sometimes, the profile of the log likelihood is not a parabola. If its shape is not symmetric around the minimum, the uncertainty band may be asymmetric, too.

In fact, when deriving the parameters from the data, they are not independent of each other and the likelihood cannot be expressed as the product of independent probabilities. There are as many relationships among data as there are estimated parameters (m), and the number of independent PDF is $N - m = \nu$. When more parameters are fitted, correlations must be taken into account.

In most practical cases, $-2 \log \mathcal{L}$ can always be well approximated with a parabola in the vicinity of its minimum, and the uncertainty about the parameters can be evaluated looking for points where $\chi^2 \to \chi^2 + \gamma$, where $\gamma = 1, 2.4, 3.5, \ldots$ depends on m. Sometimes, the minimum is not symmetric. In this case, the uncertainties can be different on each side and the result is expressed using asymmetric uncertainties, like, e.g.,

$$g = 11^{+1}_{-2} \ \mathrm{ms}^{-2} . \tag{12.24}$$

In the previous chapters, we exploited the principle of maximum likelihood, defining the χ^2 in an alternative way that, as a matter of fact, is equivalent to the one given above.

The above discussion can be regarded as a quasi-formal justification of the application of the method, within the framework of Bayesian statistics.

Summary

The reason why we believe that we are able to measure the *true* value of a quantity resides in the Bayes Theorem, according to which the probability of obtaining a quantity T from data is proportional to the probability of obtaining that data when T is its *true* value. In fact, the Bayes Theorem defines what we mean when we speak of *true* value.

When we take a measurement of a quantity, we update our knowledge about that system thanks to the Bayes Theorem. The (known) value of the measurement of something is that which we assign to the (unknown) *true* value with a given probability. We choose to assign the measured value to a parameter such that we maximise the probability of measuring it.

Measurements are repeated to increase our confidence in the results. The probability content of a measurement is always subjective, but adding information makes its estimation more reliable.

According to the maximum likelihood principle, the result of a series of measurements is the one with the highest possible probability of happening.

The product of two Gaussian PDF with respective means μ_1 and μ_2 and with the same variance is a Gaussian with mean $\frac{\mu_1+\mu_2}{2}$.

The maximum likelihood principle is equivalent to the least squares method in searching for the parameters that best describe a set of data. The probability $\mathcal{L}$ of observing a set of data is, in fact, the product of the probabilities of obtaining each single value in the given measurement. Maximising it is equivalent to minimising $-2\log\mathcal{L}$, i.e., to minimising the χ^2 if the PDF of each measurement is a Gaussian.

The χ^2, equal to $-2\log\mathcal{L}$, as a function of the free parameters, is a paraboloid in an $(m+1)$-dimensional space. For $m=1$, looking for the values of the parameters for which $\chi^2 \to \chi^2 + \nu$ allows us to find the uncertainties about the latter.

Sometimes, the profile of the log likelihood is not a parabola. If its shape is not symmetric around the minimum, the uncertainty band may be asymmetric, too.

Python

f-strings are evaluated at runtime. They are preceded by an f-character. Curly brackets contain the variable part of the string expressed as the name of the variable whose content has to be displayed.

A dictionary is a collection of key-value pairs.

FuncAnimation() repeatedly calls a function with a variable argument. That argument can be used to draw something in a figure. Cleaning the figure with clf() before redrawing it gives the impression that its graphical elements move on the screen.

2

Physics in Non-inertial Systems

Contents

© The Author(s), under exclusive license to Springer Nature
Switzerland AG 2021
G. Organtini, *Physics Experiments with Arduino and
Smartphones*, Undergraduate Texts in Physics,
https://doi.org/10.1007/978-3-030-65140-4_13

Despite appearances, Newton's dynamics is not simple at all, nor very intuitive. The three fundamental principles, for example, took centuries before they were recognised as true, as our collective experience tells us that bodies don't move unless a force is applied to them. Galileo Galilei first realised that friction is responsible for this, a fact that is not straightforward in the absence of Newton's first Law. Many conceptual difficulties in classical mechanics, such as those connected to Newton's third Law or to the usage of non-inertial reference frames, can be at least partly ascribed to the confusion between the term *force* and the term *interaction*. Forces are the result of interactions, and interactions are only effective among at least two entities. In this respect, what we call *fictitious* or apparent forces can be considered, in fact, to be as real as gravity or the elastic force. On the other hand, we can measure them, thus they exist. The confusion comes from the fact that, in the past, forces and interactions were, in fact, used as if they were synonyms, as any interaction gave rise to a force, represented as a vector in Newtonian mechanics. Today, we know that this is not always the case. The weak interaction, for example, is responsible for radioactive decays. The interaction causes the *change of state* of a system composed of an atom of a given species into another in which the state is composed of an atom of a different species, together with one or more particles (a photon, an α-particle or an electron and a neutrino). There is no place to which we can attach a vector in this case, hence there is no force, in the Newtonian sense. It would be better for us to say that interactions are responsible for the change of state of a system and the effects of interactions often can be, but are not always, represented by forces. In this chapter, we perform experiments in which we explore the physics in non-inertial frames, to dissipate doubts and to allow you to familiarise with them.

13.1 Dynamics in Non-inertial Systems

Any accelerometer will always measure accelerations within its own reference frame. If it is found in a non-inertial reference frame it experiences fictitious forces.

Consider an accelerometer at rest in a gravitational field, such as those found in smartphones or Arduino sensors. As shown in Chap. 6, irrespective of its actual shape and technology, an accelerometer can be thought of as a pair of masses m connected with a spring. Assuming a massless spring (i.e., a spring whose mass m_S is much less than m), each mass is subject to its weight $m\mathbf{g}$, directed downward.

If the accelerometer is at rest on a table, with its spring horizontal, gravity produces a force on both masses directed downward, cancelled by the normal force $\mathbf{R}$ exerted by the support, such that $\mathbf{R} = -m\mathbf{g}$ and, as a result, the acceleration $\mathbf{a}$ of both masses is null. Consequently, the distance between the masses remains constant and the accelerometer output is zero. If, on the other hand, the spring is kept vertical, while, for the mass at the bottom, $\mathbf{a} = 0$ for the same reasons, gravity is canceled on the mass at the top by the elastic force provided by the spring, such that

$$mg = -k\Delta x\,\hat{y}, \qquad (13.1)$$

$\hat{y}$ being a unit vector directed downward. In this case, $\Delta x \neq 0$ and, m and k being known, the system provides a reading of the acceleration of the system $\mathbf{g}$. If the accelerometer were in a non-inertial frame, the acceleration experienced by the system would be

$$\mathbf{a}' = \mathbf{a} - \mathbf{A} - \boldsymbol{\omega} \wedge (\boldsymbol{\omega} \wedge \mathbf{r}) - 2\boldsymbol{\omega} \wedge \mathbf{v} - \frac{d\boldsymbol{\omega}}{dt} \wedge \mathbf{r}, \qquad (13.2)$$

where $\mathbf{A}$ is the linear acceleration of the reference frame (i.e., the acceleration along a straight line), $\boldsymbol{\omega}$ its angular velocity vector, $\mathbf{r}$ the distance of the accelerometer from the rotation axis and $\mathbf{v}$ its velocity in the rotating reference frame. The various terms in this formula can be the subject of as many experiments, illustrated in the next sections.

13.2 Free-Fall

If $\boldsymbol{\omega} \simeq 0$, $\mathbf{a}' = \mathbf{a} - \mathbf{A}$. When a smartphone falls freely, its accelerometer is in an accelerated reference frame with $\mathbf{A} = \mathbf{g}$. Since it moves with acceleration $\mathbf{a} = \mathbf{g}$ in the laboratory frame, its accelerometer measures $\mathbf{a}' = 0$.

There is an experiment demonstrating this that is super easy. It is enough to start the acceleration experiment in PHYPHOX, keeping the phone at a certain height (e.g., above your head) for a few seconds, then allowing it to fall onto a soft surface, such as a bed (be sure not to damage your device: choose a large bed to perform the experiment, such that the smartphone does not fall on the floor after bouncing on the surface of the mattress).

An accelerometer at rest always measures the gravity acceleration, since gravity affects the distance between the proof masses kept separated by an elastic device.

The most general expression of the acceleration of a system includes the linear acceleration of the reference frame, the centripetal acceleration, the Coriolis term and the Euler acceleration.

The acceleration measured in a freely falling reference frame is null. This is the foundation of General Relativity.

During the fall, the absolute value of **a** is expected to be null. In fact, the result of such an experiment is shown below.

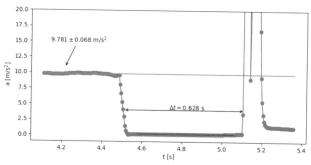

In fact, a cheap accelerometer can be affected by systematic biases, and the measured acceleration can be not null while falling. Sometimes, the acceleration measured during free-fall may not be constant, due to the fact that the device rotates around one or more axes.

At the beginning of the experiment, $a' = 9.78 \pm 0.07$ m/s^2, being that $A = 0$. When the phone falls, it drops to almost zero for $\Delta t \simeq 0.63$ s, corresponding to an height of

$$h = \frac{1}{2}g\Delta t^2 \simeq 1.94\,\text{m} . \tag{13.3}$$

Looking carefully at the data, one can see that $a' \neq 0$, and can even slightly increase with time. The average of the first experimental points after the fall has started is shown in red and can be significantly different from zero, due to mechanical stresses or electronic biases introduced by the manufacturer while mounting the phone that can often depend on temperature, too.

Such an experiment appears to be trivial, but, in fact, it was the experimental foundation for General Relativity. Remarkably enough, at the time of its formulation, the available technology prevented Einstein from actually performing the experiment, and his results are based on what he called a *gedankenexperiment*, a German word for "thought experiment".

It is worth noting that, often, the absolute acceleration of the phone during free-fall is not constant, but increases slightly with time, as shown below, where a zoomed portion of the previous graph is reported.

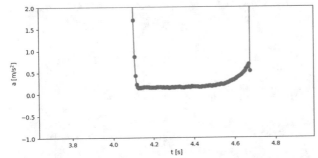

The increase is due to the fact that, while falling, the phone is subject to the drag applied by the air in which the phone is moving. It is worth noting that, in most cases, the drag can be approximated by a force proportional to the speed v of the phone, such that

$$\frac{dv}{dt} = \frac{k}{m}v,\qquad(13.4)$$

leading to a force that grows exponentially with time. A small contribution can also be ascribed to the fact that the phone often rotates while falling, leading to the development of further acceleration within its own reference frame.

13.3   Custom Experiments with PHYPHOX

One very interesting feature of PHYPHOX is its ability to collect data from more than one sensor at the same time. Custom experiments can be defined by clicking on the + on the bottom right angle of the display (see Fig. 13.1).

The simplest method for defining a custom experiment is to click on ADD SIMPLE EXPERIMENT. This opens a dialogue in which the user can select the available sensors to read during the experiment. The resulting experiment has as many tabs as the number of sensors included in it.

In an experiment that includes the accelerometer and the gyroscope, both sensors are read at (almost) the same time. In fact, the readings are made sequentially, but setting the acquisition rate to zero, sensor readings are done at the fastest possible rate, and each set of data can be considered, in many cases, as simultaneous.

PHYPHOX allows for collecting data from several sensors at the same time using custom experiments.

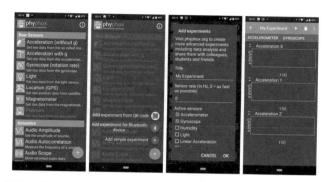

□ Fig. 13.1 Screenshots taken during the definition of a PHYPHOX custom experiment. After choosing ADD SIMPLE EXPERIMENT, the user has the ability to select the desired sensors. The experiment has as many tabs as the number of selected sensors

Once exported, data will be saved in as many files as there are sensors included in the experiment.

13.4 Centripetal and Centrifugal Accelerations

Centrifugal force is as real as other forces, because it can be measured. It is said to be fictitious because there is no source for it for the observer, while the centripetal acceleration (its counterpart in an inertial reference frame) is the result of an interaction.

Centripetal acceleration keeps a system in a rotating trajectory. Remembering that forces cause accelerations according to Newton's second Law, if a body is in an accelerated frame, as in the case of circular motion, it experiences a centrifugal force. The latter is said to be *fictitious* (apparent) or *inertial*, because there is no source for it. While the source of gravity is mass and the elastic force is exerted by a spring, a fictitious force has no source: it appears just because the system experiencing it is in a non-inertial reference frame, such that the subject can measure an acceleration a that can be interpreted as the presence of a force whose intensity is $F = ma$. The centrifugal acceleration corresponds to the term

$$\mathbf{a}_c = -\boldsymbol{\omega} \wedge (\boldsymbol{\omega} \wedge \mathbf{r}). \tag{13.5}$$

For a point-like particle rotating around an axis perpendicular to the xy-plane at distance r from it, $\boldsymbol{\omega} \wedge \mathbf{r}$ is a vector lying on the rotation plane directed as the velocity of the particle, as given by the right-hand rule. The cross-product of $\boldsymbol{\omega}$ and the latter is still a vector lying on the xy-plane, antiparallel to $\mathbf{r}$.

The computation can be done using vector coordinates as well, remembering that the cross-product of two vectors **a** and **b** can be written as the determinant of a 3×3 matrix whose elements are the unit vectors in the first row and the coordinates of the factors in the product in the second and third rows. In our case, $\boldsymbol{\omega} = (0, 0, \omega)$ and $\mathbf{r} = (r_x, r_y, 0)$, thus

To compute a cross-product numerically, it is better to use the determinant method.

$$\boldsymbol{\omega} \wedge \mathbf{r} = \begin{vmatrix} \hat{x} & \hat{y} & \hat{z} \\ 0 & 0 & \omega \\ r_x & r_y & 0 \end{vmatrix} = -\omega r_y \hat{x} + \omega r_x \hat{y} \qquad (13.6)$$

and

$$\mathbf{a}_c = \boldsymbol{\omega} \wedge (\boldsymbol{\omega} \wedge \mathbf{r}) = \begin{vmatrix} \hat{x} & \hat{y} & \hat{z} \\ 0 & 0 & \omega \\ -\omega r_y & \omega r_x & 0 \end{vmatrix} = -\omega^2 r_x \hat{x} - \omega^2 r_y \hat{y}$$

$$(13.7)$$

such that $a_c = -\omega^2 r$. Finding the cross-product this way is particularly useful when doing it with a computer. Applying this latter method, it is easy to show that the cross-product is not commutative, and that $\mathbf{a} \wedge \mathbf{b} = -\mathbf{b} \wedge \mathbf{a}$.

The cross-product does not commute.

An experiment can be set up by positioning an accelerometer and a gyroscope in a rotating frame. A smartphone with both sensors can be placed into a salad spinner, keeping its width aligned with its radius and keeping it in position with adhesive tape or cable ties. Record players or turntables like those used to support TVs and monitors are valid alternatives, as is securing the smartphone to the spokes of a bicycle wheel.

Experiments in rotating frames can be easily performed using an accelerometer and a gyroscope together.

The experiment consists in collecting the acceleration **a** and the angular velocity ω while keeping the device rotating at different speeds. As soon as the experiment starts, wait a few seconds to measure the steady state conditions, then make the device rotate, varying the speed continuously.

During these experiments, it may be necessary to correct data to take into account the non-null readings of an accelerometer at rest.

In the above configuration, the device rotates along its z-axis, such that $a_z \simeq 0$, as well as $\omega_x = \omega_y = 0$. Indeed, the support may not be perfectly horizontal and, in order to obtain the centripetal acceleration, we need to subtract the components of $\mathbf{g}$ from a_x, a_y and a_z. Using data collected during the initial seconds, when the system was at rest, we get g_x, g_y and g_z as the average of the corresponding accelerations, and subtract them from all the points.

A plot of $a = \sqrt{a_x^2 + a_y^2 + a_z^2}$ as a function of $\omega = \sqrt{\omega_x^2 + \omega_y^2 + \omega_z^2}$ is shown below (this plot is made assuming that ω is measured simultaneously with $\mathbf{a}$).

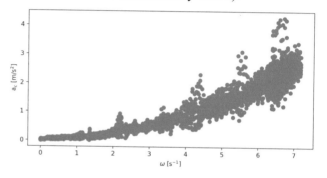

As usual, linearising the relationship between two quantities simplifies the way to find unknown parameters.

A clear parabolic dependence of the type $a \simeq r\omega^2$, as expected, can be spotted from the plot. An unweighted fit to a versus ω^2 with a straight line allows us to measure $r = 0.0526 \pm 0.0002$ m. The fit is shown below as an orange line.

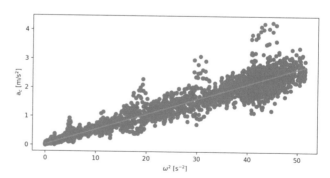

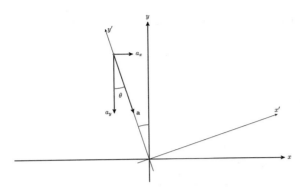

◻ Fig. 13.2 If the sensor, represented by the red dot, is not aligned with the phone's axis, its acceleration makes an angle θ with respect to it

Here, r is the distance of the accelerometer from the axis of rotation. Our result means that the device is placed at one of the points of a circumference of radius r from the rotation axis. In order to identify which of them it is, we observe that the centripetal acceleration makes an angle θ with the phone axis equal to the angle between the latter and the vector position of the accelerometer, such that

$$\theta = \tan^{-1} \frac{a_y}{a_x}, \tag{13.8}$$

as shown in Fig. 13.2.

A histogram of θ is shown below, with a Gaussian superimposed, such that its mean μ is equal to $\langle \theta \rangle$ and its width σ is equal to the standard deviation of the measurements σ_θ.

Measurements allow us to identify the point at which the accelerometer is placed inside the device. The distance from the axis of rotation defines a circumference along which the sensor is placed. To determine at which point of the latter it is, we exploit the acceleration data.

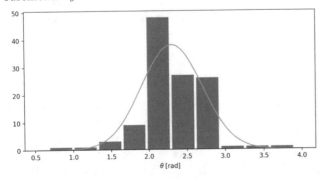

The normalisation of a Gaussian is given by its integral $N \Delta\theta$, divided by $\sqrt{2\pi}\sigma$.

The proper normalisation constant is computed as

$$C = \frac{N \Delta\theta}{\sqrt{2\pi}\sigma_\theta},$$ (13.9)

where N is the total number of events in the histogram and $\Delta\theta$ is the bin width.

13.5 Coriolis Acceleration

Gyroscopes exploit the Coriolis force exerted on a device in a non-inertial reference frame. Proof masses are forced to move such that, when the device rotates, they experience the Coriolis force. In tuning fork gyroscopes, the proof masses are the tines that vibrate at a known frequency, while the force is measured from the distance between the proof masses and fixed plates connected to them by an elastic medium.

Smartphone's accelerometers always measure the acceleration within their own reference frame. The accelerometer is clearly at rest in that frame, and it is thus difficult to measure the term $-2\boldsymbol{\omega} \wedge \mathbf{v}$, unless the phone itself is put in a rotating frame and moves within it.

It is, instead, interesting to see how gyroscopes work, simply by exploiting the Coriolis force. As seen above, inertial forces develop in accelerated frames. Fig. 13.3 shows a microphotograph of a MEMS gyroscope, known as a tuning fork gyroscope, taken using an electron microscope.

Consider a reference frame whose origin is at the center of the device, such that the x-axis is directed to the right, the y-axis points upward and the z-axis exits from the page. The four pitted wings vibrate at a frequency f like the tines of a tuning fork, driven by an external force impressed through the elastic structures attached to the central body. The velocity $\mathbf{v} = \mathbf{v}_0 \cos(2\pi f t)$ at which each wing vibrates is known: when the right wing moves to the right, the left one moves to the left; when the top wing moves upward, the bottom one moves downward, and vice versa.

If the system rotates along the x-axis, the device experiences a Coriolis force,

$$\mathbf{F}_C = -2\boldsymbol{\omega} \wedge \mathbf{v}.$$ (13.10)

Such a force is null on the left and right wings, has opposite directions for the top and bottom wings, is directed as $\hat{z}$ and has an absolute intensity equal to $F_C(t) = 2\omega_x v(t)$. The system, then, is subject to a torque and the top wings are pushed or pulled against the plate beneath it, while the bottom one tends to move in the opposite direction. Both the plate beneath the wings and the latter form a capacitor: when their distance $d \propto 2\omega_x v_0 \cos(2\pi f t)$ varies, its

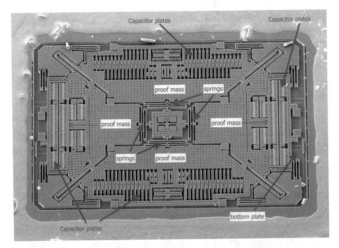

□ Fig. 13.3 Microphotograph of a tuning fork gyroscope. The x-axis is directed to the right, the y-axis upward and the z-axis is perpendicular to the page, exiting from it. Courtesy of Adam McCombs

capacitance does so as well and a current can be measured proportional to d and, in turn, to ω_x.

Similarly, if the device rotates around the y-axis, the force on the top and bottom wings is null, while the one acting on the left and right wings is again directed along $\hat{z}$. ω_y is measured from the current drawn from the capacitors made of the left and right wings, together with the plate beneath them.

Finally, if the system rotates around $\hat{z}$, the four wings experience a force. Assuming the directions of **v** as above, the top wing experiences a force to the right, the bottom one is subject to a force directed to the left, the right wing is pushed downward, and the left one is pushed upward. In this case, the device measures the variation of the capacitance of the system made of the proof masses and fixed plates visible at the four sides of the device. This way, the system measures ω_z.

In order to observe the effects of a Coriolis acceleration, the accelerometer must move (for simplicity, better to let it move along a straight line) with respect to a rotating frame as if it were, e.g., moving across a rotating carousel.

13.6 Euler Acceleration

Accelerometers can be used to experiment with accelerations that are not usually explicitly treated in physics courses, like the Euler acceleration. It is enough to measure accelerations in a frame that rotates at non-constant velocity.

The last term in Eq. (13.2),

$$\mathbf{a}_E - \frac{d\boldsymbol{\omega}}{dt} \wedge \mathbf{r}, \tag{13.11}$$

is called the Euler acceleration, named after Leonhard Euler (1707–1783), and represents the variation of the velocity with respect to time due to a change in the rotational speed of a point that rotates around an axis. This term is quite easy to understand, being analogous to the linear acceleration. Indeed, a particle moving along a circular track with non-constant speed experiences an acceleration along the tangent of its trajectory, as a particle moving along a linear track feels an acceleration directed parallel to its velocity.

In order to spot this term in experimental data, a gyroscope and an accelerometer can be put close to the inner rim of a bucket, suspended by means of two twisted ropes such that it starts to rotate. In an experiment like this, if the alignment is perfect, the angular velocity $\boldsymbol{\omega}$ will be vertical, as $\mathbf{g}$. Its orientation changes from up to down, and vice versa, when the direction of the rotation changes. The centripetal acceleration is expected to be horizontal and perpendicular to the angular velocity, while the Euler acceleration is expected to be tangent to the trajectory of the accelerometer. The relevant directions for our experiment are shown in Fig. 13.4.

Plots of the components of the acceleration, the angular velocity and the acceleration are shown below as a function of time.

3

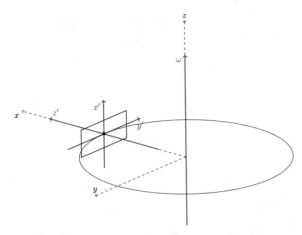

Fig. 13.4 Reference frames used in an experiment in which a device with a gyroscope and an accelerometer rotates around a vertical axis. The primed axis are those of the device, while non-primed ones are those of the inertial laboratory frame

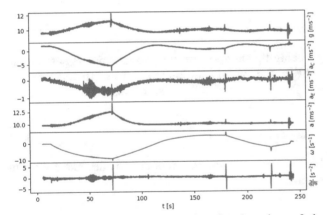

The rotation is almost uniform for the duration of the experiment (the angular acceleration $\frac{d\omega}{dt} \simeq 0$), except for when the rotation changes sign at $t \simeq 70$, 180, 220 s.

Correspondingly, we observe a jump in the acceleration's components. The centripetal acceleration a_C follows the graph of ω, being that $a_C = \omega^2 r$, with opposite sign. A similar, milder, behaviour is observed for the component tangent to the trajectory a_E. The latter is not expected, being that $a_E = \frac{d\omega}{dt} r$, however, the experiment is not perfectly aligned and the accelerometer detects a component of the centripetal acceleration along its y'-axis. The same happens for the vertical component (the one along the z'-axis).

Euler acceleration is easy to understand, however, figuring out its components from plots collected during the experiments is extremely instructive.

A plot of the acceleration along z' as a function of the angular velocity ω is shown below.

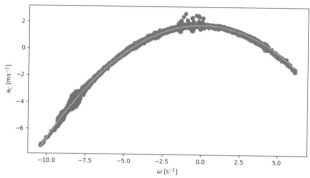

Often, results from experiments are consistent with approximations made during the analysis. Refining the analysis, however, may lead to better results or, at least, to a better understanding of possible systematics.

Data distribute along a parabola, as expected from the fact that $a_C = -\omega^2 r$. The vertex, located where $\omega = 0$, is not at $a_C = 0$, because of the imperfect alignment of the device. In fact,

$$a_{z'} = -\omega^2 r \sin\theta_{\omega,r} + g_{z'} , \qquad (13.12)$$

$g_{z'}$ being the component of $\mathbf{g}$ along z' and $\theta_{\omega,r}$ the angle between the axis of rotation and $\mathbf{r}$. The latter is close to $\frac{\pi}{2}$ and its sine is practically equal to 1, such that

$$a_{z'} \simeq -\omega^2 r + g_{z'} . \qquad (13.13)$$

Fitting data against this model provides the following results:

$$r = 8.468 \pm 0.004\,\text{cm}$$
$$g_x = 1.793 \pm 0.001\,\text{ms}^{-2} . \qquad (13.14)$$

The value of r is consistent with the diameter of the bucket $D = 2r = 16.5 \pm 0.5$ cm, even if $\theta_{\omega,r} \neq \frac{\pi}{2}$. In fact,

$$\theta_{\omega,r} = \frac{\pi}{2} + \alpha \qquad (13.15)$$

and, for small α, we can expand the sine in a Taylor series truncated at the first order such that

$$a_{z'} \simeq -\omega^2 r \left(1 + \alpha^2\right) + g_{z'} . \qquad (13.16)$$

What we measured, then, is not r, but $r\left(1 + \alpha^2\right)$.

The Euler acceleration is expected to be tangent to the trajectory, i.e., parallel to the y'-axis. Indeed, a plot of $a_{y'} = a_E$ as a function of $\frac{d\omega}{dt}$, shown below for $\left|\frac{d\omega}{dt}\right| \geq 2 \text{ s}^{-2}$, makes it clear that there is a linear relationship between the acceleration and the angular acceleration.

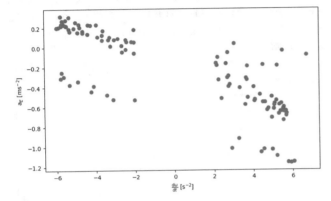

Data distribute along two parallel lines, because, during the experiment, the angular velocity changes sign a few times.

The larger angular acceleration is experienced by the device around $t \simeq 70$, 180 and 220 s. Let's look at data by zooming in the region $70 < t < 75$ s, after subtracting $a_{z'}$, as obtained above.

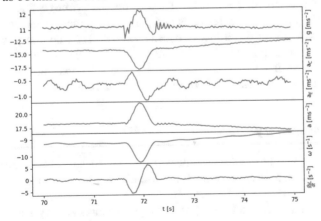

Filming the experiment may turn out to be useful in spotting features that can be hard to explain without any clues.

The ripple in $\frac{d\omega}{dt}$ corresponds to the time at which one of the ropes jumps on a node on the other one. During this phase, the rope has a shudder that is reflected in both the acceleration components and the angular rotation vector. The same ripple is clearly visible in the plot of a_E, which, in fact, is expected to be $a_E = -\frac{d\omega}{dt}r$ (in this case, the change in ω happens only in its magnitude and the angle between the angular acceleration and $\mathbf{r}$ is practically $\frac{\pi}{2}$). A plot of a_E versus $\frac{d\omega}{dt}$ is then expected to yield a linear relationship, as shown below.

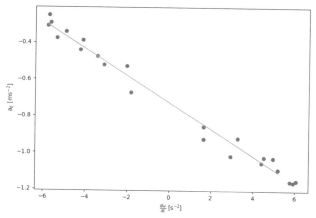

Data are fitted with a straight line, whose absolute value of the slope is the radius of the trajectory $r_E = 7.3 \pm 0.2$ cm. The intercept $q = -0.727 \pm 0.009$ m/s^2 is, as above, due to the fact that the device is not perfectly aligned. This also explains the difference between r_E and r. A similar analysis can be conducted for $t \simeq 180$ s and $t \simeq 220$ s. Averaging the results, we obtain $\langle r_E \rangle = 8 \pm 2$ cm, well consistent with the result obtained from the centripetal acceleration.

Summary

Any accelerometer will always measure accelerations within its own reference frame. If it is found in a non-inertial frame it experiences fictitious forces.

An accelerometer at rest always measures gravity acceleration, since gravity affects the distance between the proof masses kept separated by an elastic device.

The most general expression of the acceleration of a system includes the linear acceleration of the reference frame, the centripetal acceleration, the Coriolis term and the Euler acceleration.

The acceleration measured in a freely falling reference frame is null. This is the foundation of general relativity.

In fact, a cheap accelerometer can be affected by systematic biases and the measured acceleration can be not null while falling. Sometimes, the acceleration measured during free-fall may not be constant, due to air drag and the fact that the device rotates around one or more axes.

Centrifugal force is as real as other forces, because it can be measured. It is said to be fictitious because there is no source for it for the observer, while the centripetal acceleration (its counterpart in an inertial reference frame) is the result of an interaction.

The cross-product does not commute.

Experiments in rotating frames can be easily performed using an accelerometer and a gyroscope together.

During these experiments, it may be necessary to correct data to take into account the non-null readings of an accelerometer at rest.

As usual, linearising the relationship between two quantities simplifies the way to find unknown parameters.

Measurements allow us to identify the point at which the accelerometer is placed inside the device. The distance from the axis of rotation defines a circumference along which the sensor is placed. To determine at which point of the latter it is, we exploit the acceleration data.

Gyroscopes exploit the Coriolis force exerted upon a device in a non-inertial reference frame. Proof masses are forced to move such that, when the device rotates, they experience the Coriolis force. In tuning fork gyroscopes, the proof masses are the tines that vibrate at a known frequency, while the force is measured from the distance between the proof masses and fixed plates connected to them by an elastic structure.

Accelerometers can be used to experiment with accelerations that are not usually explicitly treated in physics courses, like the Euler acceleration. It is enough to measure accelerations in a frame that rotates at non-constant velocity.

Euler acceleration is easy to understand, however, figuring out its components from plots collected during the experiments is extremely instructive.

Often, results from experiments are consistent with approximations made during the analysis. Refining the analysis, however, may lead to better results or, at least, to a better understanding of possible systematics.

Filming the experiment may turn out to be useful in spotting features that can be hard to explain without any clues.

Phyphox

PHYPHOX allows for collecting data from several sensors at the same time using custom experiments.

Python

To compute a cross-product numerically, it is better to use the determinant method.

$$\mathbf{a} \wedge \mathbf{b} = \begin{vmatrix} \hat{x} & \hat{y} & \hat{z} \\ a_x & a_y & a_z \\ b_x & b_y & b_z \end{vmatrix}$$

The normalisation of a Gaussian is given by its integral $N\Delta\theta$, divided by $\sqrt{2\pi}\sigma$.

3

Dynamics of Rigid Bodies

Contents

© The Author(s), under exclusive license to Springer Nature Switzerland AG 2021
G. Organtini, *Physics Experiments with Arduino and Smartphones*, Undergraduate Texts in Physics, https://doi.org/10.1007/978-3-030-65140-4_14

Newtonian mechanics is often mistaken as being coincident with the dynamics of point-like particles. Real-life objects are certainly not point-like, however, in most cases, the physics of point-like particles is enough to describe the observations that can be made about them. This is not always true. Sometimes, the behaviour of non-point-like objects (physicists call them **rigid bodies**) exhibits effects that cannot be described by the physics of those that are point-like. Often, these effects are very surprising and largely counterintuitive. This chapter is devoted to the study of certain properties of rigid bodies, from which we derive a formal definition of a what a rigid body is and debunk beliefs arising from the naive application of Newton's Laws.

14.1 A Cylinder Rolling Along an Incline

It happens frequently that the acceleration of a rigid body rolling down an incline is mistakenly considered to be equal to that of a point-like particle. Conducting an experiment, even one that is only slightly more qualitative, is the best way to avoid such a mistake in the future.

A cylinder rolling along an incline is subject to its weight and to the normal force exerted by the plane. Naively, then, one would believe that the cylinder must fall along the plane with an acceleration $g \sin \theta$, θ being its inclination. If such a description were correct, given that $v = \omega r$, v being the speed of the cylinder, ω its angular velocity and r its radius, we could write

$$\frac{d\omega}{dt} = \frac{g}{r} \sin \theta \, . \tag{14.1}$$

Let's then set up an experiment to verify whether or not the theory is correct. In this section, we illustrate the measurement of the angular acceleration $\frac{d\omega}{dt}$ using a smartphone (see the next section for details). The measurement can be taken using Arduino, too, as described in Sect. 14.3.

The experiment is performed using a cardboard poster tube, whose diameter fits well with the width of many smartphones (see Fig. 14.1).

The inclined plane can be realised, for example, using an ironing board raised at one end. Angles are measured using a smartphone. PHYPHOX provides the angle of the device with respect to each axis with the inclination tool, which exploits the ratio between the components of the gravity acceleration measured by the accelerometer. In order to perform the experiment, we exploit a gyroscope, measuring the angular velocity around the cylinder's axis, with the

Angles can be measured using the PHYPHOX's inclinometer.

□ Fig. 14.1 Preparation of an experiment on a rolling cylinder. A smartphone is carefully inserted into a poster tube and kept in position with some paper towels

cylinder rolling down after starting from a steady state. An example of these measurements is shown below.

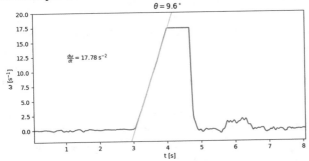

The angular velocity of a cylinder rolling down an incline linearly increases with time: its acceleration must thus be constant.

A linear fit to experimental data is shown as an orange line. From the slope of the line, one gets $\frac{d\omega}{dt}$. Data between 4.5 and 5 s appear to be constant, because the maximum range of the instrument has been reached (and the instrument is said to *saturate*). Repeating the experiment at different angles, one can collect and plot $\frac{d\omega}{dt}$ as a function of $\sin\theta$ and again fit it with a straight line, as below.

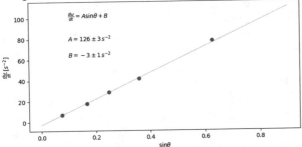

Assuming the same behaviour of a point-like particle, the slope of the line is expected to be

$$A = \frac{g}{r} \simeq 239\,\text{s}^{-2} \tag{14.2}$$

for $r = 4.1 \pm 0.1$ cm, while the intercept B should be compatible with zero. Indeed, we obtained $B = -3 \pm 1\,\text{s}^{-2}$, which is null within an interval of three standard deviations. However, $A = 126 \pm 3\,\text{s}^{-2}$ is significantly different from our expectations.

Treating an object as point-like is equivalent to neglecting any degree of freedom in excess with respect to the three translational ones. The size and shape of the object does not matter: planets and stars can often be treated as point-like, even if they are giant, while a coin has the shape of a disk even if it behaves much like a point-like object in many cases.

After checking that there were no systematic errors in our experiment, we end up with the conclusion that the theory was wrong. In fact, even if no real system is point-like, many behave as such, and we often fall into this *trap*. Indeed, treating an object as point-like is an approximation that depends on the degrees of freedom of the object. The number of degrees of freedom is the number of independent parameters needed to specify the stationary state of a system. For a point-like particle, the stationary state is defined by its position and it has three degrees of freedom. Such a number can be reduced if the particle is constrained: a particle moving along a straight line has only one degree of freedom, while, if it moves on a plane, it has two degrees of freedom.

For a rigid body, the position is not sufficient to completely specify its state. Its orientation with respect to the reference frame is important, too. For example, the axis of a cylinder can be oriented in an arbitrary way for which we need extra degrees of freedom (the coordinates of the unit vector of the cylinder axis). In some cases, non-rigid systems may have internal degrees of freedom. In a spring, for example, the length is an internal degree of freedom.

Often, the comparison of the relative size of the body r with respect to the system characteristic size R is taken as a criterion for considering the point-like approximation as valid. For example, planets are often considered to be point-like, even if their diameter r can be as large as thousands of km. The reason for that is ascribed to the fact that their distances R from the Sun are orders of magnitude larger. In fact, the reason why the point-like approximation holds is that their rotational degrees of freedom are negligible with respect to their positional ones (moreover, they are almost independent of the latter, too). A ball rolling down an incline never behaves like a point-like particle, irrespective of the ratio between its radius and the length of the path. Contrastingly, the point-like approximation for a box sliding along an incline is valid, irrespective of the size of the box, provided it does not rotate along any axis.

The need to take into account non-translational degrees of freedom depends on what we are interested in. For example, if we are interested in the time needed for a spring to fall from a given height, we might ignore its internal degree of freedom, unless the height is comparable to its length. In the latter case, the time can be much shorter, since its lower end can touch the floor much earlier its center of mass does (however, the time spent by any point of the spring while moving is equal to that of a point-like particle). Generally speaking, the point-like particle approximation is valid when either the non-translational degrees of freedom are negligible or absent or when they are decoupled from the translational ones.

The reason why the point-like dynamics is not valid in the experiment described above is that the rotational degrees of freedom are strongly coupled with translational ones. In this case, the point-like particle dynamics is not valid because, in our naive analysis, we neglected an entire class of forces acting upon each of the points of which the cylinder is made. Indeed, in order for the cylinder to keep its shape, it is necessary that any point on its surface be subject to a centripetal force. As a consequence, its motion can be described by an infinite set of equations representing the acceleration of each of its points on which, besides *external* forces like gravity and normal forces, *internal* ones maintain their relative positions. Fortunately enough, there are methods for separating the effects of internal and external forces, greatly simplifying the math.

Applying Newton's laws taking into account internal forces, a symmetric body like a cylinder, a sphere or a disk of mass M rolling down a plane rotates around its axis with an angular velocity ω, increasing with time as

> Internal forces maintaining the shape of an object must be taken into account to predict the dynamics of said object. In many textbooks, they are ruled out of equations because they sum up to zero, however, that does not mean that they are absent: they merely have consequences on the dynamics of the objects' centre of mass. Their effects are modeled throughout the moment of inertia.

$$\frac{d\omega}{dt} = \frac{Mgr}{Mr^2 + I}\sin\theta, \qquad (14.3)$$

r being its radius and I its *"moment of inertia"*. I depends on the mass distribution of the object and is defined as

$$I = \int_V \rho(\mathbf{x})r^2 dV, \qquad (14.4)$$

$\rho(\mathbf{x})$ being the mass density of the body at position $\mathbf{x}$ and r the distance of $\mathbf{x}$ from the axis of rotation, while the integral

Refer to physics textbooks for a compilation of moments of inertia.

is taken over the volume of the body V. For an empty cylinder, $I = Mr^2$ with respect to its axis.

It is worth noting that, recalling that $v = \omega r$, the acceleration $\frac{d\omega}{dt}$ obtained from the above equation substituting $I = Mr^2$ is half of the one predicted in the case of a point-like particle. This means that only half of the energy of the system is spent to make it move along the plane: the rest is spent in keeping it rolling. For a point-like object, $I = 0$, and we recover the equation of motion predicted in this case. Note that, in the case of the train car used in Chap. 10, only the moment of inertia of its wheels contributes, a contribution that is, however, negligible, because the wheels are light and small.

The acceleration of a rotating rigid body is lower than that of a point-like particle, because part of the energy gained by the object is needed to keep it rotating.

From $A = 126 \pm 3$ s^{-2}, and knowing the system mass $M = 0.606 \pm 0.001$ kg, we can then compute the moment of inertia of our device as

$$I = (9.0 \pm 0.5) \times 10^{-4} \, \text{kg m}^2. \tag{14.5}$$

This value can be compared with that which we expect from the theory. The system can be modelled as a cylinder of radius r and mass $M_C = 0.392 \pm 0.001$ kg, with a rectangle of mass $m_S = 0.200 \pm 0.001$ kg at its center and filled with a homogeneous mass $m_p = 0.014 \pm 0.001$ kg of paper, such that

$$I_{th} = \left(\frac{m_S}{3} + M_C + \frac{m_p}{2}\right) r^2 = (7.8 \pm 0.4) \times 10^{-4} \, \text{kg m}^2. \tag{14.6}$$

Despite the crude approximations, the agreement can be quantified as

$$\frac{|I - I_{th}|}{\sqrt{\sigma^2 + \sigma_{th}^2}} = \frac{90 - 78}{\sqrt{25 + 16}} \simeq 1.9. \tag{14.7}$$

It must be noted that one of the main contributions to the uncertainty comes from the measurement of the angle, whose uncertainty can be relatively large because the inclined plane often bends under its own weight and the angle is not constant along its length. This contribution is not included in the fit, in which we only take into account the uncertainty about $\frac{d\omega}{dt}$.

In principle, one can take it into account by modifying the denominator in the χ^2 definition from σ_y^2 to $\sigma_x^2 + \sigma_y^2$.

Alternatives to the poster tube are, e.g., the cardboard from a roll of packing tape or a PET bottle with its neck removed. The latter is less stable, because it is easier to distort, causing it to lose its shape.

A non-symmetrical system can be easily spotted by placing the tube on a horizontal plane. If it is symmetric, it should not oscillate around an equilibrium position, while, if the distribution of the masses is not symmetric, the heavier part tends to stay below the center and the system easily rotates spontaneously in order to reach the equilibrium position. It is very difficult to achieve a perfectly balanced system, however, if ω grows fast enough (i.e., the angle is large enough), this problem is strongly mitigated. Below, we show a measurement of the angular velocity as a function of time in the case of a poorly balanced system rolling down a plane inclined at an angle $\theta < 5°$.

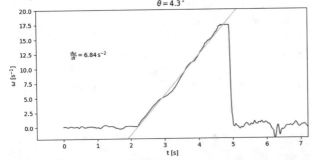

The oscillations around the mean line are a clear sign of a lack of symmetry.

Measurements can be repeated by changing the inertia of the system, adding, e.g., two identical weights on each side of the phone (e.g., two batteries). If the mass of one of the weights is m, the inertia of the body increases approximately by $2mh$, h being the distance between the axis of the cylinder and the center of mass of the weight.

14.2   Using a Smartphone's Gyroscope Remotely

Many smartphones have a three-axis gyroscope on board that can be exploited to measure the angular velocity around its three axes. This device is used in gaming applications, as well as in those related to virtual and augmented

With PHYPHOX, we can operate the phone remotely using a Wi-Fi connection. In particular, we can start, stop and configure experiments with gyroscopes.

Many public networks may be configured such that, for security reasons, they do not allow a browser to connect with PHYPHOX. In these cases, you may want to use your device as a hotspot, connecting your computer to the network managed by the phone itself.

reality, for which the device must know if and how fast it is being rotated.

PHYPHOX provides access to the raw data of the gyroscopes. Only the angular velocity around the y-axis is relevant for this measurement. It is important to insert the device into the tube such that it lies along one of its diameters and stays stable during the fall. The paper surrounding it helps keep it in position. It is also useful for manipulating the device during insertion. In fact, when the smartphone is inside the tube, it is impossible to operate its touchscreen. Cleverly enough, PHYPHOX allows us to connect to the phone using a Wi-Fi signal. Clicking on the menu icon on the top right in the experiment, a menu appears. Selecting the ALLOW REMOTE ACCESS checkbox makes it possible to control the App remotely using only a web browser, provided that it is on the same network as the smartphone. It is enough to connect to the website shown on the bottom part of the display when the option is activated to see a copy of the phone's display (Fig. 14.2).

In this way one can control the start and stop of the experiment, as well as download data to the local device. In order for the browser to connect to PHYPHOX, the App must be active. Manipulating the smartphone using the paper while inserting it into the tube avoids activating functions inadvertently.

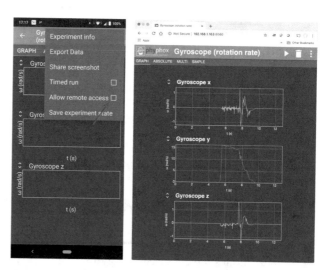

☐ **Fig. 14.2**   To allow remote access to PHYPHOX, click on the check box indicated by the red arrow in the left picture. On the right: the display of the smartphone as seen on a web browser

The experiment goes very quickly from this point on: once the connection has been set up, start data collection, leave the tube to roll down the plane and catch it when it is close to the end to avoid damaging your device through abrupt stops. Repeat the measurements a few times to evaluate the uncertainties and spot systematic errors, if any. Download the data, then change the angle and take another set of measurements.

Remember to include useful information, such as the angle at which the data were taken, in the data file name.

14.3 Arduino Gyroscopes and I2C Communications

The same experiment can be repeated using Arduino. Cost-effective gyroscopes are available on the market, often in conjunction with accelerometers in so-called IMU's (Inertial Measurement Unit).

These devices cannot work as analog devices, because they must provide a plethora of information. Moreover, the resolution required is often insufficient to be obtained with Arduino's analog-to-digital converters.

In these cases, sensors are equipped with a device able to exchange data with another device via a serial digital protocol. In other words, data are digitised on board, then sent as a sequence of bits of 0 and 1 represented as voltage pulses over a serial line. Polling the line and getting the height of the pulses allows the receiving device to collect data in the digital form. In most cases, the protocol allows for communications in both senses, such that an external device can also send information (usually to configure it) to the sensor.

One of the most popular protocols is I2C (standing for Inter Integrated Circuit), supported by Arduino. The I2C protocol, illustrated in Fig. 14.3, allows many devices on the same line to communicate with a master, thanks to two electrical lines SDA and SCL to which all the devices are attached. All I2C compatible devices thus have at least four leads: two for the power supply, to be attached, respectively, to the GND and to the 3.3V or 5V Arduino pin, and two for communication. Each Arduino flavour has its own specification about the position of the SDA and SCL lines. On an Arduino UNO, they coincide, respectively, with analog pins A4 and A5 and with the two pins close to the AREF one, on the side of digital pins. Data travels over the *Serial*

> An IMU (Inertial Measurement Unit) is a device that embeds both a gyroscope and an accelerometer to trace the state of a moving object.

> I2C is a master/slave, bidirectional, synchronised communication protocol supported by Arduino. Data travels over the SDA line, synchronised by pulses on the SCL line.

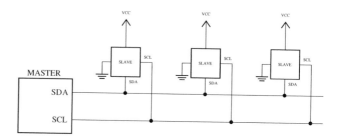

□ **Fig. 14.3** In an I2C network, two wires are used for communication between a master and many slaves. The SCL line distributes a clock signal to synchronise the data exchanged on the SDA line

DAta (SDA) line, synchronised by clock pulses that appear over the SCL (*Serial CLock*) line.

Each device on the lines can act either as a *master* or as a *slave*. Usually, Arduino behaves as the master and the sensors as slaves.

Is master/slave politically incorrect?

While we were writing this textbook, the "Black Lives Matter" movement was raising its voice to defend the rights of black people and, in general, to fight against social inequalities. Needless to say, we fully support their claims, however, there are other claims stemming from that about which we have a different opinion.

In particular, in information technology, a relatively old debate has resurfaced about the use of terms like master/slave or blacklist/whitelist, because of their racial connotation.

First of all, those connotations have little or nothing to do with black people. Blacks were only the most recent slaves forced into service in western countries. In the past, there have been plenty of slaves who were not people of color. Ancient romans, for example, used to keep slaves who were as white as them, even if they often, though not always, belonged to different ethnic groups. Similarly, the term "blacklist" most likely originated with the "black book": a list in book form compiled by King Henry VIII of people who committed crimes dating back to the XIV century [1] (the adjective "black" referring to the color of its cover).

Moreover, in particular, for the master/slave pair, its usage in information technology is appropriate exactly for its negative connotation: a slave, in fact, is a device that cannot make decisions on its own and totally depends on the master, without which it cannot (meaning it is not free to) operate. Adopting a metaphor or an alternative term would not illustrate its principles of operation as clearly. For example, one could use director/executor, primary/secondary, main/replica pairs, etc. While, in many cases, these would be appropriate (e.g., a master copy of a software repository can be called the main, while its slave copies may be called replicas), they are not as appropriate to the case that we are addressing. In particular, a director can certainly be thought of as someone who gives orders, however, executors who are not slaves are entitled to refuse them or execute them differently with respect to what the director expects.

Often, using words with negative meanings is more appropriate with respect to other, more politically correct choices and does not reflect the attitude of the speaker within different frameworks. Hiding the negative meaning of a word does not mean that the concept it represents disappears. On the contrary, carefully choosing the proper word may be of some help in making people reflect on the profound meaning of the words used. This is why, in this textbook, we prefer to continue using the traditional wording, believing that its usage will increase the awareness of its indecency in other contexts.

As usual, one can easily find the libraries needed to operate any sensor with Arduino, either using the Arduino IDE with its own library manager or just by looking for the source code on software repositories like GITHUB. And, as usual, one can choose among multiple implementations by multiple authors. The choice depends on criteria such as portability, ease of use, efficiency, etc. Libraries are meant to simplify communication management with a specific device, however, it is worth knowing, at least, the principles of native protocols: one day or another, you may need to write your own library or to modify the behaviour of a publicly available one.

> Most sensors can be operated using a library provided by its manufacturer or a third party. Libraries are useful, but knowing the native protocols boosts your reputation.

In this section, we provide an example that serves as a starting point for the realisation of experiments based on other I2C-based devices. Most sensors for Arduino have their own identifier, depending on the manufacturer: however, many of them are designed around a common architecture. As such, the sensor produced and distributed by a company can often be operated using the libraries for a different sensor based on the same electronics. Reading the documentation for the specific model usually provides such information. The measurements taken in this chapter can be done using IMU's based on the MPU6050 chip [2].

I2C-based slaves can be configured such that, for example, they provide data that is as precise as possible or as fast as possible. Many parameters, listed in the documentation, can be chosen by the user.

Often, I2C-based devices can be configured such that they can provide the best possible resolution, depending on the range to which they are sensitive. As illustrated in the previous chapters, sensors often measure quantities that are, in fact, related to the one for which they are designed. A gyroscope, for example, measures electrical currents that depend on the capacitance of the sensor that, in turn, depends on its angular velocity. The digitised current I_i measured on axis i is always a dimensionless integer number between 0 and $2^N - 1$, N being the number of bits in the ADC on board ($N = 16$ for the MPU6050). This number is then converted into an angular velocity ω_i expressed as a floating point number by multiplying it for a calibration constant C, such that

$$\omega_i = C I_i. \tag{14.8}$$

Properly setting the full scale range to which a device is sensitive improves its precision.

The current I_i, in turn, depends on the sensitivity of the device. It is, in fact, possible that the device responds differently to the same stimuli: if we need to measure only very low angular velocity, we can configure the gyroscope such that the maximum digitised current $2^N - 1$ is drawn when ω is, say, $250\,°/s \simeq 4.36$ Hz, such that the resolution of the device is

$$\Delta\omega = \frac{\omega}{2^N - 1} = \frac{250}{2^{16} - 1} \simeq 0.004\,°/s. \tag{14.9}$$

On the other hand, if we need to measure angular velocities up to $2000\,°/s \simeq 34.9$ Hz, with the same number of bits, we can achieve a resolution of

$$\Delta\omega = \frac{\omega}{2^N - 1} = \frac{2000}{2^{16} - 1} \simeq 0.03\,°/s. \tag{14.10}$$

Configuration and data are stored in the internal memory of the slave, whose locations are called registers. Each register is 8-bits long and its meaning depends on its location.

Once properly configured, I2C devices are ready to provide measurements upon request. The master asks the slave to provide them and the slave responds, transmitting as many bytes as requested to host $2^N - 1$ bits. The library running on the master transforms the integer provided by the slaves into the desired quantity, applying correction factors, if needed, and multiplying by the calibration constants.

The documentation provided by the chip's manufacturer always reports a table (register map) with the meaning of so-called *registers*. A register is a group of 8 bits internal to the chip, each with a specific function. The master in the I2C network either writes or reads from the registers, addressing the slave that provides its *address*: a unique identifier, usually provided as a two-digit hexadecimal number.

Every I2C device has a unique 16-bit address. The address is set at the factory, and the same address is shared by all the devices based on the same chip.

To write on a register (e.g., to configure the device's range), the master addresses the slave using its identifier, transmits the corresponding register's index to it, followed by its content, and then closes the communication. For example, for the MPU6050, the unique identifier is provided by the manufacturer as 0x68 (the prefix 0x tells the reader that the following digits must be interpreted as hexadecimal ones, such that the address expressed in base-10 is $6 \times 16 + 8 = 104$), while the index of the register containing the gyroscope configuration is 0x1B. According to the specifications, the gyroscope configuration register's bits 3 and 4 contain a two-binary-digit code representing the range of the device. To each range corresponds a different calibration constant. The code can be any of the four possible codes shown in the following table. In order

For better comprehension of this section, you should download a copy of the data sheet of the MPU6050 chip from the Internet.

full scale range (°/s)	hexadecimal code	binary code	sensitivity (s/°)
±250	0x00	00	131
±500	0x01	01	65.5
±1 000	0x02	10	32.8
±2 000	0x03	11	16.4

to configure the gyroscope to measure angular velocities up to 500 °/s, then, the master must transmit the following bytes over the serial data line: 0x1B, 0x08 (the binary representation of the last number is, in fact, 0000 1000, and remember that the first bit index is 0 and corresponds to the rightmost one).

To read from one or more registers, the master transmits the index of the first register to be read, followed by the number of bytes to be obtained. The slave whose address has been specified during transmission then transmits as many bytes as required that contain the content of the corresponding adjacent registers. For example, gyroscope measurements are stored in registers 67 to 72 (0x43 to 0x48). The angular velocity on each axis is represented by a 16-bit-long number: the first two registers contain the MSB (most significant byte) and the LSB (least significant byte) of the *x*-component, the next two those of

To obtain data from the slaves, the master asks them to provide N adjacent bytes starting at a given location address in the register map.

the y-component and the last two the bytes corresponding to the z-component. Then, if the master only needs the y-component of the angular velocity, it transmits the first register $0x45$ followed by $0x02$. The slave responds by transmitting two bytes (B_1 and B_0) starting from register $0x45$. The reading is then

$$I_y = B_1 \times 256 + B_0 \tag{14.11}$$

The corresponding value of ω_y can be obtained as

$$\omega_y = \frac{I_y}{65.5}. \tag{14.12}$$

In two's complement representation, binary numbers that start with a 1 are negative.

It is worth noting that, according to the device's data sheet, I_y is a signed integer represented as a two's complement number such that it represents integer numbers from $-32\,768$ ($0x8000$) to $+32\,767$ ($0x7FFF$). The calibration constant is then given by the inverse of the sensitivity. This way, when I_y contains $0x8000$, $I_y = -32\,768$ and $\omega_y = -500$.

Multiplexing

The fact that all the individual sensors that use the same I2C chip share the same address would make it impossible to use more than one device on the same network. On the other hand, it can be necessary to have, e.g., two or more IMU's on the same I2C line. In these cases, the master can address each of them using different techniques. Some manufacturers provide the possibility of changing the address of the slave, modifying the content of a register. Manifestly, the change must be done by attaching only one slave to the master. Only after the address has been changed can different devices be addressed individually.

Another possibility consists in using an extra line, often called AD0. These devices have two addresses. When AD0 is set to LOW, the device responds if it is addressed using the default address ($0x68$ for the MPU6050), while, if AD0 is HIGH, it responds if addressed with an alternative address ($0x69$ for the MPU6050). In an experiment in which there are many IMU's of the same type, one can connect the AD0 lead of each device to a different Arduino digital pin kept HIGH. When the master has to talk to one of these, it is enough to put the corresponding digital pin to LOW before starting the dialogue. Only the selected device will respond.

14.4 The Arduino Wire Library

Libraries usually provide classes from which the user can instantiate objects representing the device. One of the various libraries available for the MPU6050 IMU is (not surprisingly) MPU6050. After its inclusion, the user can instantiate an object of class MPU6050, configure it in the setup() and get data in the loop() as easily as follows:

Libraries simplify the management of external devices. On the other hand, they are not as general.

```
#include <MPU6050.h>

MPU6050 mpu;

void setup() {
  Serial.begin(9600);
  Wire.begin();
  mpu.initialize();
  mpu.setFullScaleGyroRange(MPU6050_GYRO_FS_500);
  Serial.println("IMU ready");
}

void loop() {
  int16_t ax, ay, az;
  int16_t gx, gy, gz;
  mpu.getMotion6(&ax, &ay, &az, &gx, &gy, &gz);
  Serial.print("a = ");
  Serial.print(sqrt(ax*ax + ay*ay + az*az));
  Serial.print(" w = ");
  Serial.print(sqrt(gx*gx + gy*gy + gz*gz));
}
```

In this sketch, mpu is the name assigned to the object representing the IMU. Before using it, we must initialise it using its initialize() method. The scale range of the gyroscope can be set with setFullScaleGyroRange() to which a parameter is passed in the form of an integer represented by the constant, defined in MPU6050.h, MPU6050_GYRO_FS_500. Six integer variables, each 16 bits long (int16_t), are defined in the loop() and set using the getMotion6() method of MPU6050. Those variables contain the components of the acceleration and angular velocity vectors in the form of two's complement integer numbers. In order to obtain their values in SI units, they must be multiplied by the corresponding calibration constants.

A good library defines a class to represent the device intended to be managed. The class provides methods for configuring it and accessing its data.

In the Arduino programming language, arguments are passed to functions by value, i.e., their content is assigned to formal parameters of the function working as ordinary variables. A change in the parameter, however, does not reflect a change in the content of the variable passed as an argument, because its location in the memory is different with respect to that of the parameter.

In the Arduino language, derived from the C-language, a method cannot return more than one value. In order to set two or more variables using a method, it is necessary to give to the method the address of the variables whose content has to be modified. If we passed the variable names as parameters in the method, the latter gets their values. They may, of course, change during the execution, however, that change does not reflect a change in the content of the original variable. Remember that a variable is a data container: passing it as a parameter does not imply that the content of that container changes. If a method must be able to modify the content of a variable, it must know its address in the memory. In that case, it can modify its content by overwriting the content of the corresponding memory locations. For this reason, the `getMotion6()` method, aimed at setting the values of its parameters, requires the addresses of the variables to be altered, the latter being returned by the & operator.

In the above example, we just send the modulus of both acceleration and angular velocity registered by the IMU over the serial line.

The `Wire` library allows for native communication with any I2C device.

The `Wire.h` inclusion is needed to add the functionalities of the Arduino `Wire` library, aimed at providing access to the I2C communication protocol. Communication is initialised thanks to the method

```
Wire.begin();
```

Registers are set by addressing the device, and then writing the register index and its value on the SDA line.

To configure the gyroscope configuration register, libraries implement the native communication protocol as follows:.

```
Wire.beginTransmission(0x68);
Wire.write(0x1B);
Wire.write(0x08);
Wire.endTransmission();
```

Data from registers are requested by addressing the slave and asking for N adjacent bytes starting from a given register.

The first line sets the address of the device to which the data are addressed (0x68). The next two lines write as many bytes on the serial data line. Only the addressed slave collects and interprets them accordingly, setting the gyroscope configuration register to 0x08.

`getMotion6()` must read six bytes starting from register 0x43 to obtain gyroscope data. Acceleration data are stored in six registers from 0x3B to 0x40. Setting the angular velocity components can be accomplished as follows, using native methods:

```
Wire.beginTransmission(0x68);
Wire.write(0x43);
Wire.endTransmission();
Wire.requestFrom(0x68, 6);
int Ix = Wire.read() * 256 + Wire.read();
int Iy = Wire.read() * 256 + Wire.read();
int Iz = Wire.read() * 256 + Wire.read();
```

The first group of three statements addresses register 0x43 of the device identified by 0x68. The `requestFrom()` method then asks the device whose identifier is 0x68 to transmit six bytes. These are read sequentially and returned by each `Wire.read()` statement.

A common alternative for computing the value of `Ix` from the two bytes is

```
int Ix = (Wire.read() << 0x08) | Wire.read();
```

exploiting the **shift operator** `<<`. The operator shifts the left operand to the left by the number of bits specified by its right operand. It also has a right shift operator `>>`. The "bitwise or" operator `|` performs the logical OR between all the bits of the operand on its left and those of the operand on its right. `Ix` being a 16-bit integer, if the I2C slave provides 0x31 and 0x03 as a result of a measurement (12 547), after the first reading, `Ix` reads as

$$0000\ 0000\ 0011\ 0001 \qquad\qquad (14.13)$$

(note how simple is to pass from hexadecimal to binary notation: it is enough to write down the bits separately for each hexadecimal digit). The application of the shift operator modifies the content of `Ix` into

$$0011\ 0001\ 0000\ 0000, \qquad\qquad (14.14)$$

shifting the digits on the left by eight places. The second `Wire.read()` returns 0x03, used in the logical OR operation such that, remembering that 0 or 1 = 1,

> Shift operators `<<` and `>>` shift the bits of their left operand by as many places as specified by their right operand.

> `|` and `&` are, respectively, the bitwise OR- and AND-operators. They apply the logical operation bit by bit to operands.

$$
\begin{aligned}
&0011\,0001\,0000\,0000 \text{ or} \\
&\underline{\ 0000\,0011} = \\
&0011\,0001\,0000\,0011\,,
\end{aligned}
\tag{14.15}
$$

corresponding to 0x3103, which, in turn, corresponds to decimal 12 547. The **bitwise** AND operator exists, too, represented as &.

Knowing the low level protocol described above allows you to write the code so as to communicate with every I2C compatible device, provided you have access to its documentation, usually distributed on the Internet.

14.5 Using an SD Card

In many cases, collecting data from Arduino using a computer connected via a USB cable is not possible. In these cases, data can be logged on an SD card.

To perform an experiment like the one described in this chapter, manifestly, Arduino cannot be attached to a computer with a USB cable. As a consequence, sending data over the serial line is useless. Data can instead be collected by attaching a mass storage device to Arduino, such as an SD card. This is done using a tiny SD card reader that uses the SPI (Serial Peripheral Interface) communication protocol to exchange data with the processor. The SPI protocol requires much less power with respect to the I2C, operating over shorter distances. While the latter can be a serious limitation for sensors, it is not so for a mass storage device.

The SPI communication bus is composed of four lines: MISO (Master In Slave Out), MOSI (Master Out Slave In), SCK (Serial Clock) and CS (Chip Select).

The SPI protocol uses four lines to make the communication between a master (Arduino) and a slave (the SD card reader) possible.

1. **MISO** (Master In Slave Out), also called **DO** (Data Output), brings data from slaves to the master. It must be connected to pin 12 on Arduino UNO (the pinout may be different on other Arduino platforms).
2. **MOSI** (Master Out Slave In), also called **DI** (Data Input), brings data from the master to the slaves. The corresponding Arduino pin is 11.
3. **SCK** (Serial Clock) distributes a common clock signal to the whole network of masters and slaves, via the Arduino pin 13.
4. **CS** (Chip Select) or **SS** (Slave Select) uses pin 10 on Arduino UNO, by which the master can disable communication with slaves.

As a result, any SPI-compatible device has at least six leads: two for feeding (GND and 5V) and four to participate in the communication.

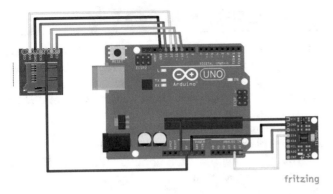

Fig. 14.4 Wiring for an experiment adopting an MPU6050-based IMU (the blue device on the right) together with an SPI SD card reader (the red one on the left)

The CS line is used to select the target of an SPI command on a network where many devices are present. If, for example, two or more SPI compatible devices share the same SPI bus, all data travelling over the MOSI and MISO lines are seen by all of them. In order to make all devices *deaf* to these data, but one, it is enough to keep all the CS lines in the HIGH state. Only the device whose CS lead is kept in the LOW state intercepts and interprets the data. By default, pin 10 on an Arduino UNO is automatically set to LOW before any transaction and to HIGH soon after. Another digital pin must be used and properly configured if a second SPI device is present in an experiment.

An SPI slave exchanges data with the master only if its CS line is kept LOW. If it is kept HIGH, the slave is disabled.

Figure 14.4 shows how to connect both an IMU based on an I2C-compatible device and an SD card reader using the SPI protocol. The wiring is complex: following conventions such as those concerning the cables' colours helps in avoiding mistakes. The MPU6050 can be supplied with a voltage between 3 V and 5 V, and we can use the 3.3v Arduino pin for it, while the 5v pin is reserved for the card reader. The power can be supplied by a common 9 V battery attached to the external black power connector on the bottom left part of the board in the figure. Care must be taken to properly align the surface of the IMU with respect to the cylinder's axis.

14.6 Using SD Cards to Store Data

The SPI protocol is relatively easy to use on Arduino that natively support reading and writing from and to peripherals. On the other hand, the need to use it as such is actually limited to very few devices. Here, we simply illustrate how to write data on a file on the SD card, being an extremely useful skill in physics data acquisition. A few details about the protocol are given in the next section.

Using a file to log data persistently is easy with an SD card reader and SPI. Just open the file in write mode and write on it using methods similar to those used to send data over the serial line.

After including the `SD.h` library, a file on the SD card can be represented by an object of the class `File`, as in

```
File dataFile;
```

The file must be opened, once the reader has been properly initialised with `SD.begin()`. To open a file in write mode, the variable representing it must be assigned with the result of the `open()` method of `SD`, usually in the `setup()` function:

```
void setup() {
    SD.begin();
    dataFile = SD.open("data.csv",
        FILE_WRITE);
}
```

The first parameter is the file name. Mode `FILE_WRITE` prepares `dataFile` to accept data to be written on the SD card. If the file does not exist, it creates it, otherwise new data are appended to existing ones. The `FILE_READ` prepares the file for reading.

Characters can be written on a file using the same `print()` and `println()` methods defined in `Serial`.

Sending data to a file works in the same way as sending data to the serial line. To write characters on a file, we can use either the `print()` or the `println()` method, analogous to those for `Serial`, as in

```
dataFile.print("omega_y = ");
dataFile.println(Iy);
```

Flushing the memory buffer so as to actually dump data on files is achieved through the `flush()` method.

Sometimes, reading the file back, it may happen that the last expected rows of a file are missing. This happens because writing is not done synchronously for reasons of efficiency. Writing on an SD card is expensive from the point of view of the time needed. The systems, then, try to keep as much information as possible in a buffer in the volatile memory and empty the buffer only when needed, actually writing its content onto a disk. To force the buffer to be emptied, the user can use the `flush()` method:

4

```
dataFile.flush()
```

The need to read data from files stored on an SD card is less common in data acquisition applications. However, it may be useful, e.g., to read a configuration stored on it before starting a new run.

Once opened for reading, the size of a file in units of bytes can be obtained using the `File.size()` method, as in

```
Serial.print("Data file size = ");
Serial.println(dataFile.size());
```

Reading a file consists in obtaining a sequence of bytes from the current position of a pointer to the file and copying them into a large enough sequence of memory locations represented by a variable in the sketch. The position of the pointer is set to the beginning of the file at opening time and it advances as reading proceeds.

When the file is no longer needed, it can be closed using

```
dataFile.close()
```

causing the memory buffer to be flushed and securing all the data on the mass storage.

Using `print()` and `println()` methods, data are written on files as if they were written to a screen, i.e., in its decimal representation. What is written in the file, in other words, are the characters representing the content of the variables or the constants passed as arguments. For example,

```
float x = 1024;
dataFile.println(x);
```

results in finding the characters `1024` in a line of the file. However, 1 024 is represented in the Arduino memory as a floating point number: while the string `1024` needs five bytes to be represented (one extra byte is the terminating one, whose ASCII code is 0), a floating point number is represented through the use of four bytes using the IEEE-754 standard.

It is not difficult to realise that its hexadecimal representation in the memory is 0x44800000. In fact, $1\,024 = 1 \times 2^{10}$. Its mantissa 1 is omitted in the IEEE-754 representation and the last 24 bits are all zero. The exponent, instead, is represented in the excess-127 notation, i.e., as $127 + 10 = 137$, whose binary representation is 1000 1001.

Writing variables containing numbers to files, the decimal representation of the corresponding numbers is recorded. Reading them back with Arduino requires converting the sequence of digits into the proper representation (`int` or `float`, for example).

Taking into account that the first bit represents the sign (0 in this case), we obtain the full representation as above.

Reading data from files is not as simple as writing them.

The above considerations are relevant when reading back data from a file, which, unfortunately, is not as simple as writing (fortunately enough, contrastingly, simple data acquisition applications only seldom require reading data from files). The `File` class has a method `read()` that accepts two arguments: a memory location, represented as the address of a character, and the number of adjacent bytes to be read. Suppose, for example, that we want to read back the `"omega_y = "` characters such that we can put them into a string of at least 11 characters:

```
void loop() {
  char string[11];
  dataFile.read(&string, 11);
}
```

will read 11 characters starting from the current pointer in the file and copy them into as many adjacent memory locations, starting at the one corresponding to the first character of string `string`, whose address is returned by `&string`. In fact, a string is just an array of characters and the `char` type represents a character variable, i.e., a 1-byte-long variable containing an ASCII code. With such a piece of code, element `string[0]` contains o, `string[1]` contains m, and so on, up to `string[9]`, which contains a blank (note that array indices run from 0 to $N - 1$, N being the size of the array).

The next characters in the file will be those needed to represent the content of the variable `Iy` in our example. What we can read, in this case, are the characters representing the number in its decimal notation, not the value of that number. It is not possible to put the value read this way into a floating point variable: in fact, its decimal representation may need anything from 1 to many characters to be written, while a floating point variable always occupies four bytes in the IEEE-754 standard. To convert a string into the content of a floating point variable, if possible, the method `toFloat()` of a `String` class exists. Its illustration is beyond the scope of this textbook, and we refer to the documentation available on the Arduino website. On the other hand, in this course, we are interested in writing data in the form of CSV files to be read on a computer.

14.7   **The Native SPI Protocol**

Even if beyond our scope, it is worth mentioning the native SPI methods, in the special case in which we need at least two devices on the bus. The SPI protocol does not need to address a specific device: the interested one is addressed by putting its CS line to LOW. Similarly to I2C, instead, each device defines a set of registers that can be either read or written by the master. The communication is always done by transferring a sequence of bits encoding both the address of the register and its content. Suppose, then, that we have two devices connected to Arduino: the CS line of device *A* is connected to pin 9, while device *B* is connected to pin 10. A command to the device is represented by, e.g., two bytes. Both pins must be configured to be output pins and both are disabled with the following code:

To write or read SPI registers, one or more bytes have to be transferred over the MOSI line. The bytes encode the addressed register, the command to be executed and its parameters, if any. The slave executes the command and returns a byte on the MISO line as a result.

```
void setup() {
  ...
  pinMode(9, OUTPUT);
  pinMode(10, OUTPUT);
  digitalWrite(9, HIGH);
  digitalWrite(10, HIGH);
}
```

(ellipses represent other statements for the configuration that may be needed). In order to address device *A*, pin 9 must be set to LOW, data are transferred on the communication lines, and then pin 9 is set back to HIGH:

```
digitalWrite(9, LOW);
recv1 = SPI.transfer(byte1);
recv2 = SPI.transfer(byte2);
digitalWrite(9, HIGH);
```

When SPI.transfer(byte1) is executed, the corresponding sequence of bits is written sequentially on the MOSI line. The device whose CS line is LOW interprets it and returns, on the MISO line, a result, if any. The returned value is put in the recv1 variable.

Summary

It happens frequently that the acceleration of a rigid body rolling down an incline is mistakenly considered to be equal to that of a point-like particle. That is a mistake, as can be easily verified by performing an experiment. Part of the energy gained in descending the incline is, in fact, converted into rotational kinetic energy and is no longer available to increase the velocity of the centre of mass.

Treating an object as point-like is equivalent to neglecting any degree of freedom in excess with respect to the three translational ones. The size and shape of the object does not matter: planets and stars can often be treated as point-like, even if they are giant, while a coin has the shape of a disk even if it behaves much like a point-like object in many cases.

The angular velocity of a cylinder rolling down an incline linearly increases with time: its acceleration must thus be constant.

Internal forces maintaining the shape of an object must be taken into account to predict the dynamics of an object. In many textbooks, they are ruled out of equations because they sum up to zero, however, that does not mean that they are absent: they merely have consequences on the dynamics of the objects' centre of mass. Their effects are modeled throughout the moment of inertia.

Phyphox

Angles can be measured using PHYPHOX's inclinometer.

With PHYPHOX, we can operate the phone remotely using a Wi-Fi connection. In particular, we can start, stop and configure experiments with gyroscopes.

Many public networks may be configured such that, for security reasons, they do not allow a browser to connect with PHYPHOX. In these cases, you may want to use your device as a hotspot, connecting your computer to the network managed by the phone itself.

Arduino

An IMU (Inertial Measurement Unit) is a device embedding both a gyroscope and an accelerometer to trace the state of a moving object.

I2C is a master/slave, bidirectional, synchronised communication protocol supported by Arduino. Data travels over the SDA line, synchronised by pulses on the SCL line.

Most sensors can be operated using a library provided by its manufacturer or a third party. Libraries are useful, but knowing the native protocols boosts your reputation.

4

I2C-based slaves can be configured such that, for example, they provide data that is as precise as possible or as fast as possible. Many parameters, listed in the documentation, can be chosen by the user.

Properly setting the full scale range to which a device is sensitive improves its precision.

Configuration and data are stored in an internal memory of the slave, whose locations are called registers. Each register is 8 bits long and its meaning depends on its location.

Every I2C device has a unique 16-bit address. The address is set at the factory and the same address is shared by all the devices based on the same chip.

To obtain data from the slaves, the master asks them to provide N adjacent bytes starting at a given location address in the register map.

Libraries simplify the management of external devices. On the other hand, they are not as general. A good library defines a class to represent the device intended to be managed. The class provides methods for configuring it and accessing its data.

In the Arduino programming language, arguments are passed to functions by value, i.e., their content is assigned to formal parameters of the function working as ordinary variables. A change in the parameter, however, does not reflect a change in the content of the variable passed as an argument, because its location in the memory is different with respect to that of the parameter.

The `Wire` library allows for native communication with any I2C device. Registers are set by addressing the device, and then writing the register index and its value on the SDA line. Data from registers are requested by addressing the slave and asking for N adjacent bytes starting from a given register.

Shift operators `<<` and `>>` shift the bits of their left operand by as many places as specified by their right operand.

`|` and `&` are, respectively, the bitwise OR- and AND-operators. They apply the logical operation bit by bit to operands.

In many cases, collecting data from Arduino using a computer connected via a USB cable is not possible. In these cases, data can be logged on an SD card.

The SPI communication bus is composed of four lines: MISO (Master In Slave Out), MOSI (Master Out Slave In), SCK (Serial Clock) and CS (Chip Select).

Using a file to log data persistently is easy with an SD card reader and SPI. Just open the file in write mode and write on it using methods similar to those used to send data over the serial line. Characters can be written on a file using the same `print()` and `println()` methods defined in `Serial`. Flushing the memory buffer so as to actually dump data on files is achieved through the `flush()` method.

Writing variables containing numbers to files, the decimal representation of the corresponding numbers is recorded. Reading them back with Arduino requires converting the sequence of digits into the proper representation (`int` or `float`, for example).

To write or read SPI registers, one or more bytes have to be transferred over the MOSI line. The bytes encode the addressed register, the command to be executed and its parameters, if any. The slave executes the command and returns a byte on the MISO line as a result.

References

1. The Merriam-Webster Dictionary, online edition 2020. ▶ www.merriam-webster.com
2. InvenSense Inc. (2020) MPU-6000 and MPU-6050 register map and descriptions. RM-MPU-6000A-00 rel. 03/09/2012. ▶ www.invensense.com

4

Wave Mechanics

Contents

The original version of this chapter was revised: Typographical errors have been corrected. The correction to this chapter is available at ▶ https://doi.org/10.1007/978-3-030-65140-4_16

Electronic supplementary material The online version of this chapter (doi:▶ 10.1007/978-3-030-65140-4_15) contains sup-. plementary material, which is available to authorized users

© The Author(s), under exclusive license to Springer Nature Switzerland AG 2021, corrected publication 2022
G. Organtini, *Physics Experiments with Arduino and Smartphones*, Undergraduate Texts in Physics,
https://doi.org/10.1007/978-3-030-65140-4_15

The physics of waves is of capital importance. It was used in the past to demonstrate that light propagates as a wave, while, nowadays, it is exploited to catch gravitational waves, one of the most elusive phenomena ever. Waves play a central role in electromagnetism, as well as in modern quantum physics, in which radiation and matter both behave like waves under certain conditions. To understand quantum mechanics, it is thus mandatory to master the physics of waves. In this chapter, we propose a few experiments that can be done using sound waves.

15.1 Making Waves

Sounds can be produced and manipulated using the `simpleaudio` module in Python.

Acoustic waves can be produced in a variety of ways. In Python, in particular, this may be accomplished by importing the `simpleaudio` module.

Sound is produced by computers driving loudspeakers with currents whose amplitude is encoded, as a function of time, as a sequence of integers, which are, in turn, represented by sequences of n bits. The sequence has several important characteristics. The "sampling rate" f_S is the number of samples in one second. Audio CDs, for example, are sampled at 44.1 kHz. In other words, each second of sound is represented as a list of f_S integers. Another important characteristic is the "resolution", often called the "bit depth". It represents the number n of bits used to represent each sample. Common values are $n = 16$, $n = 20$ and $n = 24$. Finally, a given sound can be reproduced on different "channels". A channel is a physical device through which data are sent from the source to the speaker. Each channel can transport different information. Standard audio signals can have one (mono) or two (stereo) independent channels. Professional systems may have more channels.

Audio files are characterised by their sampling rate (the number of samples per second), their bit depth or resolution (the number of bits used for each sample) and the number of channels (mono or stereo).

For the purpose of generating plain audio signals to play with waves, we need stereo sinusoidal signals sampled at a reasonable rate, such as, e.g., $f_S = 44.1$ kHz, with a relatively small depth $n = 16$.

To represent a stereo signal whose duration is t s, we need $N = t \times f_S$, n-bit integers per channel. In Python, this requires a bidimensional list (one per channel), each composed of N elements, each containing the samples of the signal, represented as an integer proportional to

5

$$y(t) = \sin{(2\pi f t)}, \qquad\qquad (15.1)$$

f being the frequency of the sound (audible signals are approximately within an interval $20 \leq f \leq 20\,000$ Hz). For a depth $n = 16$, each sample must be a 16-bit integer, whose values go from $-2^{15} = -32768$ to $2^{15} - 1 = 32767$.

The frequency of the audible signal is comprised between 20 Hz and 20 kHz.

In the previous chapters, terms like "list" and "array" were used as synonyms. In fact, they are not. There are subtle, but important differences between lists and arrays in Python. While lists can be heterogeneous, i.e., they can contain objects of different types, arrays cannot. From the practical point of view, the main difference between the two sequences is the result of the application of the * operator. Applied between a list and an integer constant N, the operator returns an extended list, whose elements are the same as the list operand repeated N times. The result of

Arrays are a special kind of list, whose elements must belong to the same type.

```
t = [1, 2, 3]
print(t)
```

is

```
[1, 2, 3]
```

The execution of the code

```
t2 = t * 3
print(t2)
```

Applying the * operator to a list and an integer extends the list.

produces

```
[1, 2, 3, 1, 2, 3, 1, 2, 3]
```

Applied to arrays, the * operator returns an array whose elements are each multiplied by the constant, e.g.,

```
a = array([1, 2, 3])
print(a)
b = a * 3
print(b)
```

The same operator applied between an array and a constant multiplies each element of the array by the constant.

produces the following output:

```
[1 2 3]
[3 6 9]
```

Note how the creation of an array implies the use of parentheses, and that the output is slightly different in the two cases. It is worth noting that multiplying an array by a non-integer number or dividing it by a constant are perfectly defined operations, while, for a list, both return a runtime error. As a matter of fact, numpy's arange() and zeros() return an array.

The samples for the left channel at $f = 440\,\text{Hz}$ of 1 s duration can then be obtained as an array as follows:

```
import numpy as np

f = 440
fs = 44100
seconds = 1
s = int(seconds * fs)
note = np.zeros((int(s, 2))
t = np.arange(0, seconds,
          seconds/len(note))
A = 2**15 - 1
x = 2 * np.pi * f * t
note[:,0] = A*np.sin(x)
```

Multidimensional lists or arrays are represented as a list of lists or as an array of arrays.

A multidimensional list (or an array) in Python can be created as a list of lists (an array of arrays). For example, for a bidimensional list,

```
aList = [[0, 1, 2, 3],
         [4, 5, 6, 7]]
```

Element aList[1] represents the second element of the list, i.e., [4, 5, 6, 7]. Its third element (6) can then be obtained as aList[1][2]. An array of N elements initialised to 0 can be obtained from numpy's numpy.zero(N, m), which returns an array of N arrays, each composed by m null elements.

To represent t seconds of a sound with a sampling frequency f_S, each sample must be computed every $\frac{t}{f_S}$ seconds.

The t array contains the times at which we have to compute the samples. The length of note is N (each element's length of note is 2, but note is an array of N arrays of length 2 each). If, then, we need to play t seconds, we need a sample every $\frac{t}{N}$ seconds. The values of the left channel of the audio sample is assigned by computing

$$y(t, 0) = \left(2^{15} - 1\right) \sin\left(2\pi f t\right) \tag{15.2}$$

such that the highest value that it attains corresponds to the maximum representable value of a 16-bit number. Note that np.sin() returns an array whose size is the same as its argument. It is assigned to the first element of each sublist of which `note[]` is made. Similarly, one can assign the values to the right channel using

```
note[:,1] = note[:, 0]
```

For a mono signal, the right (or left) channel must be set to zero, i.e.,

```
note[:,1] = [0] * len(t)
```

Note that, in this case, `[0]` is a list with a single element. Multiplied by the constant `len(t)`, it returns a list of length `len(t)` full of 0 (this is equivalent to `np.zeros(len(t))`). When the list is assigned to an array, it is automatically converted (the operation, of course, can only be possible if all the elements of the list belong to the same type, as in this case).

The corresponding sound can be played using `simpleaudio` as in

```
import simpleaudio as sa
b = note.astype(np.int16)
po = sa.play_buffer(b, 2,
                    2, fs)
po.wait_done()
```

`play_buffer()` plays the content of its first argument on the number of channels given as its second argument. The third argument represents the number of bytes per sample: 2 in the case of 16-bits samples. The last parameter is the sampling frequency. `play_buffer()` returns a playing object po, used to control the execution of the program. Its `wait_done()` method suspends the program until the object is finished playing.

Note that audio data given as the first argument of `play_buffer()` must be of a type consistent with the resolution and its third argument. For a resolution $n = 16$, each sample is composed of two bytes and `note` must contain numbers represented as 16-bit integers. The default type for its elements is `float64`, i.e., each sample is represented as a floating point number in "double precision" (using 64 bits instead of 32, 11 of which are for the exponent and 52 for the mantissa). In order to convert them into the expected type, the `astype()` method is adopted, to which the wanted type is passed as `np.int16`, representing a 16-bit integer as defined in `numpy`.

The `simpleaudio` package provides tools for interacting with loudspeakers and other peripherals intended to play audio signals. Audio data, in the form of a bidimensional array of n-bit integers, can be played using `play_buffer()`. After invoking `play_audio()`, the program continues immediately. To suspend its execution until the audio is played, the play object must wait for it.

15.2 Command Line Options

It is convenient to have the possibility of changing the characteristics of the sound produced by the script described above without changing the code. To do that, we can benefit from the possibility of passing the values of parameters from the command line. Launching the script from the command line consists in typing its name in a shell (the environment a user finds in a terminal), such as

```
./play.py
```

The command will play a sinusoidal tone with $f = 440\,\text{Hz}$ for 1 s. To play a tone with $f = 8\,200\,\text{Hz}$ for 3 s, it would be nice to write

```
./play.py -f 8200 -d 3
```

or, more explicitly,

```
./play.py --frequency 8200
          --duration 3
```

The `getopt` package allows for the use of options and long options in launching a script on the command line of a terminal. Options are preceded by a dash (–) and may or may not require arguments. They can be combined together if they don't. Long options are preceded by two dashes (--).

Here, f and d are called *options*, while `frequency` and `duration` are *long options*. Long options require two dashes to be specified. Different options may be combined into a single option if they do not require arguments, such that `ls -l -a` is equivalent to `ls -la` (`ls` is a UNIX command line used to list all the files in the current directory). The `getopt` module is intended for this. The list of possible options is given in `getopt.getopt()`. The latter accepts a list of strings consisting of the list of options and arguments given with the command, if any (in the example, this would be `--frequency 8200 --duration 3`). The list is followed by a string made of characters representing valid options (f and d), each followed by a colon if it requires an argument (e.g., `'f:d:'`). Optionally, a list of long options can be given in the format of a list of strings terminated by an equal sign if they require an argument. The first parameter of `getopt.getopt()` can be extracted from `sys.argv`, provided by the `sys` module, containing the name of the script as its first element (`sys.argv[0]`) and the list of arguments as follows, including the wanted list that is represented by `sys.argv[1:]`, as in the script below.

```python
import sys
import getopt

f = 440
```

```
seconds = 3

opts, args = getopt.getopt(sys.argv[1:], 'hf:d:',
                           ['help', 'frequency=',
                            'duration='])
for opt, arg in opts:
    if opt == '-h':
        help()
        exit(0)
    elif opt in ("-f", "--frequency"):
        f = int(arg)
    elif opt in ("-d", "--duration"):
        seconds = float(arg)
```

In this example, we added the option h (whose corresponding long option is `help`) intended to provide a help message to the user. The message is printed by the function `help()`, not shown. `getopt.getopt()` returns a list of pairs (`opts`) representing the options and their arguments, and a list of arguments passed to the command, if any (`args`, which, in this case, is empty). To understand the difference between options and arguments, consider the `tail` UNIX command, intended to display the tail of a file, i.e., its last rows.

`getopt()` returns a list of pairs representing the options with their arguments and a list of arguments.

`tail audioscope.csv` prints the last ten rows of the file `audioscope.csv`. The file name is called an argument for `tail`. If the file is growing (i.e., we are appending data to it continuously), we can monitor its content using `tail -f audioscope.csv`: it displays the last ten rows of the file when it is read the first time, but it does not exit. It keeps reading the file such that any new line appended to it will be displayed immediately. In this case, `-f` is called an option. Options may have arguments, too (not to be confused with the arguments of the command). For example, `tail -n 20 audioscope.csv` displays the last 20 lines of the file. In this case, 20 is an argument for the option `-n`.

The first object returned by `getopt()` consists of a list of pairs. Each pair is made of an option and its corresponding value. The second object is a list of arguments given in the command line.

If written in Python, `tail` would start as

```
opts, args = getopt.getopt(sys.argv[1:], 'n:f')
```

When called without arguments or options, `opts` and `args` are both equal to `[]`, i.e., they are empty lists. If the command is invoked as `tail audioscope.csv`, `opts = []`, while `args = ['audioscope.csv']`. In the case of `tail -f audioscope.csv`, `opts` is a list of pairs `[('-f', '')]`: a list with one element consisting of a tuple of two strings, the second one being empty. Finally, upon `tail -n 20 audioscope.csv`, `opts` contains `[('-n', '20')]`.

Iterating over `opts` returns each pair sequentially, decomposed into `opt` and `arg`. For each iterator, we check whether `opt` is contained in a tuple and, if so, we reassign the values of variables accordingly. Note how the construct

The `in` operator checks whether its left operand is contained in its right operand.

```
opt in ("-f", "--frequency")
```

returns true if the operand on the left (`opt`) is contained in the tuple on the right of the `in` operator.

15.3 Properties of a Wave

Almost any periodic function can be written as a series of trigonometric functions with appropriate frequency and amplitude, according to the Fourier Theorem. The latter gives the recipes for computing the amplitudes for each frequency.

Waves may have a variety of shapes, however, the Fourier Theorem states that a periodic function $f(x)$ may be expressed as a series of sine or cosine terms, at least if $f(x)$ is integrable within a finite interval of amplitude L. Terms in the Fourier series have specific amplitudes known as Fourier coefficients. In formulas,

$$f(x) = \frac{a_0}{2} + \sum_{i=1}^{\infty} a_n \cos\left(\frac{2\pi n x}{L}\right) + b_n \sin\left(\frac{2\pi n x}{L}\right).$$

(15.3)

The Fourier coefficients a_i and b_i are computed as follows:

$$a_n = \frac{2}{L} \int_L f(x) \cos\left(\frac{2\pi n x}{L}\right)$$
$$b_n = \frac{2}{L} \int_L f(x) \sin\left(\frac{2\pi n x}{L}\right).$$

(15.4)

Thanks to the Fourier Theorem, we do not need to study any possible wave: it is enough to study the dynamics of harmonic ones to predict the behaviour of any other wave. Harmonic waves are characterised by their wavelength, their period and their phase.

The Fourier Theorem guarantees that the properties of any periodic function are the same as that of sinusoidal, or harmonic, waves. For simplicity, then, we can limit ourselves to studying only the latter, since they will share their behaviour with any other wave, the latter being a sum of harmonic terms. A harmonic wave can always be written as

$$f(x,\, t) = A \cos\left(2\pi \left(\frac{x}{\lambda} \pm \frac{t}{T}\right) + \phi\right),$$

(15.5)

where λ is called its wavelength, T the period and ϕ the phase. A is called the amplitude and represents the maximum absolute value of $f(x,\, t)$, as well as having the same dimensions. For example, waves on the surface of the sea are characterised by their height with respect to the average sea level and A is measured in m. For sound waves, A may represent the local air pressure.

The inverse of the period is the frequency $f = \frac{1}{T}$. The reason why we write the wave this way is that sine and cosine both take an angle as an argument and an angle is always a multiple of 2π. Sine and cosine differ only by the relative phase. The multiple must be dimensionless: since it may depend on x and t (respectively, a length and a time), the latter must be divided by a parameter with the same nature.

The wavelength represents the distance between two successive points having the same $f(x)$ and the same $f'(x)$. The period is the interval after which the wave reaches the same $f(t)$ and $f'(t)$. These two quantities are linked by the relationship $\lambda = cT = \frac{c}{f}$, where c is called the speed of the wave and, in fact, represents the speed at which the wave propagates in space. The latter depends on the medium within which the wave propagates.

It is customary to define the angular frequency defined as $\omega = 2\pi f$ and the wave number $k = \frac{2\pi}{\lambda}$. These are useful for simplifying the wave function that becomes

$$f(x,\ t) = A \cos\left(kx \pm \omega t + \phi\right). \qquad (15.6)$$

The relationships between these quantities are easy to remember using dimensional arguments. For example, since c is a speed, it must be a length divided by a time, i.e., $c = \frac{\lambda}{T}$, while, since f is the inverse of a time, $c = \lambda f$.

To understand the characteristics of a wave, use the PHYPHOX audioscope that shows a graphical display of $f(t)$ (corresponding to the microphone position in x) in arbitrary units. A microphone works as a parallel plate capacitor: the system measures the current driven by it when the distance between the two conductors changes. The capacitance of a system of two conductors is, in fact, defined as

The properties of a wave can be explored using PHYPHOX, which shows $f(0,\ t)$, $x = 0$ corresponding to the position of the smartphone's microphone.

$$C = \frac{Q}{\Delta V} = \epsilon \frac{S}{h}, \qquad (15.7)$$

where Q is the electric charge accumulated on one of them and ΔV the voltage between the plates. S is the surface of the plates and h their distance. ϵ is a constant that depends on the insulating material between the plates. Taking the derivative of both members:

$$\frac{dQ}{dt} = -\epsilon \frac{S}{h^2} \frac{dh}{dt} \Delta V \,. \tag{15.8}$$

A microphone can be imagined as a parallel plate capacitor whose plates are separated by an elastic medium. The sound pressure causes the distance between plates to change. Consequently, an electrical current is drawn from the system.

$\frac{dQ}{dt}$ is, by definition, the current extracted from or injected into the capacitor, which then depends on the change in h. You can imagine a microphone as a parallel plate capacitor in which the two plates are connected by an insulating string. The pressure exerted on one plate by air compresses or lengthens the spring, and h changes accordingly. For microphones, it is not necessary to know the local air pressure in the proper units, thus they usually return a number proportional to it.

To properly calibrate the audioscope, we need a reference sound whose intensity is known. The intensity I of a source is proportional to A^2 and measures the power emitted from it (i.e., the energy per unit time). Sound levels are usually measured in decibels (dB). Manifestly, the power intercepted by a microphone depends on its area, and the intensity measured is then given in units of Wm^{-2}. Given that $I_0 = 10^{-12}\,\mathrm{Wm}^{-2}$, a sound intensity I in dB is defined as

$$I(\mathrm{dB}) = 10 \log_{10} \left(\frac{I}{I_0} \right) \mathrm{dB} \,. \tag{15.9}$$

I represents the energy transported by a wave that traverses a unit surface in a unit time. I_0, then, is the intensity of a sound wave that, in 1 s, carries an energy of 10^{-12} J traversing a surface of 1 m². In sound waves, the energy is that carried by the air particles.

PHYPHOX provides audio data in arbitrary units. They can be converted into absolute intensities upon calibration of the microphone, which needs a source whose intensity is known. For most experiments, absolute intensities are useless.

With the audioscope, we record the sound produced by the script described in previous sections, bringing it to the smartphone's microphone using earphones or loudspeakers. As usual, PHYPHOX exports data in columns. In the case of the audioscope, there are two columns in the file: the time t and the amplitude $y(t)$. A plot of $y(t)$ as a function of t when $f = 440\,\mathrm{Hz}$ yields

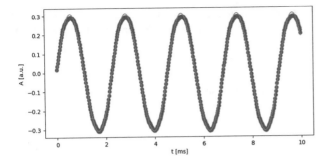

The orange line is the result of a fit with Eq. (15.6) for $x = 0$. The waveform is a bit distorted, due to the limitations of the hardware, nevertheless, the fit finds a value of $f = 440.0 \pm 0.1$ Hz. The initial parameters for the fit were estimated by taking the maximum of $f(t)$ as A and the average distance between successive peaks as T.

This is a clear example of the importance of having a good model for data. In fact, we assumed that the correct model of our data was Eq. (15.6), but, in fact, it is not. The waveform is a bit distorted and the χ^2 of the fit cannot be too close to 1. The residuals used to estimate the uncertainties measure the discrepancies between the model and the data and have little to do with the uncertainty about the period, which is a measure of how different the distances between successive peaks are.

In this case, taking the average and the standard deviation of the four differences that can be computed from the five peaks, we find $T = 2.270\,919\,352\,98 \pm 0.000\,000\,000\,02$ ms, corresponding to $f = 440.350\,291\,916 \pm 0.000\,000\,004$ Hz, which is apparently orders of magnitude better than the fit. The standard deviation of the samples is, in fact, 8 nHz and the uncertainty about the average is $\frac{\sigma}{\sqrt{N}}$, with $N = 4$. In fact, these numbers seem unreasonably good: a resolution of a few nHz seems out of reach for a smartphone.

In fact, we completely neglected the resolution of the instrument. The latter can be estimated by taking the duration of the signal (10 ms) divided by the number of samples (480) such that $\Delta_t = \frac{10}{480} \simeq 0.020$ ms. The uncertainty about the time of an event is then

$$\sigma_t = \frac{\Delta_t}{\sqrt{12}} \simeq 0.006 \text{ ms}. \tag{15.10}$$

Despite tha fact that the fits appear to be good, the model is wrong (it does not take into account distortions introduced by the hardware of both the source and the receiver). An objective quality of the fit cannot be evaluated and uncertainties do not have a precise statistical meaning.

Whenever evaluating the uncertainties, we need to take into account the limited resolution of the instruments. If it dominates over statistical fluctuations, it must be properly considered.

The total uncertainty is the sum in quadrature between the two, but the square root of the sample variance is negligible, hence $\sigma_t = 0.006\,\text{ms}$ and

$$T = 2.271 \pm 0.006\,\text{ms} \tag{15.11}$$

is the best estimate, leading to

$$f = \frac{1}{T} = 440 \pm 1\,\text{Hz} \tag{15.12}$$

for the frequency. Despite the fact that the modelling of the signal was not perfect, the final results are comparable.

15.4 The Student's t-Distribution

The uncertainty of a set of measurements is itself a measurement and, as such, its values are distributed over an interval with a PDF known as the Student's t.

In the previous section, the standard deviation s of the samples was estimated from four measurements of the period. s is itself the result of a measurement and, as such, is affected by an uncertainty that decreases with the number of samples. To assess the significance of a measurement, the ratio

$$S = \frac{x - \mu}{\sigma}, \tag{15.13}$$

where μ is the expected value of x and σ the uncertainty, is computed and the corresponding confidence interval is obtained by evaluating the integral of a Gaussian. When σ is unknown, the significance can only be assessed using an estimation of it:

$$S = \frac{x - \mu}{s}. \tag{15.14}$$

When the number N of measurements is large, $s \simeq \sigma$ and there are no significant differences. However, if N is small, the difference above does not follow a Gaussian distribution, but rather the Student's t-distribution. Such a curious name derives from the fact that it was formulated and published by William Gosset (1876–1937) using the pseudonym "a student" (according to the *legend*, the use of the pseudonym became necessary because of the policies of the company where Gossett was employed, which forbid its employees from making their names public).

The probability distribution function of the Student's t-distribution is not straightforward (nor is its derivation) and is

$$\text{PDF}_t(x) = \frac{\Gamma\left(\frac{v+1}{2}\right)}{\sqrt{v\pi}\,\Gamma\left(\frac{v}{2}\right)}\left(1+\frac{x^2}{v}\right)^{-\frac{v+1}{2}}, \tag{15.15}$$

where $v = N - 1$ is the number of degrees of freedom and $\Gamma(x)$ is the Euler Gamma function. Its shape is shown below for various v, compared to that of the normal distribution.

The PDF of a Student's t is rather complicated, but it closely resembles a Gaussian. Its width is a bit larger, but the difference reduces as the number of degrees of freedom increases, as predicted by the Law of large numbers.

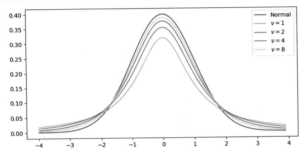

The Student's t-distribution is wider than the normal one. Its maximum is lower, while it is higher on the tails. As predicted by the Law of large numbers, the t-distribution approaches the normal one as v increases. When the average of a set of measurements differs by one standard deviation from the expected value, we assume that the probability that $|S| \le 1$ is about 68%. However, when the standard deviation is estimated from a few measurements, the probability must be estimated taking into account that S is distributed as a Student's t and

As usual, Python's scipy.stats module has functions for computing the PDF and the corresponding cumulative of many distributions, including the Student's t.

$$P\left(|S| \le 1\right) = \int_{-1}^{1} \text{PDF}_t(x)dx = \int_{-\infty}^{1} \text{PDF}_t(x)dx - \int_{-\infty}^{-1} \text{PDF}_t(x)dx \simeq 61\% \tag{15.16}$$

for $v = 3$. Such a confidence interval can be obtained in Python as simply as

```
from scipy.stats import t

cI = t.cdf(1, 3)-t.cdf(-1, 3)
print('CI: {}'.format(cI))
```

The first parameter of the t() function is x, while the second is v.

15.5 **Interference**

Interference consists in the superposition of two or more waves at a point. The resulting wave can have an amplitude ranging from the difference of their amplitudes to the sum of them.

Interference happens when two or more waves sum in the same place at the same time. Consider two equal in-phase wave sources at position $x = 0$ and $x = L$. At a point $0 \leq x \leq L$, the two waves interfere such that the resulting wave is

$$y(x) = A \cos (kx) + A \cos (k(x - L)). \tag{15.17}$$

Prosthaphaeresis formulas allow us to write the sum of two trigonometric functions as a product. In particular,

$$y(x) = 2A \cos \left(\frac{kL}{2}\right) \cos \left(k \left(x - \frac{L}{2}\right)\right) \tag{15.18}$$

(the sum of the cosines of two angles is transformed into the product of the cosines of their half-sum and half-difference). The resulting wave is null when

$$k \left(x - \frac{L}{2}\right) = (2n + 1)\frac{\pi}{2}. \tag{15.19}$$

The positions at which the amplitude of the sum of two equal waves is minimum can be easily predicted. Minima happen when the phase difference between interfering waves is an odd number of their half wavelength.

Remembering that $k = \frac{2\pi}{\lambda}$, the condition for null amplitude is

$$x = \frac{L}{2} + (2n + 1)\frac{\lambda}{4}. \tag{15.20}$$

In an experiment, we put the two speakers of a set of earphones $L = 40$ cm apart, one in front of the other, playing a sinusoidal tone with $f = 3430$ Hz and $\lambda = \frac{c}{f} = 0.1$ m $= 10$ cm. We then move a smartphone along the line connecting them at various positions x_i and record the sound using PHYPHOX's audioscope. Since the sound intensity diminishes as d^{-2}, where d is the distance from the source, in order for the two waves to sum with more or less the same amplitude, the detector must be placed close to the middle point $x \simeq 20$. For $15 < x < 30$ cm, we expect to observe a minimum for $x = 17.5, 22.5, 27.5$ cm.

There are different ways to obtain the amplitude. As in Sect. 15.3, one can fit the waveform with the amplitude as a free parameter. For this experiment, we do not need much precision, thus we can evaluate

$$A \simeq \frac{\max_i y_i - \min_i y_i}{2}. \tag{15.21}$$

In order to estimate the uncertainty σ_A, we can identify several pairs made up of a maximum and a minimum and evaluate as many A_i, such that A is given by the average $\langle A_i \rangle$ and σ_A by their standard deviation (not divided by $\sqrt{N}$, because we are interested in the spread of the values of A_i). Since minima in the waveform are expected to be symmetric to the maxima with respect to $y = 0$, the amplitude can even be estimated just as the average of the maxima recorded in the waveform, while the uncertainty is given by their standard deviation. In fact, finding the amplitude is even simpler: as shown in the next section, the standard deviation of the amplitude data is proportional to A: $\sigma_y = \frac{A}{\sqrt{2}}$. Given that A is measured in arbitrary units, taking just the standard deviation of the measurements is already a measure of the amplitude of the signal.

> The standard deviation of the samples of a waveform is proportional to the amplitude.

A plot of the sound intensity as a function of x is shown below.

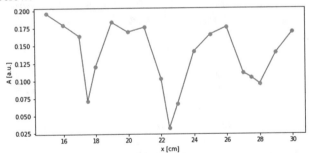

Minima are, in fact, where they are expected to be. A fit to the result cannot be particularly good, given the limitations of the setup. Indeed, apart from the limited sensitivity of a smartphone's microphone and the finite dynamic range of the earphones, during the experiment, the smartphone's microphone captures all environmental noise. Moreover, the signal's echos are not negligible, nor is the amplitude of the two waves the same (one of the sources is closer to the microphone). The sound propagates partially in the material of the surface upon which the speakers and smartphone are placed. Nevertheless, an unweighted

> In homemade experiments on waves involving sound, there are plenty of systematic effects that affect the measurements. Sound propagates in air and in solid media, it is subject to echoes and its intensity diminishes with the distance from the source. Despite these limitations, many experiments can be successfully performed.

fit to $|y(x)| + C$ brings relatively good results. Leaving λ and L as free parameters, we obtained $\lambda = 10.1 \pm 0.3\,\text{cm}$ and $L = 40.3 \pm 0.2\,\text{cm}$, perfectly consistent with expectations.

15.6 Finding the Distribution of a Random Variable

A function $y(x)$ of a random variable is a random variable itself. The PDF's $f(y)$ of y is related to that of $x, f(x)$, by $f(x)dx = f(y)dy$.

Let x be a continuous random variable whose PDF is $f(x)$. By definition, $f(x)$ is such that the probability of x lying within the interval $[x,\ x + dx)$ is $f(x)dx$. If $y = f(x)$, then y is a random variable, too, and the probability of y being in $[y,\ y + dy)$ is $f(y)dy$, $f(y)$ being the PDF of y. Clearly,

$$f(y)dy = f(x)dx , \tag{15.22}$$

because, when $x \in [x,\ x + dx)$, $y \in [y,\ y + dy)$, such that

$$f(y) = f(x)\frac{dx}{dy} . \tag{15.23}$$

While $f(x)$ and $f(y)$ are, by definition, positively defined, the derivative $\frac{dx}{dy}$ may not be such. To guarantee that, we write

$$f(y) = f(x)\left|\frac{dx}{dy}\right| , \tag{15.24}$$

authorised by the fact that $dx > 0$ and $dy > 0$, too, in the above computations.

Suppose, then, that we measure $y = A\cos(\omega t)$, with A and ω as constants. We sample t uniformly within the interval $0 \le t \le t_M$, such that its PDF is uniform and defined by

$$f(t) = \begin{cases} 0 & \text{for } t < 0 \\ c = \dfrac{1}{t_M} & \text{for } 0 \le t \le t_M \\ 0 & \text{for } t > t_M . \end{cases} \tag{15.25}$$

To find the PDF of y, we need its inverse:

$$t = \frac{1}{\omega} \cos^{-1} \frac{y}{A}, \tag{15.26}$$

from which we obtain

$$f(y) = \frac{1}{A t_M \omega \sqrt{1 - \frac{y^2}{A^2}}} . \tag{15.27}$$

In the interval $[0, \ t_M]$, the possible values of y are in $[-A, \ A]$, for $t_M \gg T$, and the appropriate normalisation factor can be found by imposing that

$$1 = \int_{-A}^{+A} f(y)dy = \frac{1}{t_M \omega} \sin^{-1}\left(\frac{y}{A}\right)\Big|_{-A}^{+A} = \frac{\pi}{t_M \omega} . \tag{15.28}$$

Finally, we can write

$$f(y) = \frac{1}{A\pi \sqrt{1 - \frac{y^2}{A^2}}} . \tag{15.29}$$

A plot of $f(y)$ is shown below, together with a histogram obtained from data generated within an interval $t \in [0, \ 70]$ in arbitrary units, and $A = 1$.

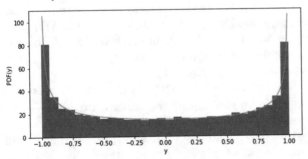

The histogram is symmetric around $y = 0$, thus the mean of the distribution of A is null, while its variance is

$$\sigma_y^2 = \int_{-A}^{+A} \frac{y^2}{A\pi \sqrt{1 - \frac{y^2}{A^2}}} dy = \frac{A^2}{2} . \tag{15.30}$$

15.7 Beats

Beats are produced by the interference of waves with different frequencies.

Beats happen when two waves of different wavelength interfere at the same point in space. Given two waves with frequency f_1 and f_2, their sum can be written as

$$y(t) = A\cos(2\pi f_1 t) + A\cos(2\pi f_2 t + \phi). \qquad (15.31)$$

The sum can be rewritten as a product using a prosthaphaeresis formula:

$$y(t) = 2A\cos\left(2\pi\frac{f_1+f_2}{2}t + \frac{\phi}{2}\right)\cos\left(2\pi\frac{f_1-f_2}{2}t - \frac{\phi}{2}\right). \qquad (15.32)$$

The latter can be thought of as a wave with a frequency equal to the average frequency between the two and an amplitude

$$A(t) = 2A\cos\left(2\pi\frac{f_1-f_2}{2}t - \frac{\phi}{2}\right). \qquad (15.33)$$

The resulting wave has a frequency equal to the average frequency between the interfering ones and an amplitude modulated with a frequency equal to half the difference of the original frequencies.

If $f_1 \simeq f_2 = f$, the resulting wave is a wave with the same frequency as the incident ones, with an amplitude modulated with a frequency $\frac{f_1-f_2}{2}$. For example, if $f_1 = 440\,\text{Hz}$ and $f_2 = 442\,\text{Hz}$, the corresponding tones are practically indistinguishable and the resulting sound has a frequency of $f = 441\,\text{Hz}$, which is again barely distinguishable from the incident ones. However, its amplitude varies from 0 to $2A$ with a frequency of 1 s.

We can easily hear beats as an intensity modulated sound whose modulation frequency is exactly $\frac{f_1-f_2}{2}$. It is enough to modify the script developed in Sect. 15.1 such that we play a sound of frequency f_1 on the right channel and a sound of frequency f_2 on the left one. The waveform for beats can be appreciated for higher frequencies. For example, for $f_2 = 2\,\text{kHz}$, $f_1 = 2.27\,\text{kHz}$, PHYPHOX's audioscope shows the following plot:

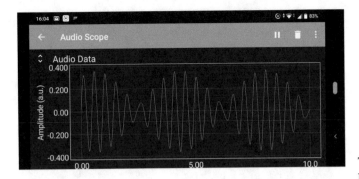

From the plot, it is easy to obtain the period of the sound wave of about

The waveform of beats is characterised by a constant frequency and a modulated amplitude. In the case of sound waves, if the frequency of the modulation is short enough, it can be heard.

$$T = \frac{2}{f_1 + f_2} \simeq 0.5\,\text{ms} \tag{15.34}$$

and the beats' period

$$T = \frac{2}{f_1 - f_2} \simeq 7.5\,\text{ms}, \tag{15.35}$$

as expected.

15.8 Collecting Audio Data with Arduino

Audio data can be captured with any microphone, including those available for Arduino. Microphones can be seen as analog sensors. They have three leads: two for feeding them and a third one for the signal, represented as a voltage proportional to the intensity of the sound. Sometimes, they have a fourth lead used as a digital pin, whose state is raised when the recorded sound exceeds a given, often adjustable, threshold.

Microphones for Arduino are analog sensors that provide a voltage proportional to the intensity of the recorded sound.

In order to record the waveform of a signal, we need to sample it at the highest possible frequency, spending as little time as possible in elaborating data. This can be achieved simply by repeatedly digitising the analog input connected to the microphone and storing the data in a data structure like an array, as in

```
void loop() {
  unsigned long t0 = micros();
  for (int i = 0; i < N; i++) {
    wave[i] = analogRead(A0);
  }
  unsigned long t1 = micros();
  float dt = (float)(t1-t0)/N;
  float t = 0.;
  for (int i = 0; i < N; i++) {
    Serial.print(t);
    Serial.print(", ");
    Serial.println(wave[i]);
    t += dt;
  }
}
```

To collect a sufficient number of samples, data acquisition must be carried out as quickly as possible.

First of all, we obtain the current time t_0, then we digitise N samples of the sound, storing them in the elements of the wave array; after reading the time t_1 at the end of data acquisition, assuming that the time needed to get each sample is constant, the time elapsed from one sample to the next is

$$dt = \frac{t_1 - t_0}{N}. \tag{15.36}$$

The division between two integers gives an integer. To obtain a rational number, one of the terms in the division must be cast into a floating point value.

Note the casting operator (float) placed in front of t1-t0. This is necessary because the numerator of the fraction is an integer, like the denominator. The result, then, would be an integer, too, unless we cast one of them into a floating point number. Only at the end of the data acquisition cycle do we send the samples over the serial line, together with the corresponding times.

N can be as large as possible. In Arduino UNO, where memory is limited, up to about 500 samples can be stored. Consequently, the range of useful frequencies is limited to around 1 000–2 000 Hz (this depends both on the memory size and the clock frequency of the board).

There are techniques, beyond the scope of this textbook, that exploit so-called *interrupts* that allow for better resolution and performance. It is worth noting that, at first sight, it might appear better to just send digitised data over the serial line, as in

Sending characters over a serial line is time-consuming and must be avoided during sampling.

```
void loop() {
  Serial.print(micros());
  Serial.print(", ");
  Serial.println(analogRead(
    A0));
}
```

In fact, sending data over the serial line is time-consuming and results are disappointing, even when removing the first two lines with the information about time.

15.9 Dimensional Analysis

Dimensional analysis is an extremely useful tool for spotting mistakes made while manipulating the equations representing physical laws. Given that each member of an equation represents a measurable quantity, both of them must have the same physical dimensions. If they do not, the equation is certainly wrong. The contrary is not always true: not all dimensionally correct equations are right. Still, a check of the dimensions of a physical quantity is always welcome and often saves lot of time.

Dimensional analysis, indeed, is even more powerful. Sometimes, it can be used to write a relationship between two or more physical quantities ignoring most of the physics behind the scenes. Here, we exploit it to determine the relationship between the speed of sound v in air and the properties of the fluid. If v is not constant, in fact, it can only depend on the properties of the medium through which the sound propagates. These are its pressure p, its temperature T, and its density ρ. The mass of the air cannot be a parameter, because it depends on the volume that we choose to consider and v cannot depend on a choice that we have made. For the same reason, the volume V is not an option. Suppose, then, that $v = f(p, T, \rho)$ and observe that the ideal gas Law establishes a relationship between p and V as

Dimensional analysis is a powerful tool for checking the correctness of a physical law. Unfortunately, its importance is often underestimated.

In fact, dimensional analysis also allows for the derivation of relationships between physical quantities. Only constants remain undetermined, but they can estimated from measurements.

$$pV = nRT , \tag{15.37}$$

where n is the amount of gas measured in units of moles and $R \simeq 8.314 \, \mathrm{J \, mol^{-1} K^{-1}}$, a universal constant called the molar gas constant. $f(p, T, \rho_{dimensionless})$ can only be a combination of the three quantities, R, and a constant C, like

$$v = f(p, T, \rho) = v_0 + C p^n T^m \rho^k R^\ell . \tag{15.38}$$

Here, v_0 represents the speed of sound in air when $T = 0 \,^\circ\mathrm{C}$ ($p = 0$ and $\rho = 0$ being non-physical). From dimensional analysis, it follows that

$$[v] = \left[LT^{-1} \right] = \left[p^n T^m \rho^k \right] . \tag{15.39}$$

Pressure is a force per unit area:

$$[p] = \left[\frac{F}{A} \right] = \left[\frac{ma}{A} \right] = \left[\frac{MLT^{-2}}{L^2} \right] = \left[ML^{-1}T^{-2} \right] \tag{15.40}$$

Density is mass per unit volume:

$$[\rho] = \left[\frac{m}{V} \right] = \left[\frac{M}{L^3} \right] = \left[ML^{-3} \right] . \tag{15.41}$$

Combining all of these together, and ignoring v_0, which already has the right dimensions, we obtain

$$[v] = \left[LT^{-1} \right] = \left[M^{n+k+\ell} L^{-n-3k+2\ell} T^{-2n-2\ell} K^{m-\ell} \right] \tag{15.42}$$

(we used K as a symbol for temperature, so as not to confuse it with time). Comparing the dimensions of the right member with those on the left, we conclude that $m = \ell = 0$, and $n = -k = 1/2$, i.e.

$$v = v_0 + C p^{\frac{1}{2}} \rho^{-\frac{1}{2}} = v_0 + C\sqrt{\frac{p}{\rho}} = v_0 + C\sqrt{\frac{nRT}{\rho V}}.$$
(15.43)

Since $n = \frac{m}{M}$, where $m = \rho V$ is the mass of the gas and M its molar mass (i.e., the mass of a mole of a substance that depends on the composition of the gas mixture),

The speed of sound in a fluid grows with the square root of the temperature of the fluid.

$$v = v_0 + C\sqrt{\frac{RT}{M}} = v_0 + c\sqrt{T},$$
(15.44)

with $c = C\sqrt{\frac{R}{M}}$. In other words, the speed of sound in a gas increases with the square root of its temperature and decreases with the mass M of the molecules of which it is composed, still as $\sqrt{M}$. Indeed, you may have seen videos in which a person speaks in a much higher tone after inhaling helium. This is because helium is much lighter than nitrogen, which makes up roughly 80% of the composition of air, thus the speed of sound in helium is higher. The frequency f of the voice is related to its wavelength λ by

$$f = \frac{v}{\lambda}$$
(15.45)

and the higher v is, the higher f is.

In order to test the validity of our result and to measure M, we need to set up an experiment in which we measure v as a function of T. Using Arduino, this is a relatively easy task. We use an ultrasonic sensor, aiming its beam of ultrasonic waves against a wall at a known distance d and measuring the time they need to come back, once reflected:

Ultrasonic signals are sound waves at high frequency, thus they propagate with the speed of sound.

$$t = \frac{2d}{v},$$
(15.46)

from which we obtain v. The temperature can be measured by means of a temperature sensor. Both sensors can be placed into a box such that d is its length. Before starting the experiment, we heat the air inside using a hair dryer, then close the box and let it thermalise, measuring both v and T at the same time.

The aim of this experiment is to verify that $v \propto \sqrt{T}$, equivalent to proving that $v^2 \propto T$. The best thing to do is thus to make a plot of v^2 as a function of T and verify that

Linearisation always simplifies the analysis of the relationship between physical quantities.

data distribute along a straight line. Alternatively, one can check that a distribution of $\frac{v^2}{T}$ is Gaussian with a width that is consistent with those expected by the ratio of two quantities.

Impress your classmates

The GPS system relies on signals emitted by a constellation of satellites that continuously send data to Earth, containing their coordinates with respect to a reference frame centred on Earth and the exact time at which the message left the satellite: (x_i, y_i, z_i, t_i). The time is provided by an extremely precise atomic clock on board. The precision needed is so high that the time of each clock must be corrected for effects of general relativity. In practice, time runs differently depending on the strength of the gravitational field in which the clock is immersed. On-board clocks orbit at a height of about 20 000 km above our heads, where the gravitational field is lower by a factor of

$$\frac{h^2}{r_\oplus^2} \simeq \left(\frac{26\,000}{6000}\right)^2 \simeq 19 \tag{15.47}$$

with respect to those on Earth, whose radius is $r_\oplus \simeq 6\,000$ km. The correction factor must be such that, when a time t elapses on Earth, on board the time is $t' \neq t$ and

$$t' = (1 - C)\,t\,, \tag{15.48}$$

where C must be dimensionless and equal to zero where the gravitational field is null. The minus sign comes from the fact that, in the presence of a gravitational field, the clock is accelerated and, if it was free, it increased its speed; in special relativity, clocks moving with respect to fixed ones run more slowly. C can only depend on the Earth's gravitational potential, which is

$$\mathcal{G} = G\frac{M}{r}\,, \tag{15.49}$$

$G \simeq 6.6 \times 10^{-11}$ being Newton's constant in SI units and M the Earth's mass. $\mathcal{G}$ has the dimensions of the square of a speed $\left[L^2T^{-2}\right]$. In order to make a dimensionless quantity, it must be divided by a quantity with the same dimensions. There is no other quantity upon which C may reasonably depend, however, there is a universal constant, the speed of light c, that has the right dimensions. The ratio between the gravitational potential $G\frac{M}{r}$ and c^2 is dimensionless, hence

$$C = C(r) = \alpha G \frac{M}{rc^2} \tag{15.50}$$

with dimensionless α. It turns out that the right correction factor computed from general relativity is, in fact, that with $\alpha = 1$. Rather impressive, in fact. Dimensional analysis can provide results that only a few people on Earth are able to compute exactly, based on one of the most technically difficult theories in physics.

A clock on the Earth's surface runs at a pace that is $\left(1 - C\left(r_{\oplus}\right)\right)$ with respect to a clock at infinite distance, while GPS clocks run such that the correction is $(1 - C(r))$. The difference between the two clocks' paces amounts to (neglecting the effects of special relativity, due to the high speed of GPS satellites, which account for a similar correction)

$$\Delta C = G\frac{M}{c^2}\left(\frac{1}{r} - \frac{1}{r_{\oplus}}\right) = 6.6 \times 10^{-11}\frac{6 \times 10^{24}}{9 \times 10^{16}}\left(\frac{1}{6 \times 10^6} - \frac{1}{26 \times 10^6}\right) \simeq 0.6 \times 10^{-9}. \tag{15.51}$$

In other words, for each second elapsed on Earth, the onboard clocks mark *only* 0.9999999994 s. This may appear to be a negligible correction, but it is a systematic correction that accumulates with time. After one year, the difference amounts to about 18 ms: during this time, light travels for about $5\,300$ km, thus the satellites appear to be farther away.

15.10 Temperature Measurements with Arduino

For the experiment described in Sect. 15.9, we can use any analog temperature sensor. It is enough to connect its GND lead to the pin with the same name on Arduino and its VCC lead with the Arduino 5V pin, measuring the voltage by connecting the third lead to one of the Arduino analog pins. The sensor data sheet contains the conversion factor from voltage to temperature.

In this case, we illustrate the use of another kind of sensor, based on a digital communication protocol known as 1-wire. Most waterproof sensors are based on this technology and it is useful to be familiar with it. As the name says, the protocol uses just one wire to run communication from Arduino and the device, and vice versa. 1-wire sensors thus have three leads: two are needed to power it, while the third

Temperature sensors can either be analog or digital. Digital thermometers often work with the 1-wire protocol. In this protocol, data between master and slaves are exchanged on the same electrical line.

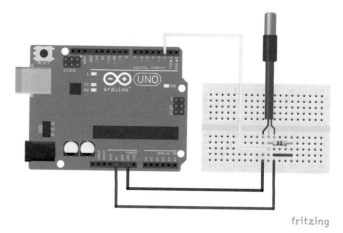

■ **Fig. 15.1**   Connecting a 1-wire temperature probe to Arduino requires the use of a *pullup* resistor of 4.7 kΩ between the signal and the VCC leads

is used to receive and send data over a serial digital line and must be connected to a digital pin. On this line, data are represented as a sequence of voltage pulses that can either be 0 V or 5 V with respect to the ground. For this reason, the signal lead must be kept at 5 V when idle. To do that, a resistor (called a *pullup resistor*) can be used to connect the signal pin to the VCC lead: its suggested value is 4.7 kΩ.

Many 1-wire slaves share the same connection, together with a pullup resistor.

Connections are illustrated in Fig. 15.1. Besides the features of the 1-wire protocol, one also has the ability to address more than one device attached to the same line, as shown below.

Here, three 1-wire devices are attached to the same line connecting them to the same Arduino pin and to the pullup resistor R. Each device has its own power inputs to be connected, respectively, to VCC and GND.

Choosing the right resistor

Resistors are electrical circuit elements whose function is to regulate the circulating current. The higher the resistance, the lower the current. The resistance is measured in Ohms (symbol Ω). In normal applications, the resistance of a conductor wire is completely negligible, and we can usually assume that it is worth zero, but there are resistors on the market that can have electrical resistance values between a few Ω and millions of Ω ($M\Omega$, pronounced *megaohm*). The resistance value is represented by a series of coloured stripes on the body of the circuit element. Each color corresponds to a number according to the following table:

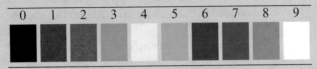

| 0 | 1 | 2 | 3 | 4 | 5 | 6 | 7 | 8 | 9 |

There are usually four strips on each resistor: three are coloured as shown in the table; the fourth indicates the accuracy of the indicated value (*tolerance*), which is 5% if the strip is gold and 10% if it is silver. Let's suppose we have a resistance with strips of the sequence red-red-brown-silver. The resistance value is obtained by writing the numbers corresponding to the first two colours (22) and multiplying them by a power of ten corresponding to the third color (1). The resistance is then $22 \times 10^1 = 22 \times 10 = 220\,\Omega$, which is a very common value. Given the tolerance of 10%, the actual value of the resistance has an uncertainty of about 22 Ω. A resistor of $4.7 \pm 0.2\,k\Omega$ is marked as yellow-purple-red-gold.

In the following, we briefly explain how pullup resistors work. You can skip directly to the next section if you are not interested in the details of the instrument, taking the above prescription as a recipe. However, soon or later, you will need to study this topic: a bit of previous knowledge about it, though qualitative, will help in understanding it.

    The output of a digital device is often driven by a transistor. In this application, a transistor can be regarded as an electronic switch. Transistors have three leads: the emitter, the base and the collector. Normally, no current flows from the emitter to the collector, unless a small current is injected into the base, which acts as a trigger for the switch. In order for the current to flow from the emitter to the collector, these two leads must be connected, respectively, to some positive voltage (e.g., 5 V) and to a ground (at 0 V), by means of a resistor: a device through which a current flows according to the Ohm Law,

The exact value of a pullup resistor is irrelevant. It must be large enough to limit the current drawn from the power supply, but low enough to not interfere with the internal circuits of the master and slaves.

$$I = \frac{V}{R}.$$

(15.52)

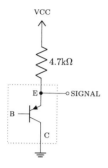

When the resistance is measured in Ohms, the current is given in ampères (symbol A). Referring to the figure on the side, in which the device is represented by the dotted rectangle and the last stage of the communication circuit is shown as a transistor, if no current is injected into the base (labelled B), the transistor works as an open switch and the SIGNAL pin is at VCC = 5 V. Upon injecting a small current into the base, a current $I = \frac{5}{4\,700} \simeq 1\,\mathrm{mA}$ flows from VCC to the ground, to which the collector (labelled C) is attached. Under these conditions, the voltage on the SIGNAL pin is 0 V. $R = 4.7\,\mathrm{k\Omega}$ is the pullup resistor, whose value is not particularly important. Its value must be large enough to keep the current flowing in the circuit small. However, if R is too large, resistors connected to the SIGNAL pin (in our case, resistors on the Arduino board internally connected to the digital pin) may dominate and current will flow from outside the circuit (from Arduino) to the ground. Usually, values from 1 kΩ to 10 kΩ are good enough.

15.11 The 1-Wire Protocol

In the initialisation phase, slaves communicate their unique identifier to the master bit by bit. A proper algorithm disentangles the signal emitted by each slave and identifies it. Once the master has collected the addresses of all the slaves, commands can be sent over the 1-wire line. They are picked up only by those slaves that are addressed.

In this protocol, a single master device (Arduino) controls one or more slaves (the sensors). Addressing a single slave on a line is made possible by assigning a unique 64-bit address to each device. Eight of these bits are reserved for identification of the *family* of the device (devices of the same type share the same family code).

The device usually completes its task (e.g., reading the temperature) upon receiving a specific *command* represented by an 8-bit code. The command is represented as a sequence of pulses over the communication line, also called the *bus*. Data are stored in an internal memory (called a *scratchpad*) that can be polled by the master to obtain them. Polling consists in issuing the reading command and reading back sequences of pulses representing the binary coded data.

Before communicating with slaves, the master must acquire their addresses. This is done by issuing a search command over the bus, to which each device responds, providing its own address. For the temperature sensor based on the DS18B20 chip, for example, the search command is 0xF0 (corresponding to decimal 240). Measuring and obtaining temperature consists in the following steps:

1. Address the device sending the proper address over the bus.
2. Issue the start conversion command, which corresponds to obtaining the temperature as a voltage and converting it into a number in the proper units. According to the device data sheet, this is represented by 0x44 corresponding to the decimal 68.
3. Wait until the conversion is complete.
4. Address the device again.
5. Send the read command (0xBE corresponding to decimal 189).
6. Read as many sequences of 8 bits (bytes) as needed. This need depends on the device: for the DS18B20, each piece of data is represented by default as a 12-bit integer, consequently, a minimum of two bytes must be read from the scratchpad.

The protocol is implemented under the default library `OneWire.h`. This library is seldom used as such. Most often, the user will install libraries dedicated to his/her specific device and use them, as they are much more user friendly. It is, however, a good idea to see how such libraries are implemented, so as to be able to do similar things in the future.

To this purpose, let's consider the waterproof temperature sensor based on the DS18B20 chip, for which we install the library provided by Adafruit, an active developer of both hardware and software that is also renowned as being certified as a Minority and Woman-owned Business Enterprise (M/WBE). The needed library `DallasTemperature.h` can be obtained via the library manager of the Arduino IDE. A very basic example follows.

As usual, libraries can be used to ignore the details of the protocol and write simple programs. Library functions encapsulate what is needed to ask slaves to perform their actions and return data collected in their own memory (also known as a scratchpad).

```
#include <OneWire.h>
#include <DallasTemperature.h>

#define BUS 2

OneWire oneWire(BUS);
DallasTemperature tSensor(&oneWire);

void setup(void) {
  Serial.begin(9600);
  tSensor.begin();
```

```
}

void loop(void)
{
  tSensor.requestTemperatures();
  float T = tSensor.getTempCByIndex(0);
  Serial.println(T);
}
```

After including `OneWire.h` and `Dallas Temperature.h`, we instantiate `oneWire` as an object of the class `OneWire`. It takes a parameter representing the pin to which the 1-wire bus is attached. `tSensor` is, instead, an object of the class `DallasTemperature` that requires the address of the `oneWire` object (obtained using the operator `&`) to communicate.

The `begin()` method of the latter is used to establish the communication protocol. With this method, the object obtains the address of the sensor on the bus. This is accomplished as follows, as one can discern looking at the source code of the library. First of all, the master (Arduino) writes 0x70 on the bus using the `write(0x70)` method of the `OneWire` library. All the slaves on the bus send back the 64th bit (b^i_{64}) of their address, followed by its complement ($\bar{b}^i_{64}$). The bus is designed such that, when more than one slave is connected, the serial line is found at a logic level corresponding to the logic AND of all the bits put on it by the slaves (let's call it b_{64}). The master gets it and responds with b_{64} such that all the slaves whose 64th bit is different from b_{64} stop responding, while the others continue the dialogue, sending b^i_{63} and $\bar{b}^i_{63}$. Again, the master responds with b_{63} and the process continues until all the devices have been discovered on the bus. It is useful to mimic this process assuming that there are three devices on the bus. For simplicity, we pretend that the address is of four bits instead of 64: $A = 0001$, $B = 1010$ and $C = 1101$. The discovery protocol works as follows (refer to Figs. 15.2 and 15.3 depicting the first two cycles):

1. The master writes 0x70 on the bus.
2. All three devices put their least significant bit on the line, respectively: 1, 0 and 1. The result is that the line is at logic level 0.
3. They then put the complement of their least significant bit on the line, respectively: 0, 1 and 0. The result is again that the line is at logic level 0.

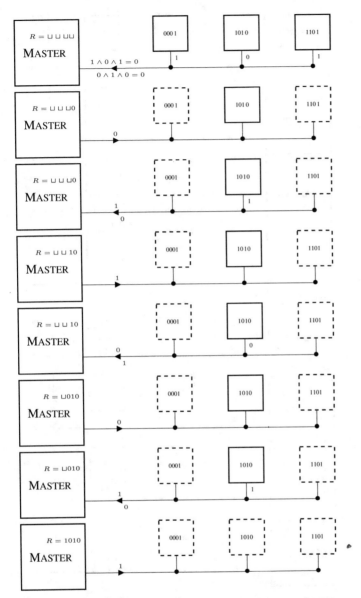

Fig. 15.2 First cycle of the search algorithm for 1-wire communication. Slaves put successive bits and their complements on the bus. The bits are read by the master and the first is sent back to the slaves. Only slaves with the same bit continue participating

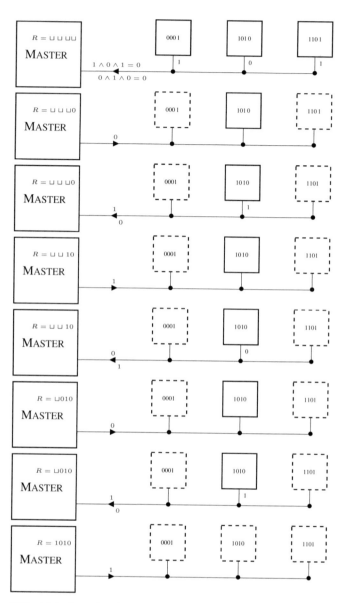

◻ Fig. 15.3 Second cycle of the search algorithm for 1-wire communication. Slaves stop participating when they have been discovered. In turn, all devices are found

4. The master puts 0 on the line (the result of the AND in step 2) and stores it as $R = 0$.

5. Slaves with 1 as the least significant bit stop (A and C).

6. The remaining slave (B) puts a 1 on the bus, followed by a 0.

7. The master gets the 1 and responds with it, storing the latter next to R at its left. R becomes $R = 10$.

8. The slave puts 0 followed by 1 on the bus.

9. The master gets the first 0 and sends it back, storing the 0 in R, which becomes $R = 010$.

10. The slave puts 1 followed by 0 on the bus.

11. The master attaches the 1 to R ($R = 1010$) and restarts the process (in the example, the address is four bits long and R is full). In the next cycle, B will not respond.

12. A and C put both 1 on the bus, which is found at logic level 1 by the master. They then put the complement of their last bit on the line, whose logic state becomes 0.

13. The master reads the first 1, such that $R = 1$, and responds with it.

14. Both devices have a 1 in their last digit and both put their third bit on the line, whose logic level becomes 0. When they transmit the complements of their third bit, the bus is found at 1.

15. The master attaches the first bit received (0) to the left of R ($R = 01$) and sends back 0 to the slaves.

16. The two slaves respond again, both having a 0 in the last transmitted bit. The response for the two slaves is 0 and 1, respectively, whose logic AND is 0. The next bit is the AND between 1 and 0, i.e., 0.

17. The master gets the first 0, attaches it to the left or R ($R = 001$) and sends back 0 to the slaves.

18. Since C transmitted 1, it stops responding. Only A responds with 0, followed by its complement 1.

19. The master attaches the first bit (0) to the left of R, obtaining $R = 0001$, and starts a new cycle.

20. In the new cycle, only one slave is present, whose address is then collected as is.

Thanks to the family bits, Arduino is able to spot all the temperature sensors on the bus and assign an index i to each of them, starting from $i = 0$, according to when they were discovered.

`tSensor.requestTemperatures()` asks to the object `tSensor` to start the measurement. This consists in only issuing the command 0x44 (convert) to all sensors.

Finally, the `getTempCByIndex()` method returns the temperature in degrees Celsius obtained by the sensor whose index is given in parentheses.

Usually, it is not worth handling the protocol by yourself; relying on the manufacturer libraries is a better option. In particular, the search algorithm is quite complicated. Nevertheless, knowing the protocol gives you a plus with respect to those who limit themselves to merely copying the software of others.

15.12 Establishing a Correlation

Once the data have been collected, the distribution of v^2 as a function of T provides a clue about the possibility that $v^2 = \alpha T + \beta$. A best fit to the data is useful for obtaining the values of α and β, while the χ^2 of the fit provides a measure of how valid the hypothesis is (in fact, $P\left(\chi^2 \geq \chi_0^2\right)$ is a measure of the probability that the observed fluctuations around the average straight line can be attributed to random fluctuations).

The Pearson correlation coefficient, defined as $\frac{\text{Cov}(x, y)}{\sigma_x \sigma_y}$, provides a means to evaluate the degree of linear correlation between two variables x and y.

Another possibility for ascertaining the existence of a linear relationship between two physical quantities x and y consists in using the "Pearson correlation coefficient", defined as

$$r_{x,y} = \frac{\text{Cov}(x, y)}{\sigma_x \sigma_y} \tag{15.53}$$

by Karl Pearson (1857–1936), also known as the "Bravais index" (after Auguste Bravais, 1811–1863). By definition, $-1 \leq r_{x,y} \leq +1$. The idea is that, when x and y are uncorrelated, their covariance is null, as their $r_{x,y}$. If, on the contrary, they are fully correlated, $\mathrm{Cov}(x, y) = \pm 1$. Two variables appear to be correlated with a strength proportional to $r_{x,y}$. It is useful to observe that such a statement is only valid for linear correlation. In particular, it is very possible that $y = f(x)$ (thus perfectly correlated with x), while $r_{x,y}$ is close to zero. Moreover, if the relationship between x and y is not linear, the result may depend on the x range.

For example, if $y = (x - 3)^2$, there is a manifest correlation between x and y, as seen in the figure below, despite $r_{x,y} = 0$.

> The coefficient is zero for uncorrelated variables. A value $+1$ indicates the maximum possible positive correlation. If x and y are negatively correlated, the coefficient is -1 when the correlation is the strongest possible.

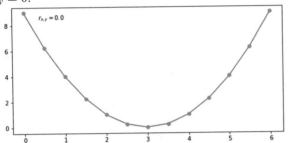

Moreover, consider $y = \sin x$, for $x_i \in [0, 2\pi]$ and $x_{i+1} - x_i = 0.5$. The correlation coefficient is not far from -1, being that $r_{x,y} \simeq -0.74$; however, when $x \in [0, 3\pi]$, $r_{x,y} \simeq 0.03$. The correlation coefficient changes, also changing the density of the points. For example, for $x_{i+1} - x_i = 1$, $r_{x,y} \simeq 0.05$.

In conclusion, the Pearson correlation coefficient is a useful tool for the rapid evaluation of the degree of correlation of two quantities, if the relationship between them is linear. If not, such a measure is of almost no use.

It can be computed directly as the off-diagonal element of the symmetric matrix R_{xy}

> The Pearson coefficient must not be used to evaluate the degree of correlation of variables, if the expected correlation is not linear.

```
R = np.corrcoef(x, y)
cov = R[0][1]
```

whose elements are defined as

$$R_{xy} = \frac{C_{xy}}{\sqrt{C_{xx}C_{yy}}}, \tag{15.54}$$

The Pearson coefficient can be evaluated using pearsonr, defined in the scipy.stats package. Besides its value, the function returns the corresponding *p*-value.

and C_{xy} is the elements of the 2×2 covariance matrix between x and y. The Pearson correlation coefficient can also be easily computed as

```
from scipy.stats import pearsonr
corr, pval = pearsonr(x, y)
```

where pval represents the *p*-value of the coefficient, indicating the probability $P(r > r_0)$ that the observed correlation coefficient r_0 has been obtained only through statistical fluctuation from uncorrelated variables. It is then close to 0 for highly correlated variables and close to 1 in the opposite case.

Summary

Audio files are characterised by their sampling rate (the number of samples per second), their bit depth or resolution (the number of bits used for each sample) and the number of channels (mono or stereo).

The frequency of an audible signal is comprised between 20 Hz and 20 kHz.

Almost any periodic function can be written as a series of trigonometric functions with appropriate frequency and amplitude, according to the Fourier Theorem. The latter gives the recipes for computing the amplitudes for each frequency. Because of that, we do not need to study any possible wave: it is enough to study the dynamics of harmonic waves to predict the behaviour of any other wave. Harmonic waves are characterised by their wavelength, their period and their phase.

Even when the fit to a set of data appears to be good, when the model is wrong, an objective quality of the fit cannot be evaluated and uncertainties do not have a precise statistical meaning.

Whenever evaluating the uncertainties, we need to take into account the limited resolution of the instruments. If it dominates over statistical fluctuations, it must be properly considered.

The uncertainty of a set of measurements is in itself a measurement and, as such, its values are distributed over an interval with a PDF known as the Student's t. It is rather complicated, but it closely resembles a Gaussian. Its width is a bit larger, but the difference reduces as the number of degrees of freedom increases, as predicted by the Law of large numbers.

A function $y(x)$ of a random variable is itself a random variable. The PDF $f(y)$ of y is related to that of x, $f(x)$, by $f(x)dx = f(y)dy$.

Interference consists in the superposition of two or more waves at a point. The resulting wave can have an amplitude ranging from the difference of their amplitudes to the sum of them.

The positions at which the amplitude of the sum of two equal waves is minimum can be easily predicted. Minima happen when the phase difference between interfering waves is an odd number of their half wavelength.

The standard deviation of the samples of a waveform is proportional to the amplitude.

In homemade experiments on waves involving sound, there are plenty of systematic effects that affect the measurements. Sound propagates in air and in solid media, it is subject to echoes and its intensity diminishes with the distance from the source. Despite these limitations, many experiments can be successfully performed.

Beats are produced by the interference of waves with different frequencies. The resulting wave has a frequency equal to the average frequency between the interfering ones and an amplitude modulated with a frequency equal to half the difference of the original frequencies.

The waveform of beats is characterised by a constant frequency and a modulated amplitude. In the case of sound waves, if the frequency of the modulation is short enough, it can be heard.

Dimensional analysis is a powerful tool for checking the correctness of a physical law. Unfortunately, its importance is often underestimated. In fact, dimensional analysis also allows for the derivation of relationships between physical quantities. Only constants remain undetermined, but they can be estimated from measurements.

The speed of sound in a fluid grows as the square root of the temperature of the fluid.

Ultrasonic signals are sound waves at high frequency, thus they propagate with the speed of sound.

Linearisation always simplifies the analysis of the relationship between physical quantities.

The Pearson correlation coefficient, defined as

$$r_{x,y} = \frac{\text{Cov}(x, y)}{\sigma_x \sigma_y}$$

provides a means to evaluate the degree of linear correlation between two variables x and y. The coefficient is zero for uncorrelated variables. A value $+1$ indicates the maximum possible positive correlation. If x and y are negatively correlated, the coefficient is -1 when the correlation is the strongest possible.

The Pearson coefficient must not be used to evaluate the degree of correlation of variables, if the expected correlation is not linear.

Arduino

A microphone can be imagined as a parallel plate capacitor whose plates are separated by an elastic medium. The sound pressure causes the distance between plates to change. Consequently, an electrical current is drawn from the system.

Microphones for Arduino are analog sensors that provide a voltage proportional to the intensity of the recorded sound.

To collect a sufficient number of samples, data acquisition must be carried out as quickly as possible. Sending characters over a serial line is time-consuming and must be avoided during sampling.

The division between two integers gives an integer. To obtain a rational number, one of the terms in the division must be cast into a floating point value.

Temperature sensors can either be analog or digital. Digital thermometers often work with the 1-wire protocol. In this protocol, data between master and slaves are exchanged on the same electrical line.

Many 1-wire slaves share the same connection, together with a pullup resistor. The exact value of a pullup resistor is irrelevant. It must be large enough to limit the current drawn from the power supply, but low enough to not interfere with the internal circuits of the master and slaves.

In the initialisation phase, slaves communicate their unique identifier to the master bit by bit. A proper algorithm disentangles the signal emitted by each slave and identifies it. Once the master has collected the addresses of all the slaves, commands can be sent over the 1-wire line. They are picked up only by those slaves that are addressed.

Libraries can be used to ignore the details of the protocol and write simple programs. Library functions encapsulate what is needed to ask slaves to perform their actions and return data collected in their own memory (also known as a scratchpad).

Phyphox

The properties of a wave can be explored using PHYPHOX, which shows $f(0, t)$, $x = 0$ corresponding to the position of the smartphone's microphone.

PHYPHOX provides audio data in arbitrary units. They can be converted into absolute intensities upon calibration of the microphone, which needs a source whose intensity is known. For most experiments, absolute intensities are useless.

Python

Sounds can be produced and manipulated using the `simpleaudio` module in Python.

Arrays are a special kind of list, whose elements must belong to the same type.

Applying the * operator to a list and an integer extends the list. The same operator applied between an array and a constant multiplies each element of the array by the constant.

Multidimensional lists or arrays are represented as a list of lists or as an array of arrays.

The `simpleaudio` package provides tools for interacting with loudspeakers and other peripherals intended to play audio signals. Audio data, in the form of a bidimensional array of n-bit integers, can be played using `play_buffer()`. After invoking `play_audio()`, the program continues immediately. To suspend its execution until the audio is played, the play object must wait for it.

The `getopt` package allows for the use of options and long options in launching a script on the command line of a terminal. Options are preceded by a dash (-) and may or not require arguments. They can be combined together if they don't. Long options are preceded by two dashes (--).

`getopt()` returns a list of pairs representing the options with their arguments, and a list of arguments. The first object returned by `getopt()` consists of a list of pairs. Each pair is made of an option and its corresponding value. The second object is a list of arguments given in the command line.

The `in` operator checks whether its left operand is contained in its right operand.

Python's `scipy.stats` module has functions for computing the PDF and the corresponding cumulative of many distributions, including the Student's t.

The Pearson coefficient can be evaluated using `pearsonr`, defined in the `scipy.stats` package. Besides its value, the function returns the corresponding p-value.

References

1. Student (1908) The probable error of a mean. Biometrika 6(1):1–25
2. Young T (1802) On the theory of light and colors. Philos Trans R Soc 92

Correction to: Physics Experiments with Arduino and Smartphones

Correction to:
G. Organtini, *Physics Experiments with Arduino and Smartphones*,
Undergraduate Texts in Physics, ▶ https://doi.org/10.1007/978-3-030-65140-4

In the original version of the book, the following belated corrections have been received post publication: Typographical corrections and amendment made in some of the equations in Chaps. 6, 7, 9, 10, and 15. The chapters have been updated with the changes.

The updated versions of these chapters can be found at
▶ https://doi.org/10.1007/978-3-030-65140-4_6
▶ https://doi.org/10.1007/978-3-030-65140-4_7
▶ https://doi.org/10.1007/978-3-030-65140-4_9
▶ https://doi.org/10.1007/978-3-030-65140-4_10
▶ https://doi.org/10.1007/978-3-030-65140-4_15

Index

Index

Symbols

Ω, 25
χ^2, 229, 232, 233, 236
&, 336, 337
*, 60, 349
+, 147
++, 60
+=, 60
//, 61, 269
:, 38, 51, 148, 151, 352
<<, 337
=, 59
>>, 337
{ }, 39
1-wire, 371, 374

A

Aachen, 26
absolute value, 47, 91, 138
absolute_sigma, 239
absorption of light, 12
abstract type, 69
acceleration, 87, 92, 228, 262, 305
accelerometer, 26, 92, 304, 309
acceptance, 159
acoustic stopwatch, 86
acoustic wave, 348
activity, 158
actuator, 27
AD0, 334
Adafruit, 375
ADC, 33, 332
address, 333, 336, 374
aesthetic, 56
air cushion, 220
air track, 220
alias, 50
Almagest, 4
α particle, 158
Ampère, 18, 374
amplitude, 354, 360
analog sensor, 32, 365
analog to digital converter, 33
analogRead(), 34, 36
angle, 227

angular frequency, 355
angular velocity, 305, 314
animation, 296
Apollo 15, 91
apparent force, 308
append(), 67
approximation, 49
arange(), 350
arbitrary units, 39
Arduino, 27
AREF, 329
argv, 352
arithmetic, 217
array, 38, 349, 365
art, 68, 128
ASCII, 58
astype(), 351
asymmetric uncertainty, 248, 300
asymmetry, 142, 159
ATmega 328, 217
atomic clock, 370
attribute, 51
audio, 350
audioscope, 355, 360, 364
augmented reality, 26, 327
auto-increment, 60
average, 46, 149, 231, 232, 236, 244
Avogadro number, 158
axiom, 3

B

balloon, 86
Banzi, Massimo, 28
bar, histogram type, 145
base, 373
baud, 36
Bayes Theorem, 120, 123, 292
Bayes, Thomas, 112
Bayesian, 112
beats, 364
beauty, 124
Becquerel, 158
Becquerel, Antoine Henri, 158
begin(), 340
beginTransmission(), 336
Bell Burnell, Jocelyn, 248

T

U

Index